DEVELOPMENTS IN
REINFORCED PLASTICS—4

CONTENTS OF VOLUMES 1 TO 3

Volume 1: Resin Matrix Aspects

1. Thermosetting Resins for Reinforced Plastics. G. PRITCHARD

2. Vinyl Ester Resins. THOMAS F. ANDERSON and VIRGINIA B. MESSICK

3. Polyester Resin Chemistry. T. HUNT

4. Phenol-Aralkyl and Related Polymers. GLYN I. HARRIS

5. Initiator Systems for Unsaturated Polyester Resins. V. R. KAMATH and R. B. GALLAGHER

6. High-temperature Properties of Thermally Stable Resins. G. J. KNIGHT

7. Structure–Property Relationships and the Environmental Sensitivity of Epoxies. ROGER J. MORGAN

8. Some Mechanical Properties of Crosslinked Polyester Resins. W. E. DOUGLAS and G. PRITCHARD

9. Crack Propagation in Thermosetting Polymers. ROBERT J. YOUNG

   Index

Volume 2: Properties of Laminates

1. Introduction: Causes of Property Variability. G. PRITCHARD

2. Imperfections in FRP Materials. G. PRITCHARD

3. Moisture Absorption in Fibre–Resin Composites. G. S. SPRINGER

4. Fatigue Behaviour of Fibre–Resin Composites. J. F. MANDELL

5. Shear Properties of Laminates. N. L. HANCOX

6. Machinability of Laminates: Punching and Drilling. R. H. PRITCHARD

7. Production and Properties of Paper-Based Laminates. D. J. NEWMAN

   Index

Volume 3

1. The Effects of Moisture on the Electrical Properties of Epoxy Composites. A. R. BUNSELL

2. Analysis of Polymer Surfaces. S. M. LEE

3. Osmosis in Resins and Laminates. J. S. GHOTRA and G. PRITCHARD

4. Stress Corrosion in Fibre-Reinforced Plastics in Aqueous Media. G. MENGES and K. LUTTERBECK

5. Thermal Degradation of Thermosetting Polyesters. G. A. SKINNER

6. Prediction of the Thermal Properties of Fibre–Resin Composites. J. KABELKA

   Index

DEVELOPMENTS IN REINFORCED PLASTICS—4

Edited by

G. PRITCHARD

*Department of Industrial, Organic and Polymer Chemistry,
Kingston Polytechnic, Kingston upon Thames, Surrey, UK*

ELSEVIER APPLIED SCIENCE PUBLISHERS
LONDON and NEW YORK

ELSEVIER APPLIED SCIENCE PUBLISHERS LTD
Crown House, Linton Road, Barking, Essex IG11 8JU, England

Sole Distributor in the USA and Canada
ELSEVIER SCIENCE PUBLISHING CO., INC.
52 Vanderbilt Avenue, New York, NY 10017, USA

British Library Cataloguing in Publication Data

Developments in reinforced plastics.—4.
—(Developments series)
1. Reinforced plastics—Periodicals
I. Series
668.4′94′05   TP1177

ISBN-13: 978-94-010-8979-1   e-ISBN-13: 978-94-009-5620-9
DOI: 10.1007/978-94-009-5620-9

WITH 96 TABLES AND 81 ILLUSTRATIONS

© ELSEVIER APPLIED SCIENCE PUBLISHERS LTD 1984

PREFACE

One of the most interesting developments in composite materials technology during the past decade has been the attempt to displace thermosetting resins from their position as the natural matrix in 'advanced' composites for such fields as aerospace. Dr McMahon gives some indication of the nature of this challenge in his chapter on fibre-reinforced thermoplastics. He acknowledges the problem of their low fibre contents, with the associated possibility of inadequate mechanical properties, especially in compression; and draws attention to the way in which the lack of suitable test methods for composites in compression has caused difficulties in assessing the latest thermoplastics composites in this respect. It is therefore of special interest that Professor Piggott deals with the whole question of compression testing of composites in Chapter 4.

On the positive side, reinforced thermoplastics seem to be much more damage-tolerant than thermosets. This is clearly an advantage that will not be lost on the aircraft industry. The subject of damage repair to FRP is one of substantial concern; the wider question of defect detection and assessment is discussed authoritatively by Reifsnider and Henneke in Chapter 3, which focuses on the technique of thermography.

Ironically, the chemistry of cured thermosetting resins, such as epoxies, is only just beginning to be understood, thanks to techniques such as Fourier transform infrared spectroscopy, which is covered by Liao and Koenig in Chapter 2 of this book. It is possible that when a better understanding has been achieved, both of the fine chemical structure of epoxies and of their supermolecular structure, there will be a better control of desirable properties such as toughness. The size and complexity of the epoxy family is illustrated by the enormous number of distinct curing agents reviewed in Scola's chapter (Chapter 5). These curing agents lead to distinct properties, and contribute to the enormous range of properties which epoxies can exhibit. This fact should not be allowed to obscure the central point that, in certain respects, all are inadequate and much remains to be done to improve organic matrix composite materials. The editor hopes that this book will help to enlarge on and clarify some of the issues raised briefly by this preface.

G. PRITCHARD

CONTENTS

Preface v

List of Contributors ix

1. Thermoplastic Carbon Fibre Composites 1
   P. E. MᶜMAHON

2. Applications of Fourier Transform Infrared Spectroscopy to the
   Study of Fibre–Resin Composites 31
   Y. T. LIAO and J. L. KOENIG

3. Thermography Applied to Reinforced Plastics . . . 89
   K. L. REIFSNIDER and E. G. HENNEKE II

4. Compressive Properties of Resins and Composites . . 131
   M. R. PIGGOTT

5. Novel Curing Agents for Epoxy Resins 165
   D. A. SCOLA

Index 267

LIST OF CONTRIBUTORS

E. G. HENNEKE II

Materials Response Group, Virginia Polytechnic Institute and State University, Blacksburg, Virginia 24061, USA

J. L. KOENIG

Department of Macromolecular Science, Case Western Reserve University, Case Institute of Technology, Cleveland, Ohio 44106, USA

Y. T. LIAO

Department of Macromolecular Science, Case Western Reserve University, Case Institute of Technology, Cleveland, Ohio 44106, USA

P. E. MCMAHON

Celanese Celion Carbon Fiber Division, Celanese Corporation, 26 Main Street, Chatham, New Jersey 07928, USA

M. R. PIGGOTT

Department of Chemical Engineering and Applied Chemistry, University of Toronto, Toronto, Ontario M5S 1A4, Canada

K. L. Reifsnider

Materials Response Group, Virginia Polytechnic Institute and State University, Blacksburg, Virginia 24061, USA

D. A. Scola

United Technologies Research Center, East Hartford, Connecticut 06108, USA

Chapter 1

THERMOPLASTIC CARBON FIBRE COMPOSITES

PAUL E. MCMAHON

Celanese Celion Carbon Fiber Division,
Chatham, New Jersey, USA

SUMMARY

The evolution of carbon fibre/thermoplastic (CF/TP) composites is being facilitated by the previous development of similar materials wherein glass fibres provided the reinforcement. CF/TP moulding compounds wherein many TP's may serve as the matrix are shown to be a rapidly developing commercial material for use in applications which require electrical grounding, improved wear, and/or dimensional stability characteristics.

The status of development of continuous CF reinforced thermoplastic preforms and potential composite applications is also reviewed. Significantly increased damage tolerance by way of improved toughness of the matrix (in comparison to thermosets) is the principal driver for development of these materials. Shorter fabrication cycles, reduced scrap and infinite shelf-life at ambient conditions are also providing impetus. With the advent of advanced TP's such as PEEK (polyether etherketone), PPS and polyamide-imides, their improved high temperature performance and solvent resistance are enabling serious development for aircraft/aerospace applications. With the development of suitable product forms and fabrication technology, cost-effective applications of CF/TPs seem certain to evolve.

1.1. INTRODUCTION

Any treatment of carbon-fibre-reinforced thermoplastics should rightfully begin with recognition of the extensive materials and manufacturing

1

experience available for glass-reinforced systems at the onset of development of high performance structural composites as metal replacements. The experience gathered in working with the fibreglass systems has been invaluable in getting carbon-fibre-reinforced plastic (CFRP) off to a fast start. Although available, glass-fibre-reinforced plastics (GFRP) lacked the stiffness required for use in high performance structures, but the technology acquired with these materials was readily transferred to carbon fibres as they were developed. Thus, light-weight composite replacements for metals have evolved based on the superior characteristics of these carbon fibres (CF), e.g. high modulus and low density.

These structural materials first moved into aircraft/aerospace components via utilisation of high performance thermoset epoxy matrix materials and the accompanying fabrication techniques developed for glassfibre. The well-known advantages of thermoplastics did not escape notice; for example,

—unlimited shelf-life
—low scrap/recyclability
—rapid fabrication cycle
—toughness/impact resistance.

However, the high cost of presses and moulds used in thermoplastic processing, combined with the relatively small quantities of each of a very large number of production items in aircraft structures, kept the industry focused on the much more flexible autoclave moulding techniques during early development.

Increasing interest in and demand for carbon fibres in aerospace applications, coupled with the growth of a secondary market in sporting goods (golf shafts, fishing rods), began to exert downward pressure on prices. When pricing and availability reached feasible levels in the early 1970s, then CF began to be used selectively to replace glass reinforcement in thermoplastic injection moulding compounds. This replacement occurred in end-use applications which could effectively utilise the significant increases in strength and stiffness at reduced density which CF provides. The properties of short-CF-reinforced nylon 6,6 are compared in Table 1.1 with those of typical glass-filled and unfilled injection moulded materials.

In this same period, aerospace companies began to consider and evaluate CF reinforced thermoplastics (CF/TPs) as CF/epoxy matrix composite replacements for selected components. Agency-funded efforts at Boeing[1]

TABLE 1.1
PROPERTY COMPARISON OF FILLED VERSUS UNFILLED NYLON 6,6
(INJECTION MOULDED LAMINATES)

Property (unit)	40% PAN carbon fibre	40% Glass fibre	Unfilled resin
Specific gravity (kg/m^3)	1 340	1 470	1 140
Flex modulus (GPa)	23·4	11·0	2·9
Flex strength (MPa)	413	331	117
Tensile strength (MPa)	276	234	83
Elongation (%)	3–4	2–3	40–80
Shear strength (MPa)	96	90	65
Fatigue endurance stress @ 10^7 cycles (MPa)	58	48	36
Notched Izod impact (J/m)	86	139	54
Coefficient of linear thermal expansion (m/m/°C × 10^{-6})	14·4	25·2	90
Heat distortion temperature (°C @ 1·82 kPa)	260	253	77
Thermal conductivity (W/m K)	1·2	0·22	—
Dynamic coefficient of friction	0·18	0·23	0·28
Volume resistivity (Ω-cm)	10^2	10^{14}	10^{14}

and General Dynamics/Convair[2] showed considerable promise in CF/TP composites using polysulphone as the matrix. (Typical mechanical properties are shown in Table 1.2.) The anticipated improvements which motivated the programmes were toughness, reduced scrap, etc., as cited above. The polysulphones were easily incorporated as the matrix in prepregs or preforms via solvent impregnation processes. Unfortunately, it was the solubilities of the polymer in certain solvents which led to the downfall of these materials and significantly delayed the appearance of CF/TP products. Parts made with CF/polysulphones are susceptible to solvent attack by cleaners and lubricants such as methylene chloride and aircraft hydraulic fluids.

When the poor fracture resistance and damage tolerance of all high performance thermoset matrix/CF composites was generally recognised in the late 1970s, the search for tougher matrix systems ensued. Unfortunately, the known routes to improve the toughness of thermosets all introduce one or more attendant weaknesses such as reduced performance temperature range and/or lower compression properties induced by lower matrix modulus. Hence, a resurgence of interest in thermoplastics as significantly tougher composite matrices occurred.

This combination of the need for greatly improved toughness, coupled with the many other advantages of thermoplastics, has resulted in a flurry

TABLE 1.2
P-1700/GRAPHITE FABRIC LAMINATE PROPERTIES

Property (unit)	Test temperature		
	21°C	80°C	150°C
Tension			
strength (MPa)	530	—	423
modulus (GPa)	68·9	—	71·7
Compression			
strength (MPa)	386	—	248
modulus (GPa)	63·4	—	73·0
Flexural (0°)			
strength (MPa)	737	634	475
modulus (GPa)	53·0	48·9	49·6
Flexural (90°)			
strength (MPa)	668	551	455
modulus (GPa)	53·0	46·9	46·2
Interlaminar shear strength (MPa)	60	—	30

of activity in the development of high performance CF/TP composites. Initial efforts have largely been expended on the development of fabrication techniques for small articles and the demonstration of acceptable properties. Wetting and/or adhesion have been identified as major obstacles to the desired performance levels. Following a brief section on short-fibre-reinforced TP moulding compounds, the major part of this chapter will discuss the development and performance of continuous CF/TP systems for aerospace applications.

1.2. SHORT FIBRE REINFORCEMENT

1.2.1. Development

A substantial technology in short-fibre-reinforced thermoplastic moulding compounds existed prior to the development of carbon fibre as a structural filler. However, little attention was initially given to CFs when the fibres were first introduced at prices of hundreds of pounds (sterling) per kilogram. With growth of both the technology and the supply of CF along with accompanying order-of-magnitude price decreases, the obvious attractions of these new reinforcements were evaluated in moulding compounds. The most pronounced advantages were:

—improved stiffness at lighter weight
—electrically conductive for grounding
—enhanced wear resistance
—dimensional stability.

CF-containing injection moulding compounds were thus developed as a high performance, speciality adjunct to the product lines of TP moulding compound suppliers.

These inherent CF attributes have been quantified. It has been demonstrated[3] that 40% by weight PAN-based† CF reinforcement in nylon 6,6 provides 110% increase in stiffness over 40% glass fibre reinforcement, Table 1.1. In addition, electrical conductivity is increased by 12 orders-of-magnitude; thus electrical grounding and EMI shielding of electronic housings, which are impossible with glass reinforcement, are now achievable. The coefficient of thermal expansion is lowered by more than 40%, which greatly enhances dimensional stability. Finally, friction is lowered by 20%, which means significant reductions in wear, and thereby improved performance in bearing applications. These improvements are verified by the data available from a variety of moulding compounders who offer CF reinforcement in a wide range of thermoplastic materials.[4] These materials include a variety of nylons, polyesters, polysulphones, polycarbonates, acetals, polyphenyl sulphide, and fluorinated polymers. The chemical inertness of CFs has also made it possible to develop CF/TP composites as filters in chemically aggressive environments where the TP matrix materials are poly-vinylidene fluoride or similarly inert polymers.[5]

1.2.2. Applications
The demonstrated performance advantages combine to provide multiple applications for CF/TP moulding compounds. Examples of these applications are shown photographically (courtesy of LNP Corporation) in Figs. 1.1(a)–(c). A paper feed sprocket shown in Fig. 1a utilises the electrical conductivity of the CF reinforcement to drain static charges and possesses enhanced mechanical strength and stiffness for lengthy trouble-free performance. The bushing of Fig. 1.1(b) capitalises on the excellent tribological characteristics of CF moulding compounds to reduce friction and wear in a bearing component. Figure 1.1(c) illustrates a typical

† Carbon fibres produced from polyacrylonitrile (PAN) precursors as opposed to those made from rayon, pitch or other starting materials.

FIG. 1.1a.　Paper feed sprocket (Thermocomp® DC-1004 EG-polycarbonate/
carbon fibre, electrical grade).

FIG. 1.1b.　Nose bushings (Thermocomp® RC-1004 EG-nylon 6,6/carbon fibre,
electrical grade).

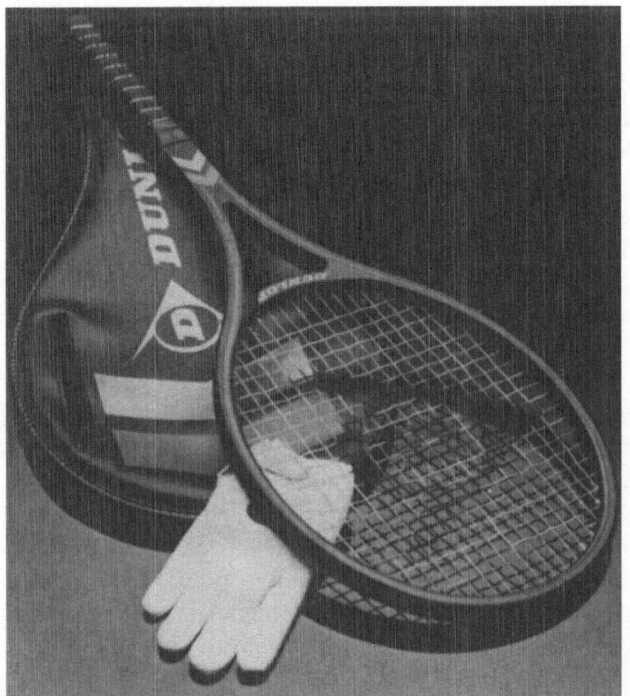

FIG. 1.1c.   Dunlop tennis racquet (Thermocomp® RC-1006-nylon 6,6 and carbon
fibre). All photographs courtesy of LNP Corporation.

sporting goods application; tennis rackets benefit from CF stiffness in
facilitating designs with larger, more uniform sweet spots.

1.3.   CONTINUOUS FIBRE REINFORCEMENT

1.3.1. Solvent Impregnation
1.3.1.1. Polymer Systems
It is of vital importance that the impregnation process used to combine
reinforcing fibre with resin matrix is conducted in such a way that each fibre
in each yarn bundle is completely wetted by the matrix. Most thermoplastic
polymers with meaningful molecular weights have melt viscosities so high
that it is extremely difficult to wet and impregnate fibres with molten
polymer. Although melt viscosities continue to decline as the melt

temperature is raised it is virtually impossible to reach a sufficiently low viscosity to allow good fibre wetting. Hydrostatic forces applied to the fibre–matrix assemblies with the aim of forcing penetration merely force the fibres within a bundle closer together, and essentially close the interstices. Thus, initial development of CF/TP prepregs focused on solution processable polymeric species. The adaptability of available coating technology and equipment used in low viscosity thermoset prepreg manufacture also provided substantial incentive to pursue solvent impregnation.

The first solvent impregnated TP prepreg system to receive serious attention was CF/P1700 polysulphone. This was followed by polyphenyl-sulphone and polyethersulphone in attempts to reduce the extreme sensitivity of the polysulphone itself to aircraft hydraulic fluids; alkyl phosphates contained in these fluids have been blamed for the solvation effects exhibited. In addition to the polysulphones both polycarbonates and chain-extendible or crosslinkable, low molecular weight, soluble polyimide-based systems have been considered. Polycarbonates received little attention since the primary interest has been in aircraft/aerospace applications and these polymers have upper use temperatures which are significantly lower than the sulphones, thus giving them limited utility.

The low molecular weight reactive systems which can undergo further polymerisation are typified by Boeing's NTS,[6] which is crosslinkable, and Amoco's poly(amide-imide),[7] which is chain-extendible; both materials are activated with thermal post-treatments. These types of low molecular weight reactive systems are still under serious scrutiny in development and evaluation programmes.[8]

The soluble polymers which have received attention, and their characteristics, are as follows.

—Polysulphone—Attacked by aircraft hydraulic fluids and paint strippers.
—Polyphenylsulphone/polyethersulphone—Resistant to hydraulic fluid, but still sensitive to paint strippers and polar solvents.
—Poly(amide-imide)—High boiling solvent required in prepreg oper-ations, chain-extending thermal post-cure required.
—Crosslinkable sulphones—Some solvent swelling, requires thermal crosslinking step.

1.3.1.2. Typical Performance
The earliest polysulphone evaluations were carried out by Maximovich[9] plus May and Goad[2] at GD/Convair and by Hoggatt and Von Volki at

TABLE 1.3

COMPOSITE MECHANICAL PROPERTIES OF RADEL POLYPHENYLSULPHONE/8H SATIN OF CLOTH[a]

Property	Condition	Strength (MPa)/Modulus (GPa)		
		Room temperature	121°C	177°C
Flex	Dry	546/28·7	—	320/21·4
(warp direction)	Wet	422/24·1		313/25·1
	Heat aged	573/32·7		452/33·1
Short beam	Dry	62	55	46
shear strength	Wet	59	54	37
In-plane shear (τ_{12}/G_{12})	Dry	63/3·3		39/3·5
±45° Tensile	Dry	125/11		79/12
Tensile	Dry	248/50		269/51
(warp direction)	Wet	300/37		274/32

[a] Data from ref. 10, courtesy of SAMPE.

Boeing.[1] These early programmes evaluated both tape and fabric prepregs coated from solution in either methylene chloride or N-methyl pyrolidone(NMP). Attractive mechanical property performance was demonstrated at both facilities; Table 1.2 illustrates the Boeing results. It must be remembered that the fibre used at this time (c. 1973) was significantly lower in strength than comparable fibres available at the time of this review (ten years later). Tension, compression and shear properties are attractive, as are their retentions at 150°C. May at GD/Convair demonstrated manufacturing concepts on the F-16 strake using a ceramic tool equipped with both heating and rapid cooling capabilities. Unfortunately, as noted in the preceding section, extreme sensitivity to hydraulic fluids and paint stripping solvents has ruled out polysulphone use in aircraft applications.

In 1978 Husman and Hartness[10] evaluated Radel™ polyphenylsulphone as a matrix for CF composites. They demonstrated good tension, flex and shear properties both at ambient temperature and at 177°C even after wet conditioning; however, solvent impregnation was much less successful than film transfer and only the latter laminate results are shown here (Table 1.3). Even though these polymers are resistant to hydraulic fluids, their

TABLE 1.4
CHEMICAL RESISTANCE OF CARBON FIBRE/POLYPHENYLSULPHONE COMPOSITES[a]

Agent	Duration				
	1 min	10 min	1 h	5 h	24 h
1. Hydraulic fluid—5606-C (Air Force usage)	ND	ND	ND	ND	ND
2. Hydraulic fluid—Skydrol 500B (Commercial usage)	ND	ND	ND	ND	ND
3. Aircraft fuel—JP-4	ND	ND	ND	ND	ND
4. Methylethyl ketone	ND	ND	Slight surface attack	Surface attack on resin down to first layer	Surface attack on resin down to first fibre layer
5. Paint stripper—Turco-5351	Slight surface attack	Increased surface attack	Progressive surface attack on resin down to first fibre layer		

[a] ND, no visible effect. Data from ref. 10, courtesy of SAMPE.

TABLE 1.5
PROPERTIES OF PES/CARBON FIBRE AT ELEVATED TEMPERATURE $(60\% \text{ VF})^a$

Property (unit)	Unidirectional carbon fibre (XAS)		Carbon fibre (T300) 5 shaft fabric	
	23°C	170°C	23°C	170°C
Flexural modulus (GPa)	94	86	64	61
Flexural strength (MPa)	1 551	975	868	577
Tensile modulus (MPa)	153	153	94	94
Tensile strength (MPa)	1 232	779	657	469
Interlaminar shear strength (MPa)	79	—	62	—

^{a} Data from ref. 11, courtesy of SAMPE.

sensitivity to solvents and paint strippers unfortunately remains (see Table 1.4). In another modification Rigby[11] has reported CF composite data for both tape and fabric with ICI 200P polyethersulphone matrix (Table 1.5). Good ambient and 170°C tensile, flex and shear properties are shown for both systems. Although general improvement in solvent resistance is claimed, no quantification is given, and sensitivity to certain solvents is admitted.

In a completely separate approach the benefits of thermoplastics in toughening CF composites are being pursued via the development of thermosetting–thermoplastic matrix systems. In this generic approach low molecular weight TP oligomers are either end-capped with thermosetting, crosslinking entities or contain reactive, chain extending end-groups. In pursuit of the first technique Boeing[6] and NASA[8] are developing crosslinkable polysulphones wherein the oligomeric sulphone moiety is of sufficient length to be able to impart TP characteristics, but not so long as to maintain solvent sensitivity after the crosslinking mechanisms have been activated by thermal cure.

Exemplative of the second method is Amoco's poly(amide-imide) (PAI) resin. This system combines soluble amide oligomers with reactive end-groups, in such a fashion as to enable solvent impregnation and lamination followed by a chain-extending post-cure. A difficulty of working with this system is the need for highly polar, low volatility solvents such as NMP. However, PAI has a pronounced elevated temperature advantage. As seen in Table 1.6(a) flexural strength drops by only 20% at 204°C. Table 1.6(b) reports data confirming excellent flex strength, and also shows well balanced tension and compression properties. Excellent

TABLE 1.6

MECHANICAL PROPERTIES OF CF/AMOCO POLY(AMIDE-IMIDE) COMPOSITES[a]

(a) *T300, 8HS fabric 0°/90° laminate*

Property (unit)	RT	149°C	204°C	260°C
Flexural				
strength (MPa)	1 000	920	790	410
modulus (GPa)	63	63	62	43
Short beam shear				
strength (MPa)	75	71	63	30

(b) *Celion 3000, 4HS fabric, 0°/90° laminate*

Property (unit)	Ambient temperature	93°C/2 week soak in	
		H₂O	Skydrol
Tensile			
strength (MPa)	680	—	—
modulus (GPa)	62	—	—
Flexural			
strength (MPa)	1 010	—	—
modulus (GPa)	59	—	—
Compression			
strength (MPa)	560	433	529
modulus (GPa)	63	—	—
Short beam shear			
strength (MPa)	100	—	—

[a] Data courtesy of Amoco Chemicals Corporation.

environmental resistance is demonstrated in the 93°C compression strength after water and Skydrol exposures.

1.3.2. Melt Impregnation

1.3.2.1. Approaches

Upon recognition that the same solvent solubility which enables ready TP impregnation of carbon fibres is also indicative of solvent sensitivity and performance problems in end-use environments, the interest in TP melt impregnation increased significantly. The formidable task of complete wetting of each individual fibre by the highly viscous matrix systems has been approached in several ways. Among these are:

(1) pre-coating with a soluble TP system;

(2)   use of sophisticated dies and metering devices;
(3)   use of wire coating dies; and
(4)   use of continuous laminating machines such as the Sandvik[12] press.

The approach of pre-coating the carbon fibre with a soluble thermoplastic system prior to introduction of the insoluble thermoplastic matrix material is the subject of a British patent.[13] This procedure has been used to prepare CF/TP composite laminates using PES[11] and TP materials in general.[14] Acceptable laminates with good properties have been claimed.

The general use of a high pressure cross-head die connected to an extruder has been patented by Hercules[15] as a process for making thermoplastic prepreg tapes and rovings by melt coating. Pultrusion through a forming orifice following passage through a TP melt reservoir was described by McMahon and Maximovich.[16] Further, moving a step closer to manufacture, Sandvik has developed a continuous belt, heated laminating press on which they have prepared continuous CF reinforced thermoplastic sheets.[12] While these methods typify the variety of ways in which melt coating fibres with a TP can be addressed, it is by no means an exhaustive list and there are certain to be proprietary alternatives either in development or actual use.

1.3.2.2. Polymer Systems

In general, any melt processable thermoplastic polymer is a suitable matrix candidate for CF reinforced composites. The early selection of model materials for developing was impacted by two simultaneous considerations:

(1)   ease of processing, i.e., temperature, pressure and thermal stability characteristics, and
(2)   performance as a matrix, i.e., temperature capability, toughness, environmental stability, etc.

In order to combat the solvent sensitivity encountered with the polysulphones the desirability of crystalline polymers was recognised. The large regions of ordered crystalline structure have significantly lower free volume than is present in amorphous polymers; this greatly restricts solvent penetration and its undesirable effects.

The desired performance characteristics chosen at the onset of TP composite development were selected to meet automotive and/or aerospace

TABLE 1.7
THERMOPLASTIC MATRICES AND THEIR TYPICAL CHARACTERISTICS[a]

Polymer	HDT (°C @ 1·82 KPa)	Flex		Notched Izod (J/m)	Crystalline or amorphous
		Strength (MPa)	Modulus (GPa)		
Nylon 6,6 (dry)	75	110	2·9	54	Semi-crystalline
PBT (Poly butylene terephthalate)	65	99	2·5	48	Semi-crystalline
Polysulphone	174	106	2·7	64	Amorphous
PPS (Poly phenylene sulphide)	135–260	96	3·8	54	Semicrystalline
PES (Poly ethersulphone)	203	128	2·6	86	Amorphous
Poly phenylsulphone	204	85	2·3	64	Amorphous
PEEK (Polyether etherketone)	159	91[b]	3·9	86	Semi-crystalline
PAI (Poly(amide-imide))	274	207	4·5	134	Amorphous

[a] Data extracted from *Modern Plastics Encyclopedia*, courtesy of McGraw-Hill Publishing Co.
[b] Tensile yield strength.

operational environments. Familiarity of the auto industry with engineering thermoplastics combined with their extreme interest in light-weighting vehicles to meet legislated fuel economy requirements in the USA influenced material suppliers to focus on commodity nylons and polyesters as the priority matrix candidates. However, following a two-year flurry of activity in the development and evaluation of CF composites for automotive components it became clear that the majority of applications under consideration were not cost-effective.

This realisation, combined with the development of new higher performance thermoplastics such as wholly aromatic polyesters and polyketones, as well as present availability of PPS, polyimides and other systems, resulted in a shift of attention to aerospace applications. Promising characterisations of these higher temperature thermoplastic candidates, coupled with the acknowledgement that significantly improved toughness was needed, led to a ground-swell of interest in CF/TP composites for aircraft components.

The thermoplastic polymers which have received attention, together with comments on the nature of the interest, are shown in Table 1.7.

1.3.2.3. Typical Performance
The first substantial quantities of performance data generated on composites made with melt impregnated tapes and rovings were reported by McMahon and Maxomivich.[16] Two matrix systems were evaluated: nylon 6,6 and polybutylene terephthalate (PBT); however, only results on the former will be discussed here, since carbon fibre/polyester composites are treated in a separate chapter of this text. Prior to the preparation of laminates for extensive property characterisation, the investigators evaluated a series of finishes on the carbon fibres. The objective was selection of a finish which would provide optimum composite performance. The sensitivity of flex and shear properties to the finishes screened was not very pronounced; nevertheless, an optimum was selected, based upon shear strength levels; a maximum shear strength of 94 MPa was achieved.

The Celion carbon fibre/nylon 6,6 composites evaluated by McMahon and Maximovich were prepared using a multi-step procedure. In the first step, a poltrusion (wire-coating) approach was used to coat Celion 6000 filament rovings with molten nylon 6,6; a dry N_2 blanket prevented oxidation and moisture regain by the hydrophilic polymer. Following the coating operation, the impregnated yarns were wound onto a large drum, and consolidated sheets were formed by hot rolling. The sheets thus produced contained 55 weight % nylon. These sheets were then stacked with the desired ply orientations (unidirectional or quasi-isotropic) and pre-consolidated using 0·34 MPa pressure at 263 °C for 5 min to force the excess resin into 181 glass bleeder cloths placed on the faces. Following the consolidation step, the bleeder plies were removed, and final lamination at 277 °C and 0·34 MPa was conducted. After a 5 min dwell the laminates were cooled under pressure to below 149 °C.

After the finish and process development steps were completed, sufficient unidirectional and quasi-isotropic laminates were prepared to enable extensive tensile flex and shear properties to be obtained. In addition to ambient, hot, and hot–wet properties, the effects of exposure to heated antifreeze and brake fluids were evaluated. The performance of the Celion/nylon 6,6 composites is surprisingly good, as shown in Tables 1.8–1.11. Room temperature (RT) tensile and flexural strengths in excess of 1200 MPa were achieved. The effects of moisture were as expected; RT flex and shear strengths are lowered by 30–40 %. This strength reduction caused by moisture is very similar to that brought about by elevated temperature (82 °C) on nylon; in both cases the polymer is above the glass transition and matrix modulus is significantly lowered. Strengths at ambient temperature after exposure to hot brake fluid and antifreeze are also reduced by

TABLE 1.8

FLEX AND SHEAR STRENGTHS (IN MPa) OF UNI-DIRECTIONAL CELION/NYLON COMPOSITES[a]

	Flexural strength		Short beam shear strength	
	(22°C)	(82°C)	(22°C)	(82°C)
Control	1 230	690	94·4	59·9
Moisture aged				
(71°C/95% RH)	680	520	68·9	52·4
Heat aged in air				
(24 h at 82°C)	1 190	720	93·7	73·0
Antifreeze				
(24 h at 110°C)	670		69·6	
Brake fluid				
(24 h at 140°C)	760		56·5	

[a] Data from ref. 16, courtesy of Pergamon Press, Inc.

30–40%. Finally, McMahon and Maximovich report flex strength and modulus of 1570 MPa and 127 GPa respectively for Celion/nylon 6,6 laminates made from thermoplastic prepreg tapes rather than the earlier wire-coated yarns.

In the same period of time Husman and Hartness[10] were experiencing difficulty in the preparation of polyphenylsulphone/CF laminates using

TABLE 1.9

TENSILE STRENGTHS (IN MPa) OF UNI-DIRECTIONAL CELION/NYLON COMPOSITES[a]

	Test temperature (°C)		
	22	82	121
Control	1 220	940	
Moisture aged			
(71°C/95% RH)	810	800	
Heat aged in air			
(24 h at 82°C)	1 140	1 060	
(24 h at 121°C)	1 180		1 000
Antifreeze			
(24 h at 110°C)	1 070		
Brake fluid			
(24 h at 140°C)	1 280		

[a] Data from Ref. 16, courtesy of Pergamon Press, Inc.

TABLE 1.10

FLEX AND SHEAR STRENGTHS (IN MPa) OF QUASI-ISOTROPIC CELION/NYLON COMPOSITES[a]

	Flexural strength		Short beam shear strength	
	(22°C)	(82°C)	(22°C)	(82°C)
Control	700	450	65·5	48·9
Moisture aged (71°C/95% RH)	500	390	53·1	41·3
Heat aged in air (24 h at 82°C)	720	500	64·1	56·5
Antifreeze (24 h at 110°C)	500		53·7	
Brake fluid (24 h at 140°C)	440		29·6	

[a] Data from ref. 16, courtesy of Pergamon Press, Inc.

solution coated fibre. They reported significantly improved laminates with attractive mechanical performance when polymer in film form was interleaved with layers of carbon fibre fabric. A distinct improvement in mouldability was observed when T-300 sized with 307 finish was used in contrast to behaviour observed with either T-300 with 309 finish or standard epoxy-sized Celion fibre. The data obtained are summarised in

TABLE 1.11

TENSILE STRENGTHS (IN MPa) OF QUASI-ISOTROPIC CELION/NYLON COMPOSITES[a]

	Test temperature (°C)		
	22	82	121
Control	400	289	
Moisture aged (71°C/95% RH)	340	310	
Heat aged in air			
(24 h at 82°C)	430	340	
(24 h at 121°C)	430		370
Antifreeze (24 h at 110°C)	350		
Brake fluid (24 h at 140°C)	440		

[a] Data from ref. 16, courtesy of Pergamon Press, Inc.

Table 1.3. The difference in fibre performance was attributed to improved wetting with 307 type finish present at the fibre-matrix interface.

Following the improved sulphone composite development effort, Hartness[17] developed and characterised continuous carbon-fibre-reinforced PPS composites. The attractiveness of PPS as a matrix lies in its potentially high temperature capability and exceptional chemical resistance. Composite processing was again carried out using the technique of interleaving thermoplastic film with carbon fibre cloth. The PPS film used was 0·03 mm thick Ryton supplied by the Phillips Petroleum Company. A repeat of the polyphenylsulphone success was obtained when the carbon fibre used contained the UC 307 finish, thus further substantiating the significance of the wetting and the interface on achievable laminate quality and performance. The fabrication process incorporated a 1 h dwell at 316 °C and 1380 KPa applied pressure. Composites contained approximately 50 % fibre by volume.

Typical mechanical performance of CF/PPS laminates evaluated in this study are shown in Table 1.12. The fabric used in the laminates was a 90 % uni-woven material, i.e., 90 % of the fibre was T-300 and was in the warp direction while the remaining 10 % of fibre present was T-400 in the filling direction (presumably with a different finish from 307). Reasonable property retention (greater than 50 %) was achieved at 121 °C, both dry and wet; however, serious decreases continue as the temperature is raised, with only 40–50 % retention at 177 °C. This is not surprising in the light of the reported T_g at 104 °C. The magnitude of the decrease in performance of the moisture-saturated PPS composites is unexpected; a drop of approximately

TABLE 1.12

COMPOSITE MECHANICAL PROPERTIES OF RYTON (PPS)/CF (90 % UNIDIRECTIONAL CLOTH)[a]

Property	Condition	Strength (MPa)/Modulus (GPa)		
		Ambient temperature	121 °C	177 °C
Flex 3 point, 0 ° direction	Dry	1 157/84	686/85	448/73
	Wet[b]	751/85	516/74	338/65
Flex 4 point, 90 ° direction	Dry	195/11	125/9·0	103/8.4
	Wet[b]	212/14·2	174/8·6	143/5·2

[a] Data from ref. 17, courtesy of SAMPE.
[b] Under water at 71 °C/Equilibrium Weight Gain = 0·6 %.

35% in wet RT flex strength was recorded with an additional drop of 25% at 121 °C wet. The flex failures are reported to shift from tension to compression mode in passing from ambient temperature to 121 °C. This moisture sensitivity is unusual in a chemically inert matrix such as PPS and further investigations will be required to provide an understanding.

Hill *et al.*[18] evaluated several thermoplastic candidates as matrix materials for continuous CF composites in the search for matrix polymers with improved environmental resistance compared to P1700 polysulphone. Candidates screened in CF film composites were Radel polyphenylsulphone, PKXAs from Union Carbide Corporation, and fluorinated polymers from Allied Corporation, CM-1 and KM-1. Mechanical properties of modified sulfones are equivalent to those of P1700, while the performance of PKXAs is lower. The Allied CM-1 and KM-1 appear to be inert to all the solvents tested, and give indications of equivalent mechanical performance to P1700, but on very limited data. Further evaluation was recommended by the investigators; however, the status of such continuing efforts, and the availability of these developmental materials are both unknown.

Many of the incentives for the development of thermoplastic matrix composites have been met, e.g. shorter fabrication times, reduced scrap, increased toughness, unlimited shelf-life. However, one property shortcoming which is common to most thermoplastics and to those discussed thus far is a performance-limiting glass transition temperature, T_g; that is, the brittle-to-ductile transition temperature of the polymers is either below or too close to the upper end-use operating temperature. This ductility translates into a matrix modulus which is too low to prevent the onset of fibre buckling at low loads. This fibre buckling at low compression loads sets the mode for and initiates compression failure.

But, significant progress in the development of high performance thermoplastic polymer systems has occurred since 1980. Two such systems are ICI's PEEK[11,19] (polyether etherketone) and Celanese's LCP (liquid crystalline polymers).[20] Since the latter materials are predominantly polyester in nature they are not discussed in this chapter. The status and performance information on carbon fibre/PEEK composites follows. Outstanding features of PEEK are its high T_g and modulus, 143 °C and 4·0 GPa, respectively. These values are typical of thermoset epoxy systems. Such characteristics, combined with the additional positive attributes discussed in preceding sections, have resulted in a high level of interest in this polymer by the aerospace community.

ICI has released extensive data on the PEEK polymer in their product brochures.[21] Some of the more noteworthy properties and their implications are given below.

High melting temperature
—High T_g and excellent elevated temperature performance
—Good creep resistance
—Difficult to process, temperatures up to 400 °C required
Semi-crystalline polymer
—Excellent solvent resistance
—Requires care to obtain stable crystallinity level and structure
Low water absorption
—Good hot/wet performance
—Resistant to thermal spikes
High resin modulus
—Good compression strength
—Good interfibre lead transfer
Good impact resistance
—tough, damage-tolerant composites

The translation of these PEEK characteristics into a very promising matrix system for continuous carbon fibre composites was broadly reported in 1982. Rigby[11] reported excellent flexural and shear strengths at ambient temperature for quasi-isotropic laminates; they are 300 and 80 MPa, respectively. In a more comprehensive evaluation Hartness[22] showed ambient flex strengths on PEEK/CF fabric composites equivalent to epoxy matrix strengths at 80–85 % of the room temperature results. Data given also confirm the very low moisture absorption levels reported by ICI. In the testing of ±45° fabric laminates in tension he demonstrated shear strengths fully equivalent to those obtained with epoxy resin systems, i.e. 91 MPa. Fibre loadings in the test laminates were 57 % by volume.

Here again Hartness used T-300 fibre with the UC 307 'thermoplastic' finish. He did note, however, that equivalent results are achievable when the finish is removed from the carbon fibre cloth prior to impregnation with TP resin. The method of impregnation and fabrication was the interleaving of CF cloth and PEEK film described previously for polyphenylsulphone and PPS.

One-week exposure to critical solvents which occur in the aircraft operating environment, e.g. hydraulic fluids and paint strippers, was shown to have little effect on PEEK composites at ambient temperature. Similar tests on both the initial polysulphones and their improved derivatives showed rupture in many instances after exposures of only a few minutes

TABLE 1.13

FLEXURAL STIFFNESS AND STRENGTH OF APC-1 PEEK/CF COMPOSITES AT 23°C[a]

Lay up	Test direction	Modulus (GN/m^2)	Strength (MN/m^2)	Strain at failure (%)
Uniaxial	Along fibre axis	120	1 670	1·4
Uniaxial	Transverse to fibre axis	10	110	1·4[b]
Quasi-isotropic	In plane	47	700	1·4

[a] Data courtesy of ICI Petrochemicals and Plastics Division.
[b] Significant yielding.

duration. This confirms PEEK's claimed excellent solvent resistance. Finally, excellent toughness is indicated in the G_{IC} measurements, which show an order-of-magnitude improvement over values obtained with state-of-the-art epoxy matrix composites.

In a series of presentations and product bulletins[21,27] issued by ICI on PEEK/carbon fibre composites, the excellent properties alluded to have been further expanded and substantiated. Uniaxial and quasi-isotropic flex properties (see Table 1.13) are at the same levels seen with the best known epoxy matrices in composites containing equivalent CFs. Further, an outstanding short beam shear strength of 106 MPa is reported. The addition of damage tolerance test results which demonstrate excellent compression strain retention after impact is illustrated in ref. 27.

The above data demonstrate the achievable performance level with ICI's APC-1 Victrex PEEK/CF composites. (APC is an acronym for Aromatic Polymer Composites.) The tapes which are available contain 54% fibre by volume; hence, reported mechanical properties compare very favourably with epoxy composite results which are usually based on 60% fibre. In support of the importance placed on the interface by Hartness, ICI also believes that the interface is critical to attainment of good wetting in impregnation, and hence, to final laminates of high quality where all fibres are wetted.[23] Figure 1.2 shows a high magnification photograph of a PEEK/CF section where fibres are at 90° and 45° to the plane of the photo. The uniform dispersion of the fibres and excellent wetting as evidenced by the lack of interfacial voids can be readily observed.

1.3.3. Potential Applications

There are no current applications where continuous (or long fibre) CF/TP composites are in service. Materials are now in an intermediate state of

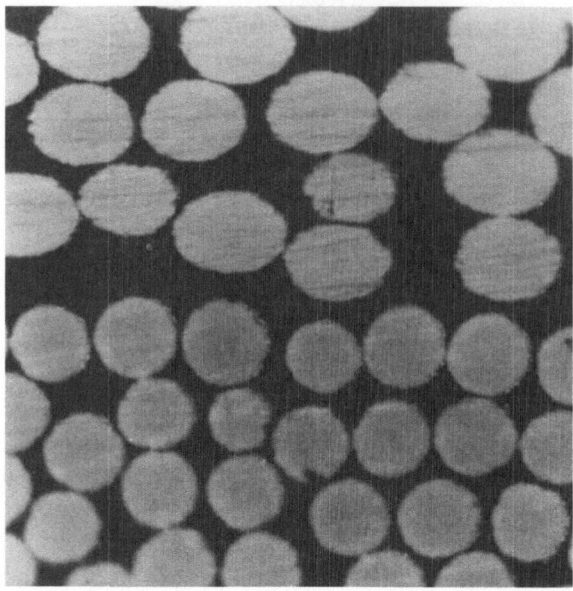

FIG. 1.2.   Photomicrograph of CF/PEEK composite section between 45° and 90° plies. Good uniformity and wetting is illustrated. (Courtesy of ICI Ltd.)

development for aircraft/aerospace applications. The high price of the materials and capital costs involved in setting up production equipment are forcing initial end-use applications to be carefully selected. Most of those under consideration are in commercial aircraft where increased toughness is essential as a counterbalance to the high liability exposure of commercial carriers. The types of components being prototyped are secondary structures such as ailerons, flaps, and access doors, along with components for space structures and orthopaedic applications.

This intense and expanding aircraft/aerospace interest in TP/CF composites is generating a variety of product forms and experience in rendering these into useful shapes typified by aircraft components. This experience will undoubtedly find its way into sporting goods and commercial product areas; the first products will most likely occur in these less-critical application areas.

Materials, fabrication approaches, and properties are being demonstrated. However, the materials and processes explored thus far are not commercial in nature. Product forms and fabrication techniques for ready conversion into parts must be developed. These will no doubt bring together elements of pultrusion, tape wrapping, and/or filament winding techniques in combination with material forms and equipment for carrying out these operations with pre-assemblies and/or preforms. Likely first generation products will be: pultruded reinforcement sections, hats, U-shapes, etc., which can be mechanically attached or thermally welded to other structures; tubular or box beam structures which can be tape or filament wound using heated transfer and placement techniques; stamped or thermoformed simple components with minimum radius of curvature and size such that they can be readily formed from stacked prepregged fabrics and sheets (ultrasonically or thermally spot-welded). Intermediate PEEK product forms and accompanying processing technology for impregnated tape lay-up and filament winding are under development.[21]

1.4. GENERAL ISSUES

1.4.1. Finishes and Fibre Wetting

Perhaps the most significant issue concerning the processing and performance of continuous carbon fibre reinforced thermoplastics is an understanding of the role which fibre finishes (e.g., add-on coatings) play both in the fabrication of high-quality laminates and in the level of achievable performance. There is certainly agreement that any third phase material, such as a coating which is incorporated into the composite system, must be compatible with both the reinforcing carbon fibre and the matrix material. This compatibility was clearly established for glass reinforced plastics, where functional silanes were developed to form linkages to the glass fibres through common —Si—O— groups and to the matrix through common chemical entities such as amino functionalities. These functionalities occur both as pendant groups on silanes and in the structure of polyamides or amine-cured epoxies. This approach has been successfully demonstrated in fibreglass thermoset and thermoplastic composites containing either continuous or chopped fibre reinforcement.

This principle of like-prefer-like, which has been amply demonstrated in glass composites, has also been successfully applied to develop optimum composites with carbon fibre reinforcement. The high quality injection moulding compounds (IMCs) described in Section 1.2 have all been prepared using fibres coated with finishes that are chemically similar to the matrix material to be used. In addition, the majority of carbon fibres used to prepare thermoset tape and fabric prepregs are also coated with compatible substances. In contrast to the glass fibre situation, where the finish also

stabilises and protects the surface involved, there is no such need for
protection of the relatively inert carbon fibre surfaces. There are a large
number of instances where sized and unsized carbon fibre have been shown
to yield equivalent behaviour in composites.

The principal reason for the use of finishes on carbon fibres has been to
facilitate handling in the processing operations to which the fibres are
subjected. These operations include: chopping and feeding during the
production of moulding compounds, where a binder is needed to increase
bulk density and/or to prevent unnecessary fibre length degradation;
prepregging and weaving, where a coherent, intact yarn bundle enables
faster, trouble-free processing; and filament winding, where fast release
from packages and ease in passing freely through guides promotes
production efficiency. Certainly where handling ease is a concern in the
preparation of CF/TP product forms, finishes can and will perform a useful
function. The question is whether finishes play an additional role in the
preparation of CF/TP preforms and fabricated parts.

The experiments reported by McMahon and Maximovich,[16] wherein a
series of finishes were screened against unfinished controls, show
ambiguous results. A fibre finish in nylon 6,6 composites does not appear to
be required, but in sharp contrast PBT polyester composite performance is
significantly enhanced with proper finishes. However, these results are
perfectly compatible with known and established polyester wetting
difficulties when nylon and polyester tyre cord performances are compared.

Hartness,[17,22] in his development of thermoplastic composites with
several resins, has repeatedly observed that one finish, UCC 307 enables
superior composites to be fabricated via the technique he has employed,
that is, alternate stacking of carbon fibre fabrics and TP films. It is also
noteworthy that he refers to equivalent results being obtained with unsized
fibre in PEEK resin.

In the ICI publications[19,21] and in private discussions[23] strong
statements as to the importance of a proper fibre finish are set forth. ICI
claims to have a proprietary finish which is in use with their APC (aligned
plastic composite) series of prepreg materials. These results are consistent
with Hartness's claims and with the claims of the NRDC[13] patent which
proposes prewetting via solvent coating with a TP material prior to melt
impregnation.

The extreme difficulty experienced in obtaining well-wetted, high quality
TP prepregs and preforms may very simply arise from a combination of
high viscosity resins and resistance to flow of resin from the outside of a
close-packed fibre bundle into the interior under pressure. The pressure

applied to induce wetting simply closes further the inter-fibre spacing thus removing all chance of good wetting.[24] It is the belief of the author that the finishes or pre-coats applied from solvent start the wetting process and prevent complete collapse of fibre bundles when highly viscous matrix resin is forced in from the exterior. For success then, the matrix must be applied to fibre monolayers; or added in-part via solution coating, such as with finish application; or added as a powder using a fluidised bed approach, such as the glass impregnation reported by Prewo and Dicus.[25] In short, new and successful approaches are evolving which will enable high quality CF/TP composites to be reproducibly manufactured.

1.4.2. Fibre Loading Level and Uniformity

When comparing the data presented on thermoplastic carbon fibre composites to the results obtained with thermoset systems, two differences stand out. The first of these is the upper level of fibre volume percent which is achievable; a limit of approximately 55% appears to exist. Second, the uniformity of fibre placement throughout the composite has been lacking. This exceptionally irregular spacing of fibres is most likely related to the difficulties in complete wet-out discussed in Section 1.4.1 and to the lower levels of fibre present. Fibre loadings in the 40–50% range are not uncommon.

The upper limit on fibre volume content achievable with TP matrix materials may well lead to an inherent performance disadvantage in comparison to thermoset matrix composites. Thermoset systems generally contain 60–65% fibre by volume. Since strength and modulus are fibre dominated properties which are readily predicted by the rule-of-mixtures then 10% less fibre corresponds to 20% less reinforcement for the TP systems. Strengths and moduli of fibre dominated properties will then be reduced by as much as 20% in the direction of fibre orientation.

There are both positive and negative aspects of this lower fibre loading. On the positive side, the increased thermoplastic resin content, which is in itself tough, will increase the toughness of the composite system. Wide separations between fibres will enable the matrix to absorb more of the impact load and further prevent fibre damage by spreading the load over a broader area via plastic deformation. Other advantages are lower thermal conductivity (where thermal protection is required) and improved load transfer between plies of multi-directional laminates where shear and transverse stresses can become very significant.

Negative features are the need for thicker laminates to achieve the same mechanical properties, and reduced dimensional stability via increased thermal expansion perpendicular to the fibre direction and through the thickness of laminates. There is also the potential for reduced electrical conductivity (important in shielding and grounding applications) and for reduced compression properties via the onset of localised buckling in larger resin domains at lower load levels.

1.4.3. Comparative Performance of Thermoplastic and Thermoset Matrices

The ultimate objective in switching from thermoset to thermoplastic matrix systems for carbon fibre composites is to increase toughness significantly without the loss of any of the attractive epoxy matrix properties. The expected and desired increases in toughness have clearly been demonstrated on PEEK composites by Hartness[22] in Mode I peel test results (see Table 1.14). In addition to the ten-fold increase in toughness, he has also shown exceptional increases in fatigue life in comparison to epoxy matrix composites. When load levels are at 60% of ultimate on ±45° tensile specimens, the cycles-to-failure with CF cloth/epoxy laminates are two orders of magnitude less than with comparable PEEK laminates. Finally, in a head-to-head comparison of ambient creep in PEEK/CF cloth and epoxy/CF cloth laminates at 70% of ultimate, the PEEK laminates showed far less creep deformation under load—0·1% versus more than 0·5% for epoxy.

However, no complete comparative evaluation of static mechanical properties for epoxy and thermoplastic matrix carbon fibre composites has been reported. The data available shows slightly lower flex and tensile strengths for CF/PES composites when normalised to the same fibre volume as with epoxy.[26] The notched Izod impact strength is much lower than would be expected on the basis of other thermoplastic/CF composite results. Slightly lower results are also obtained when the CF/nylon data of

TABLE 1.14
TOUGHNESS MODE I PEEL TEST FOR DELAMINATION RESISTANCE

Material	Critical strain energy release rate G_{IC} (J/m²)
Epoxy/GR cloth	230
Peek/GR cloth	1 990
Epoxy/unidirectional GR	140
Peek/unidirectional GR	1 400
Polysulphone/unidirectional GR	650

[a] Data from ref. 22, courtesy of SAMPE.

McMahon and Maximovich[16] are compared to standard Celion/epoxy results using fibre of the same vintage.[28] The CF/nylon data referenced here is also incomplete, since compression properties have not been reported.

The compression strength is the property most likely to be depressed for TP matrix composites in comparison to TS counterparts. This expectation is based on lower bulk moduli, which would offer less resistance to fibre buckling. Even if the initial modulus of the thermoplastic matrix is in the range of 3·4 MPa (equivalent to state-of-the-art thermosets) the thermo-plastic nature of the material would be expected to lead to a concave downwards stress–strain curve, i.e., a lowering of modulus as load is applied. Lower flex strengths reported for TP matrix composites support this rationale and supposition.

However, the more recently introduced PEEK and LCP polyester matrix systems have superior stiffness properties and may be equivalent to the majority of thermoset matrix systems. Unfortunately, early attempts to measure compression properties on unidirectional laminates have been thwarted by test specimen preparation problems. (See chapter 4.) Accepted composite test procedures require bonding of tabs to gripping surfaces, and herein arises a secondary problem. Adhesives and/or bonding techniques have not yet been developed which will enable suitable specimens and testing to be conducted.

As a result of the foregoing, it is only possible to state that very attractive TP matrix systems are evolving, but the demonstration of full mechanical equivalence awaits the development of adequate compression specimen and test procedures.

1.4.4. Large Component Manufacture

Perhaps the most formidable obstacle preventing manufacture of large, complex CF/TP composite aircraft structures is the need for a product form and fabrication technology package capable of economic production of a variety of large structural components. The current state-of-the-art is enabling production of preforms (pre-impregnated tapes and sheets) which are thin and relatively restricted in area, e.g. 1 m² or less. These sheets, which are quite inflexible in comparison to thermoset prepregs, must be assembled in layers with the desired fibre orientation, then heated, and finally pressed with pressures in the 700–7000 K Pa range in a cold tool to take advantage of the potential for rapid fabrication and to control the crystalline structure of the matrix which can affect performance. The need to assemble stiff plies and transfer the heated, fused assembly to a cold tool greatly limits the size and complexity of laminates which can be produced. An alternative is to use bi-functional tooling (such as that employed at GD/Convair[2] with polysulphone) which enables rapid heat-up and cool-down in a ceramic tool equipped with both electrical heating and water cooling capabilities. This is a very expensive approach which would require separate tooling for each part. And further, this system does not address the problems associated with pre-assembly and transfer to the tool for large part manufacture.

The option of manufacturing thermoplastic composite structures using existing autoclaves at the aircraft manufacturers needs to be addressed. Certainly the use of existing capital equipment is very attractive. However, there are several serious drawbacks. The principal obstacle is the lack of temperature capability with present claves; most have limiting upper temperatures of 200–260 °C whereas the most promising plastics, such as PEEK, LCP, Ryton, Torlon, etc., require processing in the 320–400 °C range. In addition, short process cycles cannot be used to advantage; preforming techniques need to be developed; and currently used vacuum bagging materials and sealing techniques may not be adequate.

Despite the problems cited, the inherent advantages of thermoplastic matrices for carbon fibre composites are continuing to stimulate efforts to develop the material forms and manufacturing methods which will enable economic part production. Perhaps the most promising approach being pursued is development of intermediate filament and tape products and the compatible winding equipment to enable direct filament (or tape) winding of components. This approach is under development by ICI[21] and others. It is a natural extension to the development of dry prepreg rovings, which have been developed by the major prepreggers in response to demands initiated by helicopter manufacturers such as Bell/Fort Worth. An intrinsic advantage of such impregnated preform filaments and tapes is the absolute control of fibre and matrix content in an intermediate state. The fibre content would carry over to the finished part, since no resin would be lost in successful implementation of this type of technology. Another obvious advantage of such a system would be greatly reduced scrap, which is a major cost with current thermoset materials and manufacturing technology.

1.5.   CONCLUSIONS

The use of carbon fibre in thermoplastic injection moulding compounds has been successfully implemented by the major compounders world-wide.

Tribological performance, electrical conductivity, and dimensional stability are the properties most responsible for the incorporation of these materials into end-use items such as shielded housings, bearings and seals. These moulding compounds, containing various levels of reinforcement fibres in most engineering thermoplastics are commercially available. In contrast, the commercialisation of continuous carbon fibre thermoplastic matrix composites has not yet been effected. However, the significant advantages of these materials are clearly recognised and many of them have demonstrated, in particular, toughness and damage tolerance. Further, high performance matrix materials such as PEEK, LCPs, PPS and poly(amide-imide) are emerging, and their capability to meet mechanical and environmental requirements are being demonstrated. Finally, it is fully anticipated that the significant efforts being expended to overcome product form and manufacturing technology gaps will be successful and high performance CF/TP aircraft components will become a reality. This technology will then naturally flow over into the manufacture of industrial and recreational products.

REFERENCES

1. HOGGATT, J. T. and VONVOLKI, A. D., Naval Air Systems Command Report, Contract N00017-7A-C-0026, 1975.
2. MAY, C. L. and GOAD, R., AFML-TR-75-111, 1976.
3. Celion Product Brochure GSP6, High Performance Composite Molding Compounds, Celanese Corporation, December, 1982.
4. LNP Product Bulletin 222-582 Carbon Fiber Reinforced Thermoplastic Composites, 1982.
5. ROBINSON, D. N., US Patent No. 4, 328, 151, asigned to Pennwalt, May 1982.
6. SHEPPARD, C. H., HOUSE, E. E. and STANDER, M., *36th Annual Conference SPI*, Paper 17B, February 1981.
7. COLE, W., Preliminary Data Sheet, Amoco Poly (Amide-Imide) Laminates, 1983.
8. HERGENROTHER, P. M., *J. Poly. Sci., Poly. Chem.*, **20**, 3131, (1982).
9. MAXIMOVICH, M. G., *National SAMPE Conference Series*, **19**, 262, (1974).
10. HUSMAN, G. E. and HARTNESS, J. T., *National SAMPE Conference Series*, **24**, 21, (1979).
11. RIGBY, R. B., *National SAMPE Conference Series*. **27**, 747, (1982).
12. Sandvick Conveyer International, Stuttgart, West Germany, private communication.
13. MURPHY, D. J. and PHILLIPS, L. N., British Patent No. 1,570,000, assigned to NRDC, June 1980.
14. HOGAN, P., private communication.

15. MOYER, R. L., US Patent No. 3,993,726, assigned to Hercules, Inc., November 1976.
16. McMAHON, P. E. and MAXIMOVICH, M., *ICCM-III Proceedings*, Vol. II, 1663, (1980).
17. HARTNESS, J. T., *National SAMPE Conference Series*, **25**, 376, (1980).
18. HILL, S. G., HOUSE, E. E. and HOGGATT, J. T., Naval Air Systems Command Report, Contract N00019-77-C-0561, 1979.
19. BELBIN, G. R., BREWSTER, I., COGSWELL, N., HEZZELL, D. J. and SWERDLOW, M. S., European SAMPE Conference, Stresa, 1982.
20. CALUNDANN, G. W., US Patent No. 4,161,470, assigned to Celanese Corporation, July 1979.
21. ICI Brochure, Provisional Data Sheet APC PD3, ICI Petrochemicals and Plastics Division, 1982.
22. HARTNESS, J. T., *SAMPE Quarterly*, **14**, 33, (1983).
23. COGSWELL, N., private communication.
24. CHUNG, N., private communication.
25. PREWO, K. M. and DICUS, D. L., NASA Technical Reports NASA CR165711[N81-24181/NSP], 1981.
26. McMAHON, P. E., National SAMPE Conference Series, **27**, 49, (1982).
27. ICI Brochure, Provisional Data Sheet APC PD2, ICI Petrochemicals and Plastics Division, 1982.
28. Celion Brochure.

Chapter 2

APPLICATIONS OF FOURIER TRANSFORM INFRARED SPECTROSCOPY TO THE STUDY OF FIBRE–RESIN COMPOSITES

Y. T. LIAO and J. L. KOENIG

*Department of Macromolecular Science,
Case Western Reserve University,
Cleveland, Ohio, USA*

SUMMARY

A major concern in the development of advanced composite materials technology is the problem of composite reliability. One of the sources of variations in composite performance is the differences in polymer composition and state of cure. This review will discuss the detailed character methods available for the determination of the chemical composition. structure and state of cure of the material. Particular emphasis is placed on the spectroscopic methods of Fourier transform infrared spectroscopy (FTIR) and Raman spectroscopy, with consideration also given to the thermal techniques.

2.1. INTRODUCTION

The physical, chemical, and ultimate mechanical properties of high performance glass-fibre-reinforced composites are dependent on the degree of cure and structure of epoxy matrices. Therefore, a knowledge of curing process and composition of epoxy matrices is essential for relating the properties of composites to the extent of the reaction and optimising the performance. Several methods have been developed to characterise and

31

control the curing of epoxy matrices. These methods include spectroscopy, differential scanning calorimetry, dielectric analysis, and dynamical mechanical tests. These methods can be used to characterise the curing process during or after the fabrication of the cured epoxy matrices to justify reproducibility, reliability, and durability. In general, a combination of these methods gives powerful techniques to analyse and control the quality of epoxy matrices in fibre-reinforced composites. The sensitivity, advantages, and selectivity of these techniques will be reviewed and discussed in this paper.

2.2. THE ROLE OF EPOXY MATRICES

2.2.1. Epoxy Resins

There are three major types of epoxy resins of commercial significance: (a) epichlorohydrin-Bisphenol A (conventional), (b) epoxy novolak, and (c) epoxidised polyolefin resins.[1] The most commonly used curing agents are anhydrides and amines. Miscellaneous modifications of epoxy systems are used to meet the specific performance requirements. In order to optimise the performance, several factors must be considered in choosing the epoxy resin for suitable composite matrices. These factors are based on the following considerations: (1) processing requirements, (2) economic preference, and (3) mechanical performance.

2.2.2. Processing Considerations

Processing requirements involve viscosity considerations (rheological behaviours of epoxy resins), high or low temperature cure, etc. One of the ways of decreasing the viscosity of an epoxy resin is to add a diluent.[2] Reactive diluents are liquid materials which have lower viscosity than the epoxy resin and are assimilated into the resin network during cure. Nonreactive diluents comprise materials which do not contain epoxide groups, but which are completely sorbed in the cured epoxy resin network. Nonreactive diluents may usually be removed by solvent extraction.

Since the inhomogeneities observed in cured epoxy resins by several workers appear to be related to the effectiveness of mixing the reactants, Tateosian and Royer[3] reported an improvement in impact strength by as much as 58% by use of dynamic mixing. J. P. Bell[4] studied the effect of mixing on homogeneity by using an electrical field to induce mixing of viscous fluids of different conductivity, and significantly increased the ultimate tensile strength of epoxy resins.

Another processing requirement is concerned with the temperature of cure. For advanced fibre composites, the difference in thermal expansion coefficients between the fibre and the epoxy matrices during the high-temperature-cure cycle creates a very serious problem such as visible delamination, or fibre microbuckling. On the other hand, the commonly used room-temperature-cured epoxy resins have some disadvantages, such as poor mechanical properties, and their working life is too short to be used in composite processing.

Chiao and Moore[5] used a particular amine,

$$CH_2-[-OCH_2-CH(CH_3)]_xNH_2$$
$$CH_3-CH_2-C-CH_2-[-OCH_2-CH(CH_3)]_yNH_2$$
$$CH_2-[-OCH_2-CH(CH_3)]_zNH_2$$
$$x+y+z = 5\cdot3$$

and pure diglycidyl ether of Bisphenol A (DER 332 EPOXY) to fabricate polyether amine-cured epoxy matrices. In spite of good flexibility and toughness, this resin system has fairly well balanced mechanical properties. It can be room-temperature cured, and presents no problem in filament winding, which requires a good epoxy resin of low viscosity and long working life.

2.2.3. Economic Factors

For economic considerations, sometimes it is necessary to consider energy savings as well as increases in productivity, and then a fast-curing epoxy system is required. Cordova Chemical[6] has developed accelerators that provide greater than a six-fold advantage over tertiary amines in anhydride cured epoxies. It should be emphasised that the properties obtained with a short cure (2 h) are the same as those properties previously requiring 12–24 h for the curing process with the conventional accelerators. Generally, the lower the temperature, the slower the reaction. On the other hand, the higher the temperature, the greater the risk of degradation. For production considerations, the optimum condition is the shortest cure time which will still assure a resin matrix with the desired properties.

2.2.4. Shelf-life

A current problem with epoxy resin systems used in continuous-fibre-reinforced composites is their shelf-life. Once the resin is mixed, it must be used immediately or stored at low temperatures in the form of a prepreg. Prepregs must typically be stored at $-18\,°C$ and, at this temperature, they are estimated to remain stable for six months. The ideal system from a storage point of view would be a prepreg which is stable at room temperature in the B-stage and which retains its tack and drape. Currently no such ideal system exists. However, a resin system which is cured with a sterically hindered amine is stable at room temperature in the B-stage in a glassy state. Upon mixing at room temperature, the primary amine hydrogens react to form a linear polymer. The secondary amine hydrogens do not react at room temperature because they are much less reactive due to the steric hindrance of the nearby methyl groups. The epoxy resin does not form a three-dimensional structure until the secondary amine hydrogens react upon additional heating. Buckley and Roylance[7] studied the curing kinetics of this system with FTIR. The shelf-life at room temperature of the B-staged resin system is predicted to be at least three months, based on an extrapolation of the experimental kinetic data.

On the other hand, the growing use of composite materials in commercial and military equipment has led to concern over field repair or patching of damaged composites. Field level repair has some unique materials requirements in terms of storage and curing characteristics. Since the presence of freezer storage space is not guaranteed, the resin used in the composite patch would necessarily require room-temperature stability. Also, since cure facilities are very limited, the resin system must be curable at low temperatures and short times (150 °C for 1 h). Unfortunately, commercially available prepregs typically require both freezer storage and higher temperature cures. Donnellan and Roylance[8] used 2,5-dimethyl 2,5-hexane diamine as a hardener to form a linear glassy prepolymer at room temperature. The resin was found to react to a partially cured (52 %) state at room temperature and then vitrify. Samples stored for a two-month period showed no advance to a more fully cured state. The isothermal curing behaviour was studied in a temperature range from 100 to 150 °C with FTIR.

2.2.5. Heat Resistance
Heat resistance is another requirement of the epoxy matrix. Several methods have been pursued to improve the heat-deflection-temperature. Lauze and co-workers[9] showed that cured epoxy resins based on Bisphenol S have a considerable increase in heat resistance over those based on Bisphenol A. The increased heat resistance results from the replacement of the isopropylidene group in Bisphenol A with the more thermally stable sulphone group. The improvement is indicated by the resistance to heat aging and to heat deformation as well as by retention of strength at

elevated temperature. The enhanced thermal properties of sulphone epoxy are achieved in a different way from those of the epoxy novolacs, which result from an increase in the crosslinking density. Another way of improving the high temperature performance is to incorporate a rigid structure in the backbone of epoxy resins. Polyimides have good high temperature performance and the epoxy resins possess many desirable properties of the aromatic polyimide if it contains the phthalimide moiety. Kaplan[10] has synthesised some of these epoxy/imide resins, and demonstrated good thermal stability. Another new family of resins is based on the glycidylated hydantoin ring.[11] Because of the compact, polar heterocyclic structure of the hydantoin ring, the heat distortion temperatures of these resins are significantly higher than those of comparably cured conventional epoxies. With aromatic rings replaced by heterocycles in the structure, smoke generation during combustion has been greatly reduced.

2.2.6. Load Transfer Capacity

The primary role of epoxy matrices is to transfer stress from the fibres to the finished composites. Currently prevailing epoxy resins are designed for glass fibre in fabricating composites. Since graphite fibres have a higher modulus ($400\,GN\,m^{-2}$) than glass fibre ($70\,GN\,m^{-2}$), it is important to use high strength matrices to maximise the efficient transfer of the fibre strength to the composites. Holler and co-workers[12] synthesised a series of pure epoxy resins of the following structural types: diepoxy Bisphenols, glycidyl esters, diepoxy cycloaliphatics, acyclic Bisphenol, diepoxides and glycidyl amines, and then related the mechanical properties to the Bisphenol diepoxide structure. The glycidyl glycidate resins had a tensile strength of over $140\,NM\,m^{-2}$; this is one of the highest values ever reported for a bulk polymer. These kinds of resins can be used in the fabrication of high modulus graphite-fibre-reinforced composites. Another serious problem concerned with epoxy matrices in composites is the brittle nature of the fully cured resin. F. J. McGarry *et al.*[13] have shown that the impact strength of aromatic epoxy resins can be improved by incorporating a specific carboxyl terminated elastomer in concentrations of up to 10 parts per hundred parts of resin. Soldatos and Burhans[14] extended this investigation to cycloaliphatic epoxides, which have many outstanding properties, and got toughened epoxy resins without significantly degrading the heat distortion temperature (HDT). Liquid rubber can be used to toughen or flexibilise epoxy resins. The toughened epoxy resins show improved crack resistance and improved impact strength, with a minimum loss of

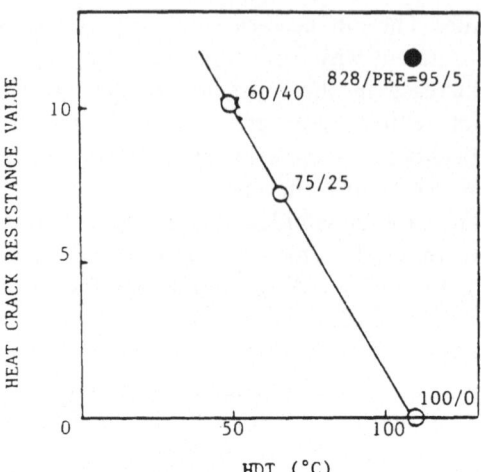

FIG. 2.1.   Heat distortion temperature vs. thermal shock resistance value.

mechanical and thermal properties (e.g. heat distortion temperature). On the other hand, the flexibilised systems are those where the liquid rubber has reacted with the epoxy resin and they are single phase systems. The flexibilised system provides epoxy matrices with high impact strength, but always accompanied by a significant loss in thermal–mechanical properties. The first criterion for a toughened system is that it should contain a dispersed second phase, and this phase must have particles greater than about 100 nm in diameter. The non-functional liquid rubber that contains no reactive groups to react with epoxy resins will have a higher fracture energy but will not improve the impact strength. It is generally accepted[15] that three criteria are necessary for toughening:

(1)   proper size of dispersed phase;
(2)   bonding between the matrices and the dispersed rubber;
(3)   elastic character to the rubber particles.

A. R. Siebert et al.[15] showed that the toughest epoxy systems are those where the second phase exists as a bimodal distribution of rubbery particles that contain small particles with 100 nm diameters and large particles with 1–5 μm diameters. H. Samejima[16] reported a new elastomeric modified epoxy which gave a cured product having superior impact resistance. In Fig. 2.1, the thermal shock resistance is plotted versus the heat deflection temperature for polyetherester (PEE) modified epoxy resin, and epoxy resins blended with the flexible epoxy. It is obvious that flexible epoxy resins will decrease the HDT, but improve the heat crack resistance value. On the other hand, the toughened epoxy resins improve the heat crack resistance value without sacrifice of HDT.

2.3.   CALORIMETRIC ANALYSIS

2.3.1. Introduction

The processing of epoxy resins, such as curing/crosslinking, involves the exposure of these materials to various levels of heat treatment. The physical and mechanical performance and the quality of the cured articles are largely determined by the extent of cure, the control of temperature distribution, and the rate of temperature rise during processing. Moreover, temperature variations during cure, which determine the degree of cure in an epoxy resin system, depend to a large extent not only on the heat of the reaction but also on the specific heat and the thermal conductivity of the material at different stages of the cure cycle. These parameters can be characterised by differential scanning calorimetry, and are essential for optimising product quality and processing condition considering the heat transfer.

2.3.2. Differential Scanning Calorimetry (DSC)

Differential scanning calorimetry measures the difference in the rates of heat absorption or evolution by a sample with repect to an inert reference as the temperature is raised at a constant rate. On the other hand, differential thermal analysis (DTA) measures the differential temperature caused by heat changes in the sample. DSC can be used to characterise the curing reaction of epoxy resins in the presence of filler or without it. The basic assumption made is that the heat of reaction is proportional to the extent of the reaction. Moreover, it is also assumed that the specific heat of the material either stays constant or varies linearly with scanning temperature during a scan while both the temperature and degree of cure change simultaneously. These assumptions are valid for simple reactions, but not obviously valid for the complicated crosslinking reactions which occur in the cure of epoxy resins. There are three ways[17] of measuring the extent of curing in epoxy resin. This can be achieved by: (1) isothermal operation, (2) analysis of thermograms with different scan rates, and (3) scans on partly cured resins. For isothermal operation, because a short time is needed for the samples and the test cell to heat up to the desired temperature, the beginning portion of the exotherm is lost. Therefore, isothermal scans

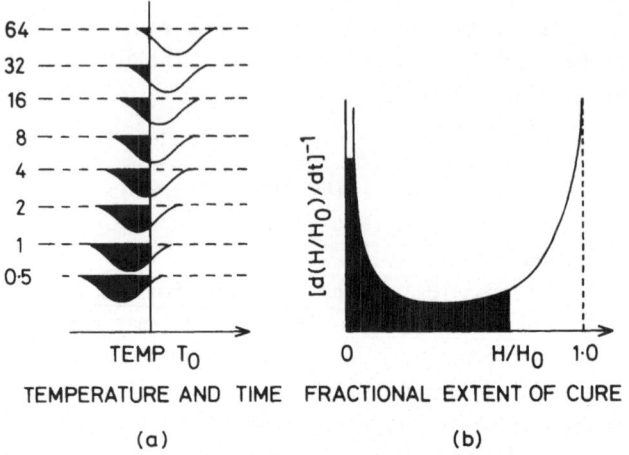

TEMPERATURE AND TIME   FRACTIONAL EXTENT OF CURE

(a)   (b)

FIG. 2.2.   (a) Set of displaced thermograms; (b) curve deduced from (a) and used to obtain isothermal cure curve at temperature T_0. (Reproduced from Fava, R. A., Differential scanning calorimetry of epoxy resins, *Polymer*, **9**, 137 (1968), by permission of the publishers, Butterworth & Co. (Publishers) Ltd ©.)

become unreliable for very fast cures. This problem can be solved by analysis of thermograms with different scan rates. Figure 2.2 shows a series of thermograms with different scan rates on a temperature axis.

An isothermal cure curve can be drawn using these curves. An ordinate at T is drawn in each thermogram. The state of the resin is described by three parameters; temperature (T), heat generation rate (dH/dt) and heat of reaction (H). H is given by the shaded area, and the total heat of reaction H_0 is given by the area enclosed by the complete curve. The eight states shown in Fig. 2.2(a) are plotted in reduced forms as shown in Fig. 2.2(b). The integral of the curve from zero to H/H_0, is equal to the time to reach degree of cure H/H_0 at temperature T_0. Therefore, the isothermal cure curve of H/H_0 versus t can be constructed. This process can be repeated at any temperature except at temperatures lower than T_0 in Fig. 2.2(a) since the final part of the cure curve will be missing. Thus, at low temperature the isothermal method gives reliable results while the scan method gives insufficient data from which to construct a reliable curve. On the other hand, for fast cure or when the curing temperature is too high, the thermograms with different scan rates must be used. The third method is to scan only partly cured resins. Each partly cured sample is scanned and the area of the thermogram gives the residual heat of reaction $(H_0 - H)$ from which H could be derived. This method becomes useful when the rate of heat evolution is too small for isothermal detection.

2.3.3. Degree of Cure

On the other hand, characterisation of the kinetics of epoxy resin chemistry is not only essential for a better understanding of structure–property relationships but is also required for determining the time–temperature dependence of the degree of cure. DSC has been used by several workers both to monitor the state of cure and to determine the kinetic parameters of cure of epoxy resins both in isothermal and dynamic modes. There are two interrelated methods for analysing DSC curves of the crosslinking reaction of epoxy resins. The first method utilises a single DSC thermogram to evaluate the kinetic parameters, such as activation energy, E, kinetic order of reaction, n, and the total heat of reaction, H_0, by detailed differential–integral analysis of the DSC thermogram. The second method uses the multiple DSC scan, obtained at various rates of heating.

The first method was proposed by Ellerstein[61] and is based on the calculation of Borchardt and Daniels.[62] The general kinetic equation which is of Arrhenius type is given by the following expression:

$$\mathrm{d}x/\mathrm{d}t = k(1 - x)^n$$

where $\mathrm{d}x/\mathrm{d}t$ is the rate of reaction, k is the reaction rate constant, and x is the fractional extent of reaction (extent of cure).

Since $k = A \exp(-E/RT)$ we can rewrite the equation as follows:

$$\mathrm{d}x/\mathrm{d}t = A \exp(-E/RT)(1 - x)^n$$

where A is the frequency factor, E is the activation energy, R is the gas constant, and T is the absolute temperature. From the assumption that the heat evolved is proportional to the extent of reaction:

$$x = H/H_0$$

and

$$\mathrm{d}x/\mathrm{d}t = \mathrm{d}H/\mathrm{d}H_0$$

where H is the partial heat of reaction, which varies in direct proportion to the fraction reacted, x, and H_0 is the total heat of reaction, which is equal to the total area under the curve.

Combining the above equations gives:

$$\mathrm{d}H/\mathrm{d}t = A(H_0) \exp(-E/RT)(1 - H/H_0)^n$$

In a dynamic DSC run, the heating rate (or scan rate) $Q = \mathrm{d}T/\mathrm{d}t$, so that the above equation can be written as:

$$\mathrm{d}H/\mathrm{d}T = A(H_0/Q) \exp(-E/RT)(1 - H/H_0)^n$$

This equation shows the change of degree of cure with temperature. These calculations were extended by Crane, Dynes and Kaelble[63] to a study of the curing of epoxy resin. They derived the following equation to calculate the kinetic parameters of the system they studied.

$$\left[\frac{d^2H/dT^2}{dH/dT}\right]T^2 = \left[\frac{E}{R} - \frac{nT^2}{(1-\alpha)H_0}\left(\frac{dH}{dT}\right)\right]$$

dH/dt is the ordinate scale of a DSC curve, and it is converted through the scan rate $Q = dT/dt$ to become curve height $h = dH/Qdt = dH/dT$. The term d^2H/dT^2 becomes the slope of the curve, S, at temperature T. Furthermore, from the definition of $x = H/H_0$, the term $(1 - x)H_0$ becomes equal to $H_0 - H$, which is the remaining area, A, under the DSC curve. By making these substitutions in the above equations, the following equation is obtained:

$$(S/h)T^2 = E/R - nT^2(h/A)$$

After evaluating the values of the slope, S, curve height, h, and remaining area A at various temperatures, a plot of $(S/h)T^2$ versus $T^2(h/A)$ will be a straight line whose slope defines the order of reaction, n, while the activation energy, E, is obtained from the intercept at $T^2(h/A) = 0$.

2.3.4. Kissinger's Method

The second method which utilises multiple DSC curves obtained at various scan rates, is due to Kissinger.[64] The method assumes that the reaction rate is a maximum at the peak temperature, T, of a DSC curve. The quantity $h = dH/dT$ attains its maximum value at the peak temperature and the slope $S = d^2H/(dT^2) = 0$. From the same assumptions stated above, Kissinger obtained the following equation.

$$\frac{d(\ln Q/T^2)}{d(1/T)} = -\frac{E}{R}$$

or alternatively

$$\frac{d \ln Q}{d(1/T)} = -\frac{E}{R} - 2T$$

where Q is the scan rate and T is the peak temperature of the DSC curve. A plot of $\ln Q$ versus $1/T$ from several DSC curves should be linear and the activation energy, E, is obtained from the slope when $E/R \gg 2T$. It should be noted that the activation energy is determined by using the above equation regardless of order of reaction, which is assumed to remain constant throughout the reaction.

It is obvious that the kinetic order, n, activation energy, E, and heat of reaction of a curing system can all be defined using a single DSC curve at a constant scan rate by using the above equations. The great advantage of the Kissinger method is to evaluate the effects of scan rate on the cure kinetics. Therefore, a combined analysis using these two interrelated methods provides a more comprehensive evaluation of the curing kinetics in epoxy resin systems.

2.3.5. Filled Resins

Most research on the cure kinetics of epoxy resin systems was without fillers.[18-25] Very little DSC information has been reported on epoxy systems containing fibres. Pappalardo[26] used DSC to determine the activation energy and reaction rate constants for some epoxy–glass-fibre composites. The effect of filler on the curing behaviour was found to depend on the specific filler and polymer system. Dutta and Ryan[27] used DSC to investigate the effect of carbon black and a silica filler on the cure kinetics of a model epoxy resin cured stoichiometrically with an aromatic diamine. They found that the filler does not significantly affect the order of the reaction but does influence the reaction rate. It appears that the carbon black fillers affect kinetic rate constants through the Arrhenius frequency factor, whereas the surface-treated silicas influence the kinetic rate constants through the activation energy. Parker and Smith[28] used DSC to evaluate cure and shelf-life of epoxy prepregs, and to correlate the total heat of reaction to resin flow, so as to predict the processing characteristics.

2.3.6. Baseline Problems

The most serious problem[29] with DSC is the requirement of an accurate knowledge of the baseline. The baselines are always taken at locations preceding and following the exotherm. However, the baselines should take into account the change in heat capacity of the cured resin compared to starting materials. Schneider *et al.*[29] have studied the curing behaviours of two DICY-containing epoxy resins. The DSC scan of epoxy resins that consist of DGEBA (diglycidyl-ether of bisphenol A), Novolac, and amine as curing agent are shown in Fig. 2.3. The baseline, which represents the scan of the cured resin, is lower than the uncured resin at the starting point, then undergoes an upward endothermal shift at the glass transition temperature (107 °C) and stays above the initial scan for the higher temperature.

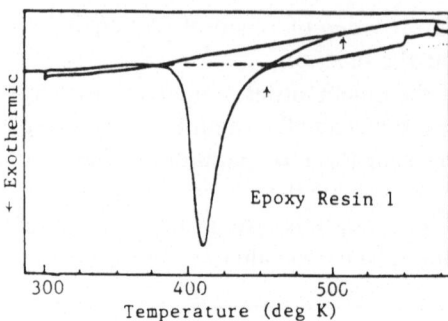

FIG. 2.3. DSC scan of resin 1 showing cure curve and various baselines. Full curves, scan of cured and uncured resin; dotted curve, scan with empty sample cells; chain curve, tangent to cure peak. Arrows indicate extrapolation of cure curve to baseline represented by scan of cured resin.

Therefore, it is impossible to produce an accurate extrapolation from the cure curve to get the correct baseline. If the baseline is taken as the tangent to the DSC curve at a location preceding and following the exotherm, then the heat of reaction is 0.29 J/kg. On the other hand, if the extrapolation is done as shown, by the portion of the curve between the arrows, the value obtained is 0.38 J/kg, which is very close to the result obtained by isothermal curing runs at higher temperature. It is obvious that the results are quite different with different baselines. Since it is impossible to get a correct baseline, only an approximation of the heat of reaction can be obtained from the scanning runs. In other cases the scan shows complex behaviour, such as that shown in Fig. 2.4. The resin used for Fig. 2.4 is a mixture of DGEBA, Novolac, and TGMDA (tetraglycidyl methylene-dianiline). It shows a major exotherm at 141 °C, a broad and smaller

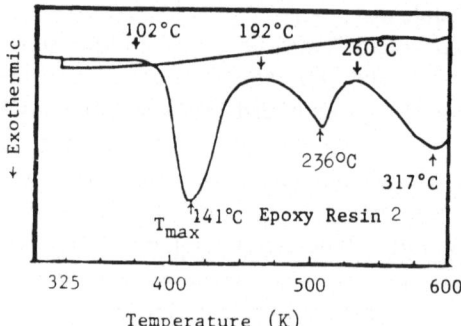

FIG. 2.4. DSC scans of resin 2. Heating rate 5 °C/min, 320 to 600 K, atten. 10 mcal/s, sample weight 13·1 mg.

exotherm peaking at 236 °C, and a third peak starting at 260 °C. The exotherm above 260 °C arises from sample degradation. It is obvious that any determination based on the tangent baselines rather than from the baseline produced by re-scan of the cured epoxy resin will be in error. The problem of estimating the conversion is even more complicated in this case. It involves the resolution of the two lower temperature exotherms. J. F. Sprouse estimated the extent of reaction occurring in the second curing region. The value obtained is too high compared with FTIR results (38 vs. 24 %). Since the infrared measurement is considered more reliable than the value extracted from DSC, Sprouse and co-workers[29] used 24 % conversion from FTIR results and calculated the total heat of reaction from the lower limit, and got 84 J/mol. Compared with a result from the literature, which has a value of 109 kJ/mol, it implies that the area from the DSC curve for calculating the percentage conversion could lead to a result that is 30 % too low. On the other hand, the deviation of the DSC results is about 1 %.

2.3.7. Differential Thermal Analysis (DTA)

Compared with DSC, DTA[30] is seldom used in the study of epoxy resins. It was shown that there is a good correlation between the gelation time and the temperature corresponding to the peak of the exotherm on DTA curves. The DTA method permitted a rapid and sufficiently accurate estimate of gelation time of epoxy resin in a broad temperature range.

Isothermal DSC provides the heat output as a function of time, representing directly the rate of the reaction. However, the information provided by this method gives us little insight into the chemical mechanism of the curing process and the chemistry o cure.

2.4. DYNAMICAL MECHANICAL TESTS

2.4.1. Gelation

One of the most important factors affecting the processing of the epoxy resins is gelation. When gelation begins, the viscosity of the system rises exponentially and has a remarkable effect on the processing such as compression moulding or autoclaving. A typical method for determining time to gelation is by a standard ASTM test,[31] which is based on steady-state viscosity. A schematic representation for the cure behaviour of an epoxy system is shown in Fig. 2.5. Steady-state viscosity measurements only characterise the liquid state. Characterisation of the rubbery and glassy states can be made with dynamic mechanical techniques.

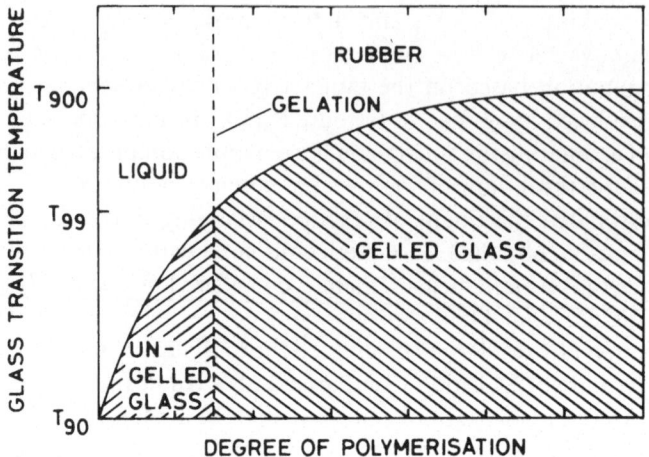

FIG. 2.5. Glass transition temperature vs. degree of polymerisation for a thermosetting system.

2.4.2. Torsional Braid Analysis (TBA)

Torsional braid analysis (TBA), a dynamical mechanical technique,[30] involving an adaptation of the torsional pendulum with a free hanging composite specimen, consists of a multifilamented glass braid and the epoxy system. With this technique, we can measure the change in rigidity and damping in the reacting system throughout the cure, and study transitions during cure, cure kinetics, and activation energy.

Two physical transitions are usually observable with TBA as cure proceeds isothermally. The first, gelation, corresponds to a transition from linear or branched molecules to an infinite network. The second, vitrification, corresponds to a transition from a rubbery to a glassy state. Each of these phenomena is accompanied by an increase in rigidity and by a maximum in mechanical damping. There exist two critical isothermal temperatures. These are $T_{g\prime}$ (the maximum glass transition temperature) and T_{gg} (the glass transition temperature at its gel point). As shown by J. K. Gillham:

    (i)   gelation only is observed, if $T_{cure} > T_{g\prime}$,

    (ii)   both gelation and vitrification are observed if $T_{g\prime} > T_{cure} > T_{gg}$,

    (iii)   vitrification only is observed, if $T_{cure} < T_{gg}$.

This information is related to the rheological changes which occur during the complete cure of the epoxy resin. A typical TBA interpretation is given in Fig. 2.6.

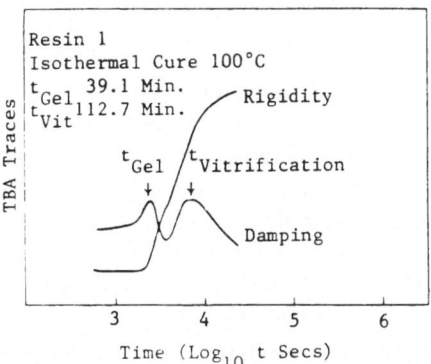

FIG. 2.6.   Relative rigidity and damping curves vs. time for the reaction of resin 1
at 100 °C.

A number of studies[29,32-34] have been reported on the curing behaviour of epoxy resins employing TBA. Quite a few of the studies are concerned with mixtures of fibre and epoxy. Schneider and Gillham[35] used TBA to study the prepreg material consisting of epoxy resin on glass fibre, carbon fibre, or aramid fibre, and measured gelation and vitrification times as a function of temperature. The study has shown that substantially all of the information concerning the curing behaviour of epoxy resins, which has been obtained by TBA, can also be obtained on the prepreg material directly. The only difference is a marked weakening of the gelation mechanical damping peak in the prepreg material. In spite of its weak intensity, the gel points can still be constructed very accurately to determine the activation energy by plotting an Arrhenius equation over a wide range of temperature. It is obvious that the filler does not affect the behaviour significantly, nor does it interfere with the results under study.

2.4.3.   Other Dynamic Techniques using Resin Supports

In addition to torsional braid analysis, there are other similar techniques. W. J. Macknight and co-workers[36] used dynamic spring analysis to study the curing behaviour of two commercially formulated epoxy resins. It is demonstrated that this supported viscoelastic technique is suitable for the determination of the onset of gelation but the method is not useful for studying later stages of reaction when the resins become more rigid. Goldfarb and Lee[37] studied the curing behaviour with torsion impregnated cloth analysis. This method used glass cloth as a resin support, monitoring the dynamic mechanical properties of resins during the curing process. In spite of the similarity between TBA and this technique there are some differences between these two techniques. Torsional impregnated cloth analysis (TICA) has some advantages over TBA. Its frequency of measurement is constant, and the frequency can also be varied to study the frequency effect on transition. In TBA, the resin impregnated braid is set in free oscillatory motion and its damping decay characteristic is recorded. The frequency of oscillation can be related to the storage modulus (G') of the resin while the log of the amplitude decrement is proportional to $\tan \theta$ (where θ is the phase angle between stress and strain). On the other hand, in the TICA experiment the cloth undergoes an oscillatory strain at a fixed frequency, and the resultant torque is analysed by a frequency response analyser to give both the in-phase and out-of-phase component amplitudes. The results obtained from TICA are not absolute values, and have an advantage over TBA in that the frequency of measurement is constant. The frequency can be varied to study the frequency effect on the transitions. Another advantage is that in TBA the rigidity may be influenced by substrate–resin interactions,[38] while in TICA the in-phase and out-of-phase components are measured directly, so the loss of modulus transition can be identified more confidently.

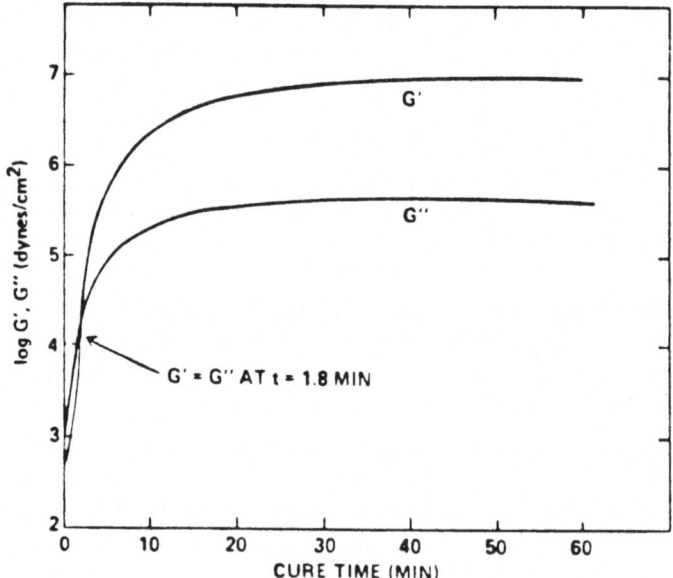

FIG. 2.7.　Dynamic viscoelastic characterisation of the 171 °C isothermal curing of a DICY cured epoxy resin at 10 rad/s and 10 % strain.

2.4.4. Determination of Gel Time by a Viscoelastic Method

Tung and Dynes[39] used a viscoelastic test of a kind by which gelation is not clearly observed. They described a method for determining the gel time of epoxy resins from dynamic viscoelastic data, which is based on the crossover of the dynamic storage G' and loss G'' moduli measured during isothermal curing. Figure 2.7 shows the dynamic mechanical properties during the 171 °C isothermal cure of an epoxy resin, crosslinked with dicyandiamide. The two modulus curves intersect, i.e., ($G'' - G'$ or $\tan \theta = 1$) at $t = 1 \cdot 8$ min. It was found, however, that the time to reach the modulus crossover point ($\tan \theta = 1$) coincides with the gel time as measured by the standard gel time test. Further examples of this correlation are shown in Fig. 2.8 for a variety of crosslinking systems varying widely in gel time. This correlation suggests a loss tangent ($\tan \theta$) of unity at the gel point. Loss tangent, being the ratio of energy lost to energy stored in a cyclic deformation ($\tan \theta = G''/G'$), measures the relative contributions of elasticity and viscosity of an epoxy system. Tan θ of a viscous liquid should

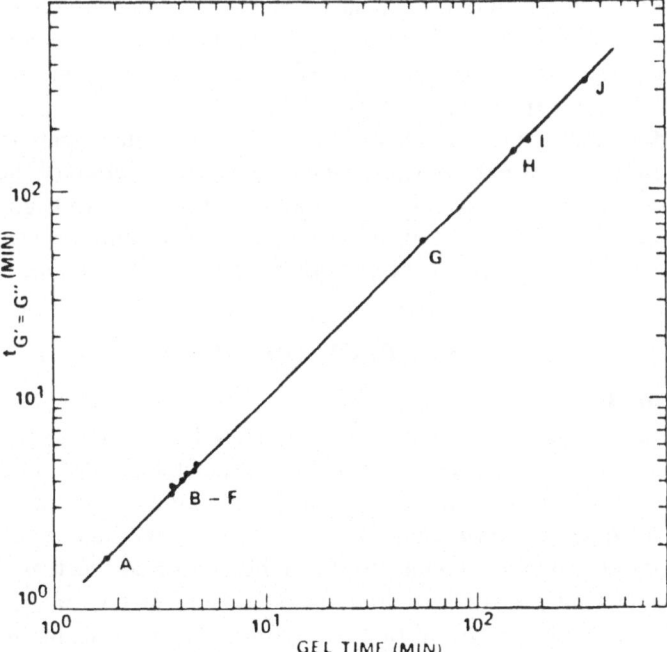

FIG. 2.8.   Correlation of gel time and time of modulus crossover of thermosetting resins. (A) brominated epoxy, $T_{cure} = 171$ °C; (B)–(F) epoxy, $T_{cure} = 171$ °C; (G) ATS, $T_{cure} = 150$ °C; (H)–(J) silicones, $T_{cure} = 25$ °C.

therefore be greater than 1, while that of an elastic solid should be less than 1. When the cure temperature is above T_{gg}, resin systems proceed from the viscous liquid state through gelation to elastic solids. Under these conditions, tan θ of gelation, the transition state between viscous liquid and elastic solid, would be expected to be equal to 1. As demonstrated by Tung and Dynes, this method is more precise and free of operation error than the conventional method in the determination of gel times of epoxy resin.

2.4.5. Other Applications of Dynamic Mechanical Tests

Another application of dynamic mechanical testing is the characterisation of cured composites. For example the glass transition temperature can be determined. As demonstrated by Katherine E. Reed,[40] the most interesting observation was the appearance of an additional damping peak above the glass transition temperature of the matrix resin. Also, as the fibre direction was changed from transverse (90°) to longitudinal (0°), the transition region broadened, and the change in frequency over this temperature range exhibited a maximum at intermediate angles. This phenomenon is associated with the unique resin layer in the interface which has been estimated to comprise roughly 0·1 % of the total resin in the composite, assuming thickness on the order of 100 Å. The structure of this interfacial region is different from that farther away.

Dynamic mechanical tests can be used to detect gelation and vitrification, which are the key parameters in any description of the curing behaviour. They are useful in prepreg technology and composite processing. However, dynamical mechanical tests cannot provide the necessary information about the curing chemistry and curing mechanism.

2.5. SPECTROSCOPY

2.5.1. Introduction

Infrared spectroscopy has long been recognised as a powerful tool for monitoring the curing process, and for characterising the cured epoxy composites. It is based on the absorption of radiation in the infrared frequency range due to the molecular vibrations of the functional groups contained in the polymer chain. Prior to FTIR, infrared spectroscopy was carried out using a dispersive instrument utilising prisms or gratings to geometrically disperse the infrared radiation. Using a scanning mechanism, the dispersed radiation is passed over a slit system which isolates the frequency range falling on the detector. In this manner the spectrum, that is, the energy transmitted through a sample as a function of frequency, is obtained. This infrared method is highly limited in sensitivity because most of the available energy is being thrown away, i.e. it does not fall on the open slits. For polymer analysis, where the bands are generally broad and weak, this energy limitation is particularly severe. Hence, to improve the sensitivity of infrared spectroscopy, FTIR was developed, which allows the examination of all the transmitted energy, all of the time.

2.5.2. Fourier Transform Infrared Spectroscopy—A Description of the Method

Fourier transform spectroscopy uses the Michelson interferometer rather than the conventional grating instruments. The Michelson interferometer has two mutually perpendicular arms. One arm of the interferometer contains a stationary, plane mirror; the other arm contains a movable mirror. Bisecting the two arms is a beamsplitter which splits the source beam into two equal beams. These two light beams travel down their respective arms of the interferometer and are reflected back to the beam splitter and on to the detector. The two reunited beams will interfere constructively or destructively, depending on the relationship between their path difference (x) and the wavelengths of light. When the movable mirror and the stationary mirror are positioned the same distance from the beam splitter in their respective arms of the interferometer, the paths of the light beams are identical. Under these conditions all wavelengths of the radiation striking the beam splitter after reflection add coherently to produce a maximum flux at the detector and generate what is known as the centre burst. As the movable mirror is displaced from this point, the path length in that arm of the interferometer is changed. This difference in path length causes each wavelength of source radiation to destructively interfere with itself at the beam splitter. The resulting flux at the detector, which is the sum of the fluxes for each of the individual wavelengths, rapidly decreases with mirror displacement. By sampling the flux at the detector, one obtains an interferogram. For a monochromatic source of frequency v, the interferogram is given by the expression:

$$I(x) = 2R(v)T(v)I(v)(1 + \cos 2\pi vx)$$

where $R(v)$ is the reflectance of the beam splitter, $T(v)$ is the transmittance of the beam splitter, $I(v)$ is the input energy at frequency v, and x is the path difference. The interferogram consists of two parts: a constant (DC) component equal to $2R(v)T(v)I(v)$ and a modulated (AC) component. The AC component is called the interferogram and is given by

$$I(x) = 2R(v)T(v)I(v)\cos 2\pi vx$$

An infrared detector and AC-amplifier converts this flux into an electrical signal

$$v(x) = \operatorname{Re} I(x) \, \text{volts}$$

where Re is the response of the detector and amplifier.

For highest accuracy in the digitised signal, the maximum intensity in an interferogram should match as closely as possible the maximum input voltage of the analog/digital converter (ADC). The noise must also be given at least four or five units, so the computer word length in FTIR spectrometers is 16 (or 32 in double precision), 20 and 24 bits. Griffiths[41] gives the example of measuring the spectrum of a continuous source whose intensity is uniform from 4000 to 400 cm and zero outside this band. The signal-to-noise ratio of the spectrum $(S/N)_S$ is related to the signal-to-noise ratio of the interferogram $(S/N)_I$ by

$$(S/N)_I = M^{1/2}(S/N)_S$$

where M is the number of resolution elements. So if we want to measure the spectrum with a $(S/N)_S = 500$ at resolution 1 cm ($M = 3600$) in a single scan, it can readily be seen that $(S/N)_I = 3 \times 10^4$ and the full dynamic range of a 15-bit ADC (215) is only just large enough to digitise the signal adequately. If $(S/N)_I$ is any greater than 30 000 the noise level in the digitised interferogram is set by the least significant bit of the ADC rather than by detector noise. For this reason, minicomputers with a word length as large as 20 bits are required so that a reasonable number of scans can be signal-averaged.

The interferogram for a polychromatic source $A(v)$ is given by

$$I(x) = \sum^{+\infty} A(v)\{1 + \cos 2(\pi v x)\} \, dv$$

The methods of evaluating these integrals involve a determination of the values at zero path length and very long or infinite path length. At zero difference

$$I(x) = 2 \int^{+\infty} A(v) \, dv$$

and for large path differences

$$I(x) = \int_{-\infty}^{+\infty} A(v) = I(0)/2$$

so the actual interferogram $F(x)$ is

$$F(x) = I(x) - I(\infty) = \int_{-\infty}^{+\infty} A(v) \cos 2(\pi vx)\, dv$$

From Fourier transform theory

$$A(v) = 2 \int_{-\infty}^{+\infty} F(x) \cos (2\pi vx)\, dx$$

This Fourier transform process was well known to Michelson and his peers but the computational difficulty of making the transformation limited the application of this powerful interferometric technique to spectroscopy. An important advance was made with the discovery of the fast Fourier transform (FFT) algorithm by Cooley and Tukey.[42] The use of the FFT revived the field of spectroscopy using interferometers by allowing the calculation of the Fourier transform to be carried out rapidly. The essence of the technique is the reduction in the number of computer multiplications and additions for n data points. The normal computer evaluation requires $n(n-1)$ additions and multiplications whereas the FFT method requires $(n \log_2 n)$ additions and multiplications. If we have a 4096-point array to Fourier transform, it would require $(4096)^2$ or $16 \cdot 7$ million multiplications. The FFT allows us to reduce this to $(4096) \times \log_2$ (4096) or 49152 multiplications, a saving of a factor of 341 in time. The advantage of the FFT increases with the number of data points. As computers have improved, the time required for a Fourier transfirm has reduced until currently the transformation can be carried out in less than a second with fast array processors, thus the spectra can be calculated during the return of the moving mirror if desired.

However, it is to be noted that the Fourier transform integrals have infinite limits, while the optical path differences are finite, so modifications or approximations must be made. We will use the approximation of the limits between $-L$ and L, where L is the maximum distance of the mirror drive. So

$$S(v) = 2 \int_{-L}^{+L} F(x) \cos 2\pi vx\, dx$$

where $S(v)$ is used to indicate that we are approximating the Fourier transform. It is of interest to examine the effect of this finite optical path length approximation on the $S(v_K)$ of an incident monochromatic source of wavelength v_K. The interferogram for this source is

$$F(x) = A(v_K) \cos 2\pi v_K x$$

where $A(v_K)$ is the amplitude of the light intensity. Making the substitution, we obtain:

$$S(v_K) = 2 \int A(v_K) \cos(2\pi v_K x) \cos(2\pi v x) \, dx$$

and after the transformation one obtains:

$$S(v_K) = A(v_K) 2L \operatorname{sinc} 2\pi(v_K - v)L$$

where the sinc function of y is $(\sin y)/y$. This function represents the instrument line shape of a Michelson interferogram. Obviously, this instrument line shape is not satisfactory because of the strong side lobes. These side lobes can be removed by apodisation. Apodisation is carried out by multiplying the interferogram by a function $H(x)$, before the Fourier interferogram is transformed. Thus:

$$S(v) = 2 \int_{-L}^{+L} F(x)H(x) \cos(2\pi v x) \, dx$$

A variety of apodisation functions[43,44] have been examined. Triangular apodisation has the form $H(x) = 1 - |x/L|$ if $x \leq L$ and zero otherwise.

Again, for a monochromatic source, we have

$$S(v) = 2 \int_{-\alpha}^{+\alpha} A(v_K) \left(1 - \frac{X}{L} \right) \cos(2\pi v_K x) \cos(2\pi v x) \, dx$$

Note that the integration limits could now be changed to plus and minus infinity without changing the result, because the integrand is zero outside the range $-L \leq x \leq L$; integration yields:

$$S(v) = A(v_K) L \sin^2(v_K - V)L$$

A summary of the effects of apodisation has been given and recommendations[45] made for the appropriate apodisation function for quantitative infrared measurements. There are a number of other problems such as phase correction arising from the fact that, in practice, the radiation undergoes phase shifts due to beamsplitter characteristics, signal processing delays, refraction effects in materials and so forth. Several techniques have been developed which allow appropriate corrections to be made for these effects.[41]

2.5.3. Advantages of FTIR

The advantages of FTIR over grating infrared arises from several sources. For measurements taken at equal resolution and for equal measurement time with the same detector and optical throughput, the signal-to-noise

(S/N) of spectra from FTIR instrument will be $M^{1/2}$ times greater than on a grating instrument, where M is the number of resolution elements being examined during the measurement. Alternatively, for a given observation time, it is possible to repeat the FTIR measurement M times, which increases the signal by a factor of M and the noise by a factor of $M^{1/2}$ to achieve an S/N enhancement of a factor of $M^{1/2}$. This advantage arises from the fact that the FTIR spectrometer examines the entire spectrum in the same period of time required for a dispersive instrument to examine a single spectral element. Theoretically, an FTIR spectrometer can acquire the spectrum with the same S/N from 0 to 4000 cm with 1-cm resolution 4000 times faster than a dispersive instrument. Or from another point of view, for the same measurement time a factor of approximately 63 increase in S/N can be achieved on the FTIR instrument. Therefore, when there is a limited time for measurement, there is a definite time advantage for the FTIR. When time of measurement is not an important consideration, the time can be used to multiscan with an FTIR to signal-average and increase the S/N. Of course, there is also the inherent time advantage associated with rapid scanning FTIR since it requires a very short time to obtain the complete spectrum (1·5 sec). This time advantage of the FTIR has been particularly important for the study of the curing reaction of epoxy resins, the degradation process, and other time-dependent processes of epoxy resins.

Another advantage of FTIR comes from the fact that the frequencies of an FTIR instrument are internally calibrated by a laser whereas a conventional IR instrument exhibits drifts when changes in alignment occur. This advantage is particularly useful for coaddition of spectra to signal-average, since the frequency accuracy is a requirement in this case. For the absorbance subtraction technique to be useful for epoxy resins examined over a period of time, such as months or years, long term frequency accuracy must be maintained. Applications such as quality control and long term aging and weathering require the reproducibility of the frequency that can be achieved by an FTIR instrument over the long term.

The overall simplicity of an FTIR compared to a dispersive instrument is also an advantage. For example, a single instrument can be easily converted to study the near-, mid- or far-infrared frequency regions, whereas with the dispersive method, three totally different instruments are required. To improve resolution with an FTIR instrument, the basic design is only slightly modified, while for a dispersive instrument different optical components are required. This is important, since in cases where interfering absorptions are present in the epoxy resins, the near-IR may be very useful to characterise epoxy content at $4535\,\mathrm{cm}^{-1}$ and hydroxyl content at $6784\,\mathrm{cm}^{-1}$.

2.5.4. Quantitative Infrared Measurements

Quantitative IR methods are based on the Beer–Lambert law. $A = EC$, where A is the absorbance for unit path length, E is the extinction coefficient, and C is the concentration. The analysis of any multicomponent resin or composite is greatly facilitated if the spectrum of that material may be expressed by a linear combination of a finite set of pure component spectra. The entire process may be separated into three steps:

  (i)   calculation of the number of species present,
  (ii)   identification of each of those species, and
  (iii) a suitable fitting of the spectra of those species to the spectra of composites.

The factor analysis methods are useful in determining the number of spectroscopically distinguishable components in the composites.

2.5.5. Factor Analysis of Mixtures

Factor analysis is based upon expressing a property as a linear sum of terms called factors, and the technique has found wide application to a variety of multidimensional problems. The technique has been applied to infrared and Raman spectra and most recently to FTIR spectra.

The Beer–Lambert law can be written for a number of components over a wavelength range as:

$$A_i = \sum_{j}^{n} \varepsilon_j c_{ij}$$

where A_i is the absorbance spectrum of mixture i, ε_j is the absorptivity for the jth component, and c_{ij} is the concentration of component jth mixture i. Factor analysis is concerned with a matrix of data points. So in matrix notation we can write the absorbance spectra of a number of solutions as $\mathbf{A} = \mathbf{EC}$, where $\mathbf{A}$ is a normalised absorbance matrix which is rectangular $(r \times p)$ in form, having columns containing the r absorbance at each wavenumber recorded and p rows corresponding to p different mixtures being studied. The $\mathbf{A}$ matrix could thus be 400 by 10 corresponding to a measurement range of 400 wavenumbers at one wavenumber resolution for 10 different mixtures or solutions. $\mathbf{E}$ is the molar absorption coefficient matrix ($r \times n$) and conforms with the **A** matrix for the wavelength region, but only has n rows corresponding to the number of absorbing components. **C** is the concentration matrix ($n \times p$) and has dimensions of the number of components, n, by the number of mixtures or solutions, p, being studied. Of course, we do not know **E** and **C** or we would not have a spectroscopic problem since **E** and **C** contain all of the information required to interpret **A**. Factor analysis can be used to generate **E** and **C**, which allows a complete analysis of a series of mixtures containing the same components in differing amounts.

There are two basic assumptions in factor analysis. First, that the individual spectra of the components are not linear combinations of those of the other components, and secondly, that the concentration of one or more species cannot be expressed as a constant ratio of another species. It is the different relative concentrations of the components in the mixtures that provides the additional information necessary to deconvolute the spectra.

What factor analysis allows initially is a determination of the number of components n in the p mixtures required to reproduce the absorbance or data matrix **A**. In factor analysis we find the rank of the matrix **A**, and the rank of **A** can be interpreted as being equal to the number of absorbing components. To find the rank of **A**, the matrix **M** = **AA**′ is formed where **A**′ is the transpose of **A**. This matrix termed the covariance or second moment matrix has the same rank as **A**, but has the advantage of being a square matrix ($p \times p$) with the dimensions corresponding to the number of solutions or mixtures being examined. In the absence of noise, the rank of **A** is given by the number of nonzero eigenvalues of **M**.

2.5.6. Least-squares Quantitative Analysis of Infrared Spectra

Since the set of pure components spectra determined are linearly independent, they may be unambiguously fitted by a least-squares criterion to the spectrum of the composite. The least-squares algorithm is derived from the criterion of minimisation of the sum of squared differences between the experimental and the fitted spectra. The algorithm employed is a simple linear regression model for gamma-spectra:

$$\sum_{i=1}^{N} R_{i,k} = \sum_{j=1}^{M} \times \left(\sum_{i=1}^{N} W_i R_{i,j} R_{i,k} \right) X_j$$

In this equation N = number of data points in each spectrum; M = number of standard spectra; $R_{i,k}$ = datum in the ith channel of the kth standard spectrum; S_i = datum in the ith channel of the composite spectrum; $W = 1/S_i$ = weighting factor; and X_j = number to be determined such that the sum of all standard (pure component) spectra, when multiplied by their X_j values, gives a best least-squares fit to the composite spectrum. The least-squares coefficients, as derived, are useful in determining proper scaling factors for absorbance subtraction techniques in FTIR. If the pure component spectra are properly scaled, then the least-squares coefficients yield the correct volume fractions (or weight or mole fractions, if desired) for a composite consisting of any mixture of those components. Besides, if the quality of the least-squares fit within random error, then a statistical error analysis may be performed to determine confidence intervals for each least-squares coefficient.

2.5.7. Absorbance Subtraction Method

One of the spectral processing operations most widely used in polymer analysis is the digital subtraction of absorbance spectra, in order to reveal or emphasise subtle differences between two samples or between a sample and a reference material. Spectral subtraction with FTIR is a powerful method of extracting structural information about components of composite spectra. When the epoxy resin is examined before and after a chemical or physical treatment, and the original spectrum is subtracted from the final spectrum, positive absorbances reflect the structures that are formed during the treatment, and negative absorbances reflect the loss of structures. The advantage of FTIR difference spectra lies in their ability to compensate for differences in the thickness of the two solid samples. The balance of thicknesses allows small spectral differences to be associated with structural changes and not to be outweighed by the difference in the amount of sample in the beams. Additionally, with properly compensated thickness, the differences in absorbances can be magnified through computer scale expansion to reveal small details of the spectral differences. The scaling parameter, k, is chosen such that:

$$(A_1 - kA_2) = 0$$

where A_1 and A_2 correspond to the absorbances of the internal thickness bands of samples one and two. Multiplication of the absorbance spectrum of sample two by k will yield a new spectrum having the same optical thickness as sample one. One may use the peak absorbance, integrated peak areas, or a least-squares curve fitting method to calculate the scaling factor k.

E. M. Pearce[46] demonstrated the subtraction method of FTIR to compare the functional group stability of epoxy resins. This is based on the following considerations.

If two functional groups x and y decrease their IR spectral absorbance at the characteristic frequency v_x and v_y, then the difference absorbance of each functional group can be expressed as:

$$A_s^{v_x} = A_2^{v_x} - k' A_1^{v_x}$$
$$A_s^{v_y} = A_2^{v_y} - k' A_1^{v_y}$$

In order to remove functional group x from the difference spectrum, the k' parameter can be obtained:

$$k' = A_2^{v_x}/A_1^{v_x} = C_2/C_1$$

where C is the concentration of the functional group x. Three situations can occur in relation to the absorbance of the functional y:

$$A_s^{v_y} - k' A_1^{v_y} = e_y b(C_2 - k'C_1) \lessgtr 0$$

where e_y is the extinction coefficient.

If the value is more than zero, the group y is more stable than the group x; if as stable, the expressions above equal zero; and if group y is less stable than group x, they are less than zero, under specific degradations.

2.6. SAMPLING TECHNIQUES FOR FTIR

Transmission spectroscopy, diffuse reflectance spectroscopy, and internal-reflection spectroscopy can be used to characterise the epoxy resin in composites.

The optics of the sampling chambers of commercial FTIR instruments are the same as those of the traditional dispersive instruments, so the accessories which are generally available commercially can be used. The main difference between the two types of instrumental optics is that the beam is round, and larger at the focus, for FTIR. Thus, some of the sampling accessories may block some of the beam energy in FTIR experiments. When energy is a limiting factor, the accessories can be modified to accommodate the larger beam. However, the improved sensitivity of FTIR allows one to obtain better sensitivity using the conventional sampling accessories and expand the range of sampling techniques.

In internal reflection spectroscopy (IRS) the sample is in optical contact with another material (e.g. a prism). The prism is optically denser than the sample; the incoming light forms a standing wave pattern at the interface within the dense prism medium, whereas in the rare medium the amplitude of the electric field falls off exponentially with the distance from the phase boundary. If the rare medium exhibits absorption, the penetrating wave becomes attenuated, so the reflectance can be written:

$$R = 1 - \mathrm{kd_c}$$

where d_c is the effective layer thickness. The resulting energy loss in the reflected wave is referred to as attenuated total reflection (ATR). When multiple reflections are used to increase the sensitivity, the technique is often called multiple internal reflection (MIR). Thus, qualitatively, an IRS spectrum resembles a transmission spectrum. There are two adverse effects arising from the wavelength dependence of IRS. First, the long wavelength side of an absorption band tends to be distorted, and secondly, bands of longer wavelengths appear relatively strong. With FTIR spectrometers, one does not achieve the same improvement in IRS as in transmission compared with dispersion instruments because the ATR attachments have not been redesigned for the larger, round beam. However, the signal-averaging capability and speed have increased the utility of IRS for polymers, particularly for surface studies. In IRS, the infrared beam penetrates the surface of the composite between a few tenths of a micron to a few microns depending on the type of reflection plate, the angle of incidence, and the wavelength of the infrared beam. The depth of beam penetration can be reduced by placing a thin barrier film between the trapezoidal reflection plate and the epoxy resin under study. Hirschfeld has generated the algorithms which are necessary to use IRS to determine the optical constants of a sample from a pair of independent reflectivity measurements at each frequency. The optimum method appears to be to determine the total reflectance at two polarisations at the same incidence angle. Another important sampling technique is diffuse reflectance measurements. When light is directed onto a sample it may either be transmitted or reflected. Hence, one can obtain the spectra by either transmission or reflection. Since some of the light is absorbed and the remainder is reflected, study of the diffuse reflected light can be used to measure the amount absorbed. However, the low efficiency of this diffuse reflectance process makes it extremely difficult to measure, and it was initially speculated that infrared diffuse reflectance measurements would be futile. Initially, an integrating sphere was used to capture all of the reflected light, but more recently improved diffuse reflectance cells have been designed which allow the measurement of diffuse reflectance spectra using FTIR instrumentation.

The requirement for reflectance to be diffuse is that the intensity of reflected light is isotropic but for a solid sample both scattering and absorption occur and since the scattered radiation is angularly distributed, it is by no means isotropic. However, with a large number of particles, as is found in a powder, an isotropic scattering distribution can be achieved, so the emerging light will still be diffuse. Kubelka and Munk used an empirical theory to relate the absorption coefficient (k) and the scattering coefficient (s).

$$f(r_\alpha) = (1 - r_\alpha)^{1/r_\infty} = k/s$$

where r_α is the absolute reflectance of an infinitely thick layer. In practice a standard is used and the following ratio is calculated:

$$r'_\infty = r'_\alpha(\text{sample})/r'(\text{standard})$$

The principal problem with diffuse reflectance is that the specular component of the reflected radiation, i.e. that which does not penetrate the sample, is measured along with the diffuse reflected light which penetrates the sample. Generally, the change in specular reflection with frequency is small, except in regions of strong absorption bands where the anomalous dispersion leads to restrahlen bands in the specular reflection spectrum. When the restrahlen bands are observed, the absorption bands can appear inverted at their centre. This effect makes quantitative measurements on samples with strong absorptivity very difficult.

For powdered samples, diffuse reflectance offers considerable advantages, particularly since no sample preparation is required which could change the morphology of the sample.

2.7. BAND ASSIGNMENTS FOR EPOXY RESINS

Some general assignments for common epoxy systems can be found in the *Handbook of Epoxy Resins* by Lee and Neville.[2] The literature contains several reports assigning characteristic absorptions to the epoxide group in small molecules. The infrared and Raman assignments for EPON 828 and NMA (nadic methyl anhydride) by Antoon[47] are shown in Table 2.1. Absorption decreases due to the reacting species during the crosslinking reaction may be observed at 3008 cm^{-1} (assigned to v_s (CH$_2$) of the epoxide ring), at 1858 and 1780 cm^{-1} (v_s (C=O) and v_{as} (C=O) of the anhydride

TABLE 2.1(a)
INFRARED AND RAMAN BAND ASSIGNMENTS FOR EPON 828

IR	Raman	Assignment	IR	Raman	Assignment
~3500		ν(OH)	1133	1134	
	3210w		1120w		
	3151w			1113s	
3123vw	3125w		1108		
3098w			1086	1087	$\delta(\phi$—H) in-plane
	3067s	$\nu(\phi$—H)	1076sh		
3057		ν_{as}(CH$_2$) epoxy		1055w	
3038		$\nu(\phi$—H)	1036	1040w	$\nu_s(\phi$—O—C)
	3006	$\nu(\phi$—H)	1012	1015w	$\delta(\phi$—H) in-plane
2998sh		ν_s(CH$_2$) epoxy		992w	
2968	2969	ν_{as}(CH$_3$)+	971		
		ν_{as}(OCH$_2$)?	936w	938	$\delta(\phi$—H) out-of-plane
2929	2927	ν(CH) epoxy?	916	917	epoxy ring
	2910		~906sh		
~2890sh			863	863	epoxy ring
2874	2871	ν_s(CH$_3$)		836	
2836	2833w	ν_s(OCH$_2$)	831		$\delta(\phi$—H) out-of-plane
2805vw				824s	$\delta(\phi)$ out-of-plane
2756vw	2755w		808sh	809	
	2710w		772		
~2064w		ϕ, disubstituted		765	
1891w		ϕ, disubstituted	758		
~1766w		ϕ, disubstituted	737w	737	
1608	1607s	ν(C=C)ϕ	727w		
1583	1583	ν(C=C)ϕ		677sh	
1511s	1510w	ν(C=C)ϕ		667	
	1481w	δ(CH$_2$) epoxy		654	
1470sh				650	
1457	1462	ν(C=C)ϕ +	639w		
		δ_{as}(CH$_3$)	604w		
1431w	1429	δ(OCH$_2$)?	586w		
1414w	1414w	δ(CH) epoxy?	575	580w	
1385	1385vw	δ_s(CH$_3$) gem-dimethyl	557		
1363	1360vw	δ_s(CH$_3$) gem-dimethyl	504w	498w	
1347	1348	δ(CH) epoxy	450	455w	
1312sh	1310sh			395	$\delta(\phi)$ out-of-plane
1298	1298	ν(C—O)+ν(C—C)?		374sh	
	1261	epoxy ring?		350w	
	1254	epoxy ring?		318w	
1248s		$\nu(\phi$—O)		277sh	
1230sh	1231	$\delta(\phi$—H) in-plane	240		
1185	1187	$\delta(\phi$—H) in-plane		213sh	
1157w	1154	ϕ		175w	

s = strong, w = weak, vw = very weak, sh = shoulder, ν = stretching, δ = deformation.

TABLE 2.1(b)

INFRARED AND RAMAN BAND ASSIGNMENTS FOR NADIC METHYL ANHYDRIDE

IR	Raman	Assignment	IR	Raman	Assignment
	3 077		1 039w		
3 072vw			1 015w	1 017w	
3 058w	3 060		1 003vw	1 006w	
3 018sh		$\nu(=C-H)$	990vw	989	
	2 993			970w	
2 981	2 985s	$\nu_{as}(CH_3)$	943s		anhydride ring
2 945		$\nu_{as}(CH_2)$	929	934s	anhydride ring
	2 924	$\nu(CH)$ tertiary?	916s	920	anhydride ring
2 917	2 916	$\nu_s(CH_3)$	899s	902	anhydride ring
2 879	2 882	$\nu_s(CH_2)$	868w	870w	anhydride ring
2 857sh	2 857		853vw		
2 828vw			843	844w	
	2 744w		816sh	819	
1 858s	1 859	$\nu_s(C=O)$	798	799w	$\delta(=CH)$ out-of-plane
	1 852		786sh		
~1 820?	1 834?		764sh		
1 780vs	1 781	$\nu_{as}(C=O)$	758w	759w	
1 704w			736w		
1 626w	1 627s	$\nu(C=C)$	712	717w	
	1 576		696w		
1 465w	1 465sh	$\delta(CH_2)$	673sh	672	
1 445	1 449	$\delta_{as}(CH_3)$	649sh		
1 382w	1 381	$\delta_s(CH_3)$	635w		
1 345vw		$\delta(CH)$	622	622s	
1 326w	1 329w	$\delta(CH)$ in-plane		599	
1 312vw			589w		
1 299			573w	576w	
1 289	1 290w		536w	539w	
1 277vw			506vw		
1 267w			495w	498w	
1 256vw			461w	465w	
	1 246w		447w		
~1 240			431w	434w	
1 228s	1 230w	$\nu(C-O)$		427w	
1 216			419vw		
1 194w			408vw	412sh	
1 180vw				376w	
1 137w	1 141			360w	
1 124vw	1 125			330w	
1 106w	1 107			242w	
1 083s		anhydride ring		211w	
	1 074w			188w	
1 053w	1 054w			154w	

vs = very strong, s = strong, w = weak, vw = very weak, sh = shoulder, ν = stretching, δ = deformation.

TABLE 2.1(c)

INFRARED SPECTRAL CHANGES DURING THE CROSSLINKING OF A STOICHIOMETRIC NADIC METHYL ANHYDRIDE/EPON 828 MIXTURE CATALYSED BY 2·0% WEIGHT BENZYLDIMETHYLAMINE

Intensity decrease	Assignment	Intensity increase	Assignment
∼3 060	$v_{as}(CH_2)$ epoxy		
3 010	$v_s(CH_2)$ epoxy		
		2 963	$v_{as}(CH_2)$
		∼2 904	$v(CH)$
		2 863	$v_s(CH_2)$
1 858	$v_s(C{=}O)$ anhydride		
1 780	$v_{as}(C{=}O)$ anhydride		
		1 743	$v(C{=}O)$ ester
		1 454	$\delta(CH_2)$
		1 398	$\omega(CH_2)$
		1 361	
		1 332	
		1 267	$v(C{-}O) + (C{-}C)$
1 228	$v(C{-}O)$ anhydride		
		1 178	$v(C{-}O)$ ester
		1 155	$v(C{-}O)$ ester
		1 127	$v(C{-}C)$? ester
		1 112	
1 083	anhydride ring		
		1 056	
		1 012	
942	anhydride		
928	anhydride		
915	anhydride + epoxy ring		
898	anhydride		
865	anhydride + epoxy ring?		
842			
798			
713			

v = stretching, δ = deformation, ω = wagging.

ring), and 916 cm^{-1} (epoxide ring). Intensity increases after the crosslinking reaction are at 2963 cm^{-1} (v_{as} (CH$_2$) adjacent to the ester group), at 1743 cm^{-1} (v (C=O) ester), and at 1778 cm^{-1} (v (C—O) of the ester). Table 2.2 shows the IR spectra of three epoxy resins after curing and before degradation by Pearce.[46] The 3550 cm^{-1} band for the resins was assigned to stretching the —OH group. The absorptions at 3060 and 3036 cm^{-1} are due to C—H stretching of the —CH$_3$ group. C—H

TABLE 2.2

TENTATIVE INFRARED ABSORPTION ASSIGNMENTS FOR THE THREE CURED EPOXY RESINS AND THE ABSORPTION VARIATIONS DURING DEGRADATION

Wavenumber (cm^{-1})	DGEBA IR[a]	DGEBA TD[b]	DGEBA TO[c]	DGEBA PO[d]	DGEPP IR	DGEPP TD	DGEPP TO	DGEPP PO	DGEPF IR	DGEPF TD	DGEPF TO	DGEPF PO	Functional group	Vibration mode
3 570	×[e]								×				⎫ R—OH	v(O—H)
3 550		+[f]			×								⎬	v(O—H)
3 525			+			+	+				+		ArOH	v(O—H)
3 430						+	+						O=ArCOH	v(O—H)
3 350								+				+	⎫ R—OOH	v(O—H)
3 300								+					⎬	v(O—H)
3 200				+								+	Ar—OOH	v(O—H)
3 068													⎫	v(C—H)
3 060														
3 052	×	−[g]			×	−	−	−	×	−	−	−	⎬ Arylene	v(C—H)
3 038	×	−	−		×	−	−			−	−	−		
3 034	×	−	−	−	×	−	−			−	−	−		
2 970	×	−	−	−						−	−	−	⎫ Methyl	v(C—H)
2 935	×	−												
2 933					×	−	−	−						
2 980										−	−	−	⎫	v(C—H)
2 880		−							×				⎬ Methylene	
2 876	×	−			×	−			×	−	−	−		

(continued)

TABLE 2.2—contd.

Wavenumber (cm⁻¹)	DGEBA IRᵃ	TDᵇ	TOᶜ	POᵈ	DGEPP IR	TD	TO	PO	DGEPF IR	TD	TO	PO	Functional group	Vibration mode
1808		+	+	+		+	+			+	+	+	RC(=O)—O—C(=O)R (anhydride)	$\nu(C{=}O)$
1790														$\nu(C{=}O)$
1784			+	+			+			+	+	+	RCOOH	$\nu(C{=}O)$
1782	×				×			−					Lactone	
1775						−								$\nu(C{=}O)$
1765	×	+	+	+		+	+			+	+	+	RCOOR	
1761	×				×	−	−						Lactone	
1745		+	+	+						+	+	+	RCOR	
1732	×	+	+	+					×	+	+	+	RCHO	
1725						+	+	+					(o-hydroxybenzaldehyde-type ring, COH, C=O)	
1715	×	+	+	+	×	+	+	+	×	+	+	+	RCOH	$\nu(C{=}O)$

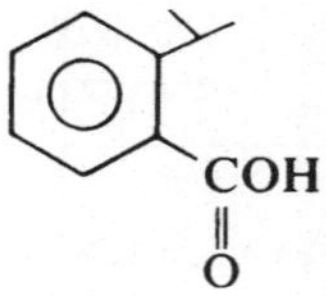

	Quadrant stretching				Semicircle stretching						
	Phenoxy	Phenylene	Phenylene								
	1665	1610	1578	1510	1505	1480	1465	1448	1440	1425	1412
	+	−	−	−	+			−		+	−
	+	−	−	−		+		−		+	−
		−	−	−				−		+	−
		×	×	×				×			×
	+	−	−	−		+	−		+		−
	+	−	−	−		+	−			+	−
	+	−	−	−			−		+		−
		×	×	×			×				×
	+	−	−	−		+					−
	+	−	−	−		+				+	−
	+	−	−	−		+				+	−
		×	×	×							×

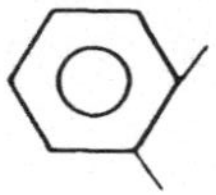

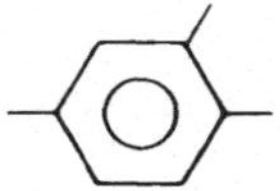

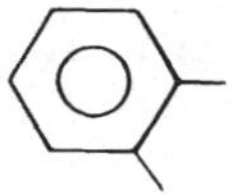

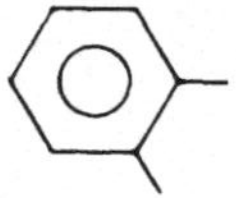

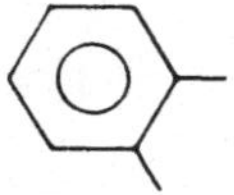

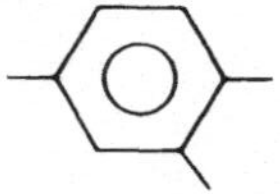

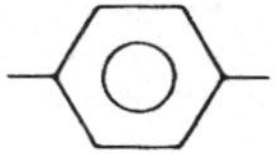

(continued)

TABLE 2.2—contd.

Wavenumber (cm^{-1})	DGEBA IR[a]	TD[b]	TO[c]	PO[d]	DGEPP IR	TD	TO	PO	DGEPF IR	TD	TO	PO	Functional group	Vibration mode
1 288	×			—	×	—	—	—					Lactone Ester or acid	ν(C—O—C) ν(C—O—C) or ν(C—O)
1 286				+		+		+						
1 280			+	+				+			+	+	⬡—OH	ν(C—O)
1 280			+	+				+			+	+	⬡—OH	ν(C—O)
1 255	×	—	—	—	×	—	—	—		—	—	—	Ar—O—R Ar—OH	ν(C—O—C) ν(C—O—C)
1 245				+				+	×			+		
1 202		—		—				—		—	—	—	Ar—C—Ar C	ν(C—C)
1 184	×			—	×			—		—	—	—	Ar—C—Ar C	ν(C—C)
1 180									×				Ar—C—Ar C	ν(C—C)
1 170		+	+	+		+		+		+	+	+		ν(C—C)
1 120	×	—	—	—	×	—	—	—	×	—	—	—	Aliphatic chain	ν(C—C)

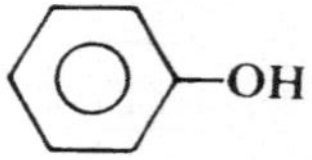

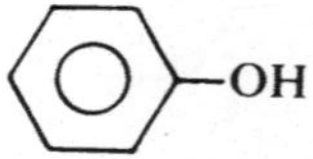

Wavenumber (cm⁻¹)	Assignment		IR / TD / TO / PO observations (left → right)
1096	Aliphatic ether	ν(C—O—C)	− − − × × × × × ×
1086			− − − × × × ×
1035			− − − × × × ×
1015	Lactone	ν(C—O—C)	− − − × × ×
1010			− − − × × ×
965			×
940			×
925			×
890	Peroxide	ν(O—O)	+ + + ×
885			− − −
840			
830	p-Phenylene	In-phase out-of-plane hydrogen wagging	+ + + × × ×
825			× × ×
822			×
775	o-Phenylene	In-phase out-of-plane hydrogen wagging	+ + + ×
754			− − −
746			+ + +
740			× ×
736			+
725			−
715			
692	o-Phenylene	Out-of-plane sextant ring bending	+ + + − ×

[a] IR: Original IR spectrum of epoxy resin.
[b] TD: Thermal degradation.
[c] TO: Thermooxidative degradation.
[d] PO: Photooxidative degradation; stretching.
[e] ×: Absorption present in the original spectrum.
[f] +: Absorbance increase during degradation.
[g] −: Absorbance decrease during degradation.
[h] The absorption formed possibly by decreasing peak, 1782 cm⁻¹ and increasing peak, 1784 cm⁻¹.

stretching of the —CH_2 group can be assigned to 2933 and 2878 cm^{-1}. The absorption at 1732 cm^{-1} may be assigned to stretching of the carbonyl groups in DGEBA and DGEBF. Bands at about 1610 and 1578 cm^{-1} are derived from quadrant stretching of the benzene rings.

2.8. INFRARED STUDIES OF CURING OF EPOXY RESINS

The studies of curing of epoxy resins with infrared spectroscopy are based on the absorption intensity of the epoxy, anhydride and hydroxyl functional groups that appear at 913, 1858 and 345 cm^{-1}, respectively. In other cases the near-IR spectra of the epoxy functional group are used. Since glass fibres have a strong absorption in the mid-IR region, it is very difficult to measure the IR spectra of epoxy matrices with dispersive spectrometers for glass-fibre-reinforced composites. Perkinson[48] developed a method of separating the epoxy resin from the cured fibreglass–epoxy composites. The method is based on the concept of differential flotation, which involves the separation of two materials of different specific gravity with a liquid of intermediate specific gravity. Infrared spectra of three differently prepared specimens of the same lot of prepreg composite were compared, and the results demonstrate the feasibility of this approach for cured composites. The low signal-to-noise resolution limit and the time needed for scanning the entire spectrum with a dispersive instrument, limit any application of conventional IR to the study of reactions with short half lives. Since FTIR can solve this kind of problem with rapid scanning (1 sec), Buckley and Roylance[7] demonstrated this technique by studying the kinetics of a sterically hindered amine-cured epoxy resin system. Variation in epoxide absorbance caused by differences in specimen thickness can be eliminated by normalising the epoxide peak height to an internal reference peak, which appears at 1510 cm^{-1} due to the phenyl groups.

$$f_{915} = \frac{A_{915,t}}{A_{1510,t}} \times \frac{A_{1510,0}}{A_{915,0}}$$

where f_{915} = fraction unreacted epoxide at time t; $A_{915,t}$ = specimen absorbance at 915 cm^{-1} due to epoxide at time t; $A_{1510,t}$ = specimen absorbance at 1510 cm^{-1} at time t; $A_{915,0}$ = initial absorbance at 915 cm^{-1}; and $A_{1510,0}$ = initial absorbance at 1510 cm^{-1}.

Dannenberg[49] used near-IR to study the curing reaction of epoxy resins, and demonstrated the advantages of this technique, which is very sensitive due to the fact that the epoxide group has a strong overtone absorption frequency in the near-IR region. Several studies[7,8,50-52] have been reported on the epoxy curing process by FTIR. Appropriate programs for data processing are available, and are easily modified to any specific reaction needs. These techniques have been found very helpful in interpreting the complicated crosslinking reactions of epoxy resins. Sprouse, Halpin and Sacher[53] used FTIR spectroscopy to measure the extent of cure in fibre-reinforced-epoxy composites by two different techniques. The two sampling techniques are:

(1) thin films of neat resin held between salt plates, and
(2) internal reflectance spectroscopy.

Thin films were cured in the same program as used in the corresponding composite fabrication. Infrared spectra were recorded at short time intervals throughout the cure cycle. Internal reflectance measurements were performed on the corresponding composites and results were compared with thin film measurements. Antoon, Starkey and Koenig[54] demonstrated the utility of FTIR difference spectra for investigating the composition of neat epoxy resin, hardener, and catalyst, as well as the composition and degree of crosslinking of the cured matrices. Improved precision is achieved by using a least-squares curve-fitting program for the determination of the composition of the uncured and cured epoxy matrices mixtures.

2.9. QUALITY CONTROL OF EPOXY MATRICES IN FIBRE-REINFORCED COMPOSITES

Fibre-reinforced composite structures have gained a significant position as materials of construction for the aerospace industry and transportation industry. It thus follows that large expenditures of funds, to say nothing of human lives, are dependent on the reliability of these products. An important step in gaining confidence in a product is knowing that the starting prepreg has the same chemical formulation, and each batch of material has been processed in the same manner.

Antoon, Zehner and Koenig[55] developed a general and convenient method for determining simultaneously the initial resin composition and the extent of crosslinking of epoxy matrices. This technique can be used as a quality control method to determine that the starting prepreg has the same

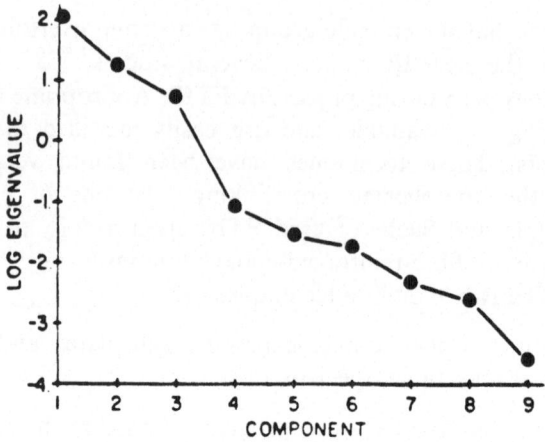

FIG. 2.9.   Factor analysis of crosslinked epoxy/anhydride resin in the 2000 to 1400 cm^{-1} region. Tertiary amine-catalysed reactions.

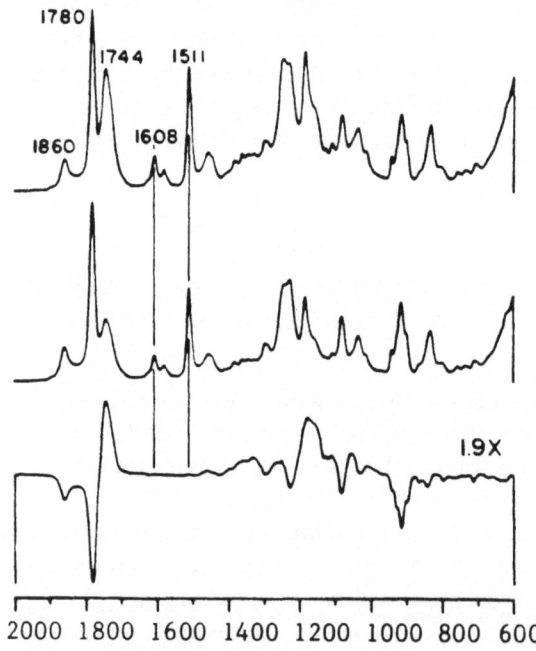

FIG. 2.10.   Generation of difference spectrum characteristic of 80 °C crosslinking of stoichiometric mixture of NMA/EPON 828. Top: crosslinked 83 min. Middle: crosslinked 37 min. Bottom: difference spectrum (top minus bottom).

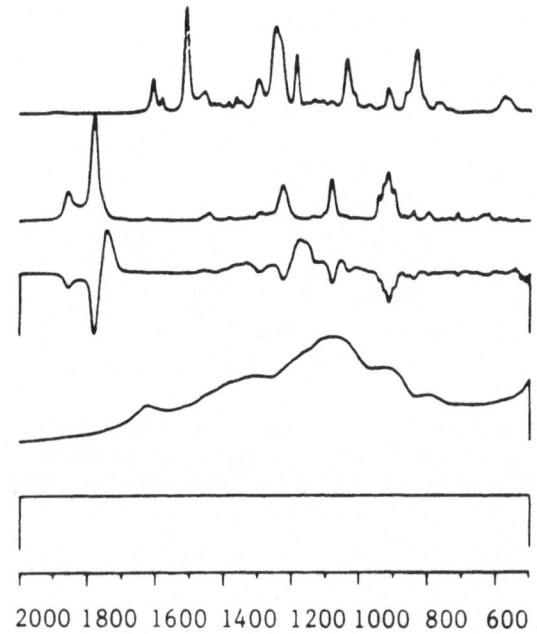

2000 1800 1600 1400 1200 1000 800 600

FIG. 2.11. Spectral components in a crosslinked epoxy/anhydride (polyester) matrix filled with S-glass. From top: EPON 828 epoxy resin, nadic methyl anhydride, crosslinking difference spectrum (from Fig. 2.2), S-glass and boxcar function.

chemical formulation and that each batch has been processed in the same manner.

Figure 2.9 shows the results of the factor analysis of crosslinked epoxy/anhydride resins in the 2000–1400 cm^{-1} region. The plot of log (eigenvalue) in descending order reveals a break in the eigenvalue magnitude between the third and fourth eigenvalues. As indicated by Antoon and Koenig, the number of these larger eigenvalues is equal to the number of components in the system. Therefore, the spectrum of the crosslinked epoxy matrix may be approximated by a linear combination of only three linearly independent component spectra. The three spectra chosen to represent the components were those of pure NMA, pure EPON 828, and a difference spectrum characteristic of the crosslinking reaction. The difference spectrum was calculated by subtracting the spectrum of a stoichiometric mixture of NMA and EPON 828 crosslinked for 37 min at 80 °C from the spectrum of the same reactant mixture crosslinked for 83 minutes at 80 °C. The procedure is illustrated in Fig. 2.10. Figure 2.11 shows the absorbance spectra employed for the analysis of the crosslinked

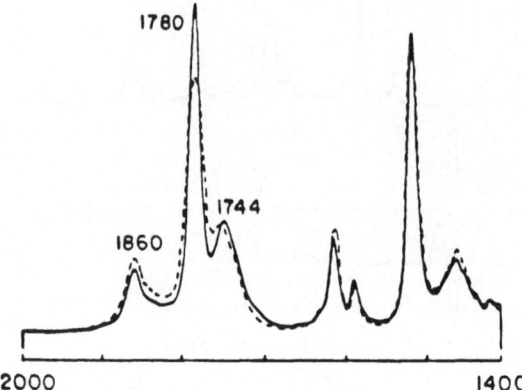

FIG. 2.12. Least-squares fit of the spectra EPON 828 and nadic methyl anhydride, the crosslinking difference spectrum and the boxcar function to the spectrum of a partially crosslinked polyester matrix. Solid line, experimental matrix spectrum; and dotted line, fitted spectrum.

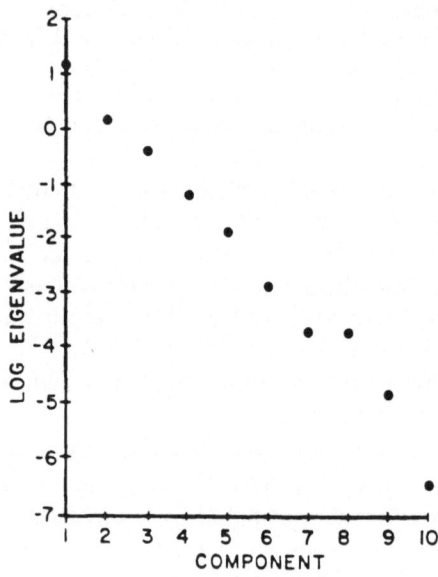

FIG. 2.13. Factor analysis of the spectra of crosslinked polyester composites in the 1550 to 850 cm^{-1} region.

epoxy matrix. A boxcar function is included as a fourth component in the least-squares analysis of the resin, in order to prevent errors in the background level from affecting the accuracy of the least-squares coefficients. The spectrum of S-glass, also shown in Fig. 2.11, is included as a fifth component in the least-squares curve-fit analysis for S-glass reinforced composites. After performing the least-squares curve-fitting, the quality of the fit between the experimental resin spectrum and the least-squares fitted spectrum is illustrated in Fig. 2.12. The least-squares coefficient corresponding to the difference spectrum is a measure of the extent of crosslinking, and the coefficients corresponding to the pure NMA and pure epoxy are a measure of the amount of reactants.

Figure 2.13 shows the factor analysis of glass-fibre-reinforced epoxy composite. Since there is no obvious separation between the nonzero and error eigenvalues, it is an arbitrary decision to determine the number of components that contribute to the composite spectrum. Antoon and Koenig attribute this problem to systematic errors in the spectra themselves Figure 2.14 shows a least-squares curve-fitting of five component spectra to the spectrum of a composite. As can be seen, the fitted curve and the experimental result coincide satisfactorily. The results for nonreinforced matrices consisting of stoichiometric mixtures of epoxy and anhydride crosslinked to various extents are given in Table 2.3. The weight percent

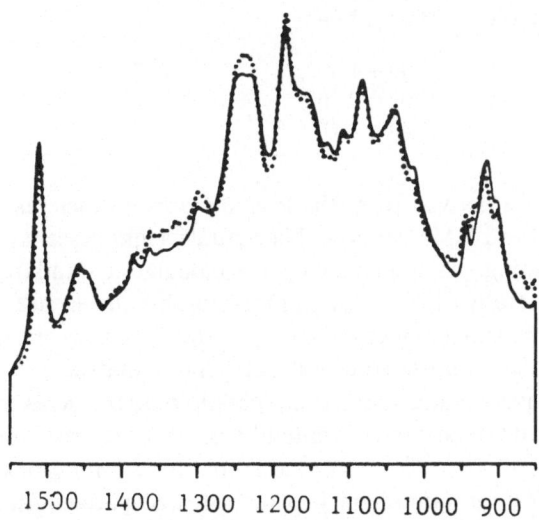

1500 1400 1300 1200 1100 1000 900

FIG. 2.14.   Least-squares fit of five component spectra (from Fig. 2.3) to the spectrum of a composite in the 1550 to 850 cm^{-1} region.

TABLE 2.3

LEAST-SQUARES ANALYSIS 2000–1400 cm^{-1} OF 1:1 STOICHIOMETRY ANHYDRIDE:EPOXIDE MIXTURES (0.5% WEIGHT BDMA CATALYST) CROSSLINKED AT 90°C

Uncured	Least-squares calculation (%)	Composition by weighing	Extent crosslinked from I(1860)/I(1608)
Uncured			
EPON 828	51·99 ± 1·89 (wt %)	50·77	
NMA	48·01 ± 0·96 (wt %)	49·23	
Extent crosslinked	−1·65 ± 1·37		0·0
30 min at 90°C			
EPON 828	52·06 ± 1·97	50·77	
NMA	47·94 ± 1·00	49·23	
Extent crosslinked	5·97 ± 1·42		6·6
60 min at 90°C			
EPON 828	51·62 ± 2·11	50·77	
NMA	48·38 ± 1·07	49·23	
Extent crosslinked	12·42 ± 1·51		13·2
90 min at 90°C			
EPON 828	51·58 ± 2·24	50·77	
NMA	48·42 ± 1·14	49·23	
Extent crosslinked	18·47 ± 1·60		18·8
120 min at 90°C			
EPON 828	51·11 ± 2·31	50·77	
NMA	48·89 ± 1·17	49·23	
Extent crosslinked	28·14 ± 1·64		31·0
6 h at 90°C			
EPON 828	51·99 ± 1·89	50·77	
NMA	48·01 ± 0·96	49·23	
Extent crosslinked	65·35 ± 1·36		70·2

compositions are derived from the least-squares coefficients after scaling the EPON 828 and NMA spectra. The actual weight percent compositions are from weighing. As indicated by the calculated standard errors, the accuracy is generally within 2% and most notably the accuracy is retained even at very high extents of crosslinking. Table 2.4 shows the least-squares analysis of S-glass reinforced crosslinked epoxy matrix.

The least-squares analysis of an epoxy matrix yields reproducible information, with an accuracy limited by several spectroscopic problems. A source of error comes from sample preparation limitations. The wedge effect occurs from nonuniformity in the infrared pathlength through the sample, which results in a deviation from Beer–Lambert law and causes a weakening of the stronger absorption bands in highly absorbing materials.

TABLE 2.4

LEAST-SQUARES ANALYSIS OF S-GLASS REINFORCED CROSSLINKED EPOXY MATRIX

Temperature, time of curing	E (predicted)	$2\,000$–$1\,400\,cm^{-1}$		$1\,550$–$850\,cm^{-1}$	
		R	E	R	E
80 °C, 70 min	31·1	1·2	51	0·93	45
80 °C, 90 min	39·3	1·1	59	0·88	55
80 °C, 110 min	47·5	1·1	57	0·90	59
80 °C, 130 min	55·6	1·0	62	0·92	62
80 °C, 180 min	63·7	1·0	70	0·96	67
160 °C, 30 min	83·5	1·0	82	0·93	80

Another source of inaccuracy is the change in the spectra of NMA and EPON 828 when the molecules are in electronic environments different from those of the pure liquids. Finally, since the refractive indices of KBr, crosslinked resin, and S-glass are substantially different, significant nonlinearity in spectra occurs by the scattering of the infrared beams. Although there exist some spectroscopic problems, Antoon and Koenig[55] suggest some methods of improvement. Statistical methods of determining the number of components present may be helpful. Multiple difference spectra, each representing the crosslinking at each stage, can improve the sensitivity of least-squares methods.

2.10. INFRARED STUDIES OF DEGRADATION OF EPOXY RESINS

Another important application of infrared spectroscopy in the characterisation of cured epoxy composites is to determine the effect of in-service exposure conditions (degradation, hydrolysis, weathering, aging) on composites. The surface of a glass-fibre-reinforced epoxy composite is degraded rapidly upon outdoor exposure unless it is protected by a UV-absorber or a paint. The degradation phenomena are difficult to study because of uncertainty about the epoxy resin composition and the intractable nature of the cured composite. These difficulties prevent the use of conventional techniques of polymer analysis. Although DSC can be used to evaluate the thermal stability of epoxy composites, it cannot provide enough information to interpret the degradation process. Several techniques, such as gas chromatography, chemical analysis, and mass spectroscopy have been developed to study degradation phenomena. These

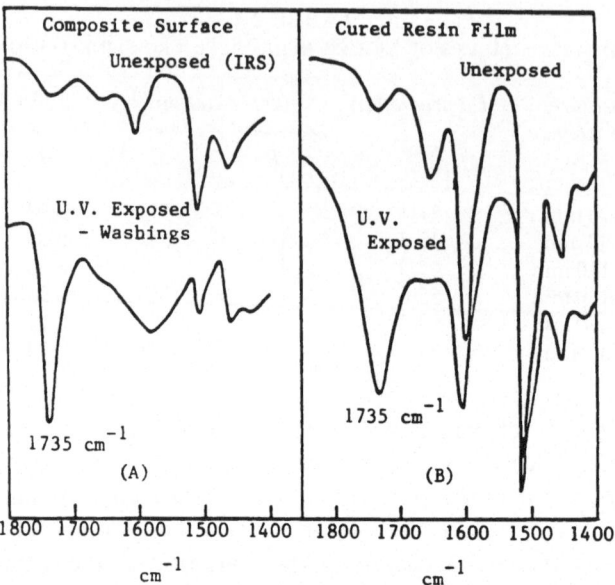

FIG. 2.15. (A) IRS–IR spectrum of unexposed 1009-26 composite surface and transmission IR of surface washings from composite surface exposed for 4000 h to sunlamp. (B) Transmission IR of air-cured 1009 resin film before and after 55 h of exposure to sunlamp.

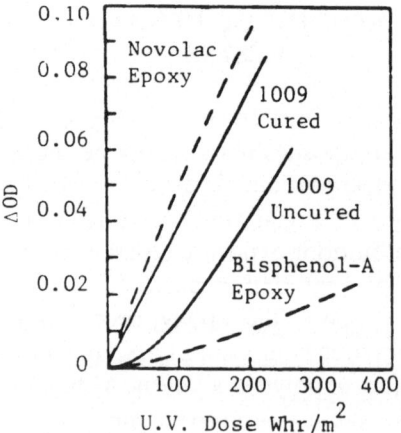

FIG. 2.16. Change in aliphatic carbonyl concentration measured by absorption intensity at $1735\,cm^{-1}$ (ΔOD) with total UV dose from 300 to 350 nm for resin systems indicated.

methods are based on the analysis of the degradation products derived from the original polymer. However, these methods are complicated, since the degradation of materials at high temperatures may cause rearrangement of the degradation fragments, or other reactions. Therefore, degradation product analysis may lead to a false conclusion about the mechanism. On the other hand, FTIR can eliminate this problem. George, Sacher and Sprouse[56] investigated the photo-protection of the surface resin of a glassfibre-reinforced composite with FTIR using a single-pass internal reflectance attachment for the surface study. As shown in Fig. 2.15, the IR spectrum shows a strong ester carbonyl band at $1735\,cm^{-1}$, which is used as an index of photo-oxidation under exposure to the sunlamp. This band can be used to measure the photo-oxidation rates of different epoxy resins, as shown in Fig. 2.15. The 1009 resin is a mixture of novolac epoxy and Bisphenol A epoxy.

Figure 2.16 shows the oxidation rates for different epoxy resins. The observed oxidation rate for the epoxy novolac is eight times that of the Bisphenol A epoxy. The high photo-oxidation rate of the cured epoxy novolac is related to the cure process itself.

The uncured resin showed negligible absorption of solar radiation (Fig. 2.17). When heated at 165 °C for a short time, an intense absorbing chromophore is formed. If the cure is carried out under vacuum, chromophore formation can be minimised. It was found that novolac

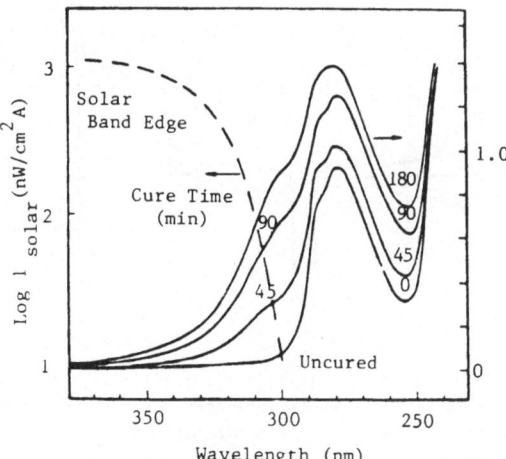

FIG. 2.17.   Changes in the UV absorption spectrum of $1\cdot5\,\mu m$ film of 1009 resin during cure in air at 165 °C for the time shown. The reported solar spectrum in this region is also shown.

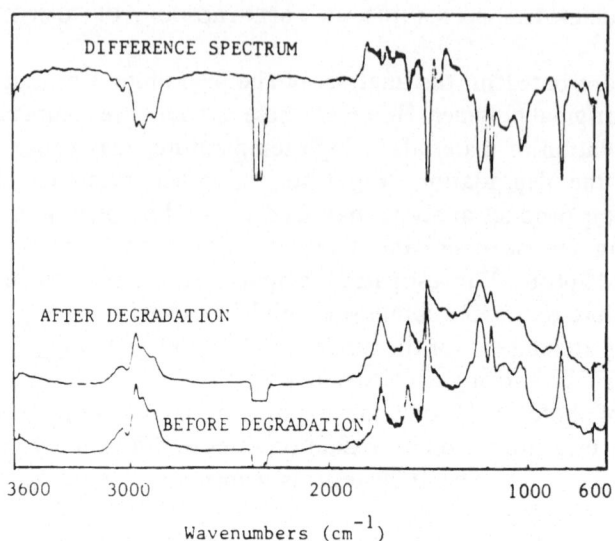

FIG. 2.18.   IR and difference spectra of cured DGEBA before and after thermal degradation at 300 °C.

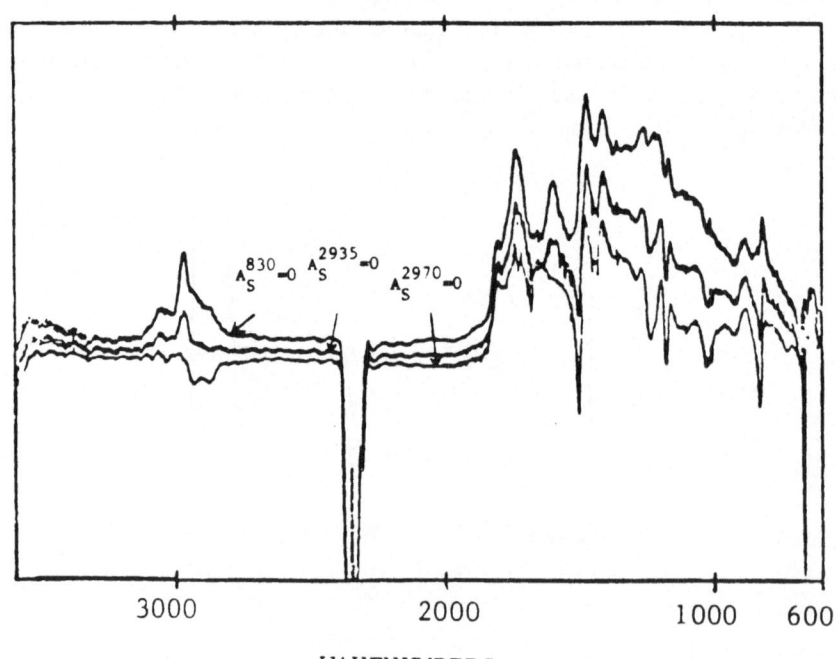

FIG. 2.19.   Variations in the thermal degradation difference spectrum of cured DGEBA during cancelling process.

epoxy shows the chromophore formation which explains why the photo-oxidation rate of the novolac epoxy is higher. On the other hand, from the changes in the IR spectrum during cure, a strong carbonyl group absorption was found, due to oxidation, and this absorption does not occur if the cure is under vacuum. Therefore, the weathering stability of an epoxy resin can be affected by the conditions of cure.

Pearce used the subtraction of absorbance spectra to compare the functional group stability of epoxy resin. Figure 2.18 shows the IR spectra and difference spectra of cured DGEBA, before and after thermal degradation. The absorption bands of the cured resin have decreased in intensity and new bands have appeared. To understand the relative stability of the functional groups in the resin, as shown in Fig. 2.19, for DGEBA, the subtracted spectra were obtained by changing the k parameter. After the band at $830\,\mathrm{cm}^{-1}$ (p-phenylene) is cancelled ($A_s^{830} = 0$) it shows that the difference absorbances A_s^{3052}, A_s^{3034}, A_s^{2970}, A_s^{2935} and A_s^{2876} change from negative to positive. These changes indicate that the p-phenylene group is not as stable as the —CH$_3$ and —CH$_2$— groups and may rearrange to a more stable form of substituted benzene species. After cancelling the —CH$_3$ group absorption, the C—H stretching frequencies almost disappear, while those of the —CH$_2$— group show negative difference absorbances. The result indicates that the —CH$_3$ group has similar stability to the benzene ring and higher stability than the —CH$_2$— group. Repeating this method, Pearce established the order of functional group stability as total methyl group ~ total benzene group > methylene > p-phenylene > ether linkage > isopropylidene. Based on FTIR analysis, it is proposed that initially the isopropylidene group degrades, releasing the first methyl group and retaining the second methyl group until the latter stages of degradation. The p-phenylene group undergoes a Claisen rearrangement and forms a 1,2,4-trisubstituted benzene. Other possible initial degradation steps are proposed. The oxidative, thermal, and photo-degradations were found to be related to the auto-oxidative degradation processes for aliphatic hydrocarbons. The Wieland rearrangement and Norrish type reactions, as well as other possible oxidation degradation mechanisms, were also suggested.

2.11. STUDY OF THE EFFECTS OF MOISTURE

On the other hand, Antoon and Koenig utilised FTIR to identify the irreversible chemical effects of moisture on an anhydride-cured epoxy

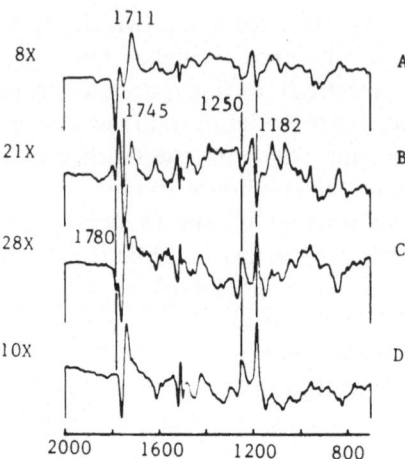

FIG. 2.20. Difference spectra showing the progression of spectral changes occurring in epoxy resin in 80 °C distilled water. Same sample as in Figure 2.1. (A) 4.4 days minus zero days exposure. (B) 35 days minus 12 days. (C) 50 days minus 35 days. (D) 224 days minus 98 days.

resin.[57] Exposure of a crosslinked epoxy film to 80 °C liquid water environment caused a hydrolysis of the unreacted anhydride groups (approximately 5 % of the initial anhydrides) to diacids. This effect, dominant during the early stages of moisture exposure, is illustrated in Fig. 2.20. The decrease in anhydride concentration is indicated by the intensity decrease of NMA vibrations at 1860 and 1080 cm^{-1}. Figure 2.20 also shows the effect of long-term exposure of the same epoxy film to 80 °C liquid water. The short-term effect is hydrolysis and leaching of unreacted NMA molecules; the longer-term effect rules out matrix hydrolysis and may be due to subtle structural changes such as perfection of the hydrogen bonding of polar groups.

The effect of stress on moisture stability is shown in Fig. 2.21. The intensity decrease at 1744 cm^{-1} suggests a loss of ester group; intensity increases at 3500 and 1720 cm^{-1} may be assigned to the formation of alcohol groups and carbonyl groups, respectively. Such a hydrolysis reaction, though very slow, is expected to be important in initiating irreversible matrix degradation. As shown by Antoon and Koenig, the degradation effect is slightly more rapid when the epoxy resin is under stress. In Fig. 2.22 the relative intensity of the ester carbonyl peak at 1744 cm^{-1} is plotted versus exposure time in pH 11.9 water (80 °C) for epoxy films with a range of applied tensile stress levels. High stress dramatically increases the hydrolysis rate.

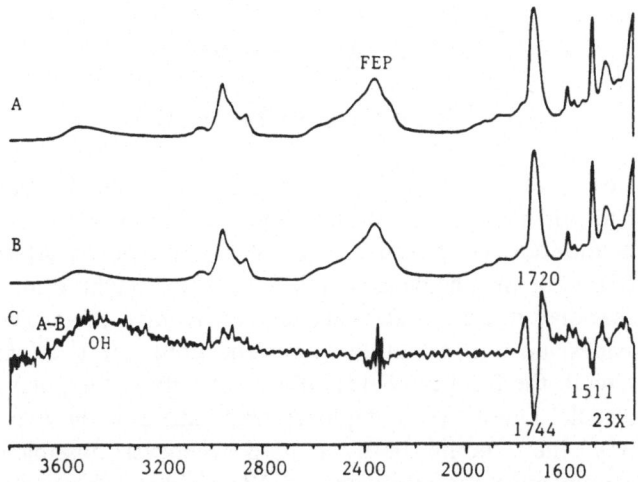

FIG. 2.21. Exposure of epoxy resin to 100 % relative humidity air at 80 °C while under $2 \cdot 6 \times 10^8$ dyn/cm^2 tensile stress. (A) Absorbance spectrum after 155 days exposure; (B) absorbance spectrum after 71 days exposure; (C) difference spectrum, A minus B.

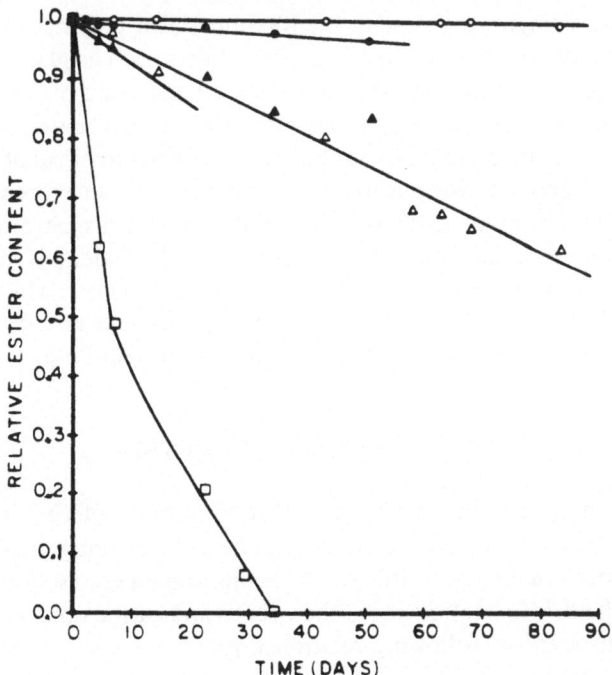

FIG. 2.22. Relative ester content (1744 cm^{-1} intensity) versus time for epoxy resins under varying tensile loads. Environment is pH $11 \cdot 9$, water at 80 °C. Stress (dyn/cm$^2 \times 10^{-8}$): $\bigcirc$, $0 \cdot 0$; $\bullet$, $1 \cdot 9$; $\triangle$, $2 \cdot 6$; $\blacktriangle$, $4 \cdot 0$; $\square$, $5 \cdot 4$.

2.12.   RAMAN SPECTROSCOPY

Since the advent of lasers, Raman spectroscopy has become an important analytical tool in polymer research. Raman scattering occurs for those vibrational motions which produce a polarisation or distortion of the electron charge of the chemical bonds. Thus the stretching motion of a homonuclear diatomic molecule is active in the Raman effect. The Raman effect provides more information about the nonpolar portions of the molecules, while the IR effect yields information about the polar portions of the molecule. Due to the complementary nature of the two types of spectroscopy, they should both be used whenever possible; Raman spectroscopy enhances the effectiveness of IR for solving chemical structure problems, and vice versa.

Since the Raman effect is a light scattering rather than a light transmission process, the transparency, size, and shape of the samples are relatively unimportant. Thus, one can run large samples or extremely small samples with comparative ease in the Raman. Filled polymer composites present difficulty for IR investigations, since fillers such as glass, clay, and silica are strong IR absorbers that block the IR spectrum of the polymer. These particulate fillers (glass, silica) are poor Raman scatterers, so the Raman spectrum of the polymer is obtainable without removal of the filler.

Only a few reports have appeared in the literature. Lu and Koenig[58] used Raman spectroscopy to study the curing of epoxy resins. Strong Raman lines which are characteristic of epoxy resin and independent of the state of cure are found at 640, 823, 1114, 1188, 1232, and 1608 cm^{-1}, which are due to the Bisphenol A skeleton. The epoxy group has lines at 768, 809, and 1156 cm^{-1}, which are sensitive to the degree of crosslinking.

2.13.   DIELECTRIC ANALYSIS

Dielectric analysis is based on measuring the ability of the dipole in a system to align with an oscillating electric field. In aligning the dipoles, a certain amount of energy is utilised. By subjecting an epoxy system to the oscillating field, information about the viscous and elastic properties can be obtained through the following relationships:

$$E = E_0 \exp iwt$$
$$D = D_0 \exp (iwt + \delta)$$

where E is the amplitude of the electric field, D is the displacement, w is the

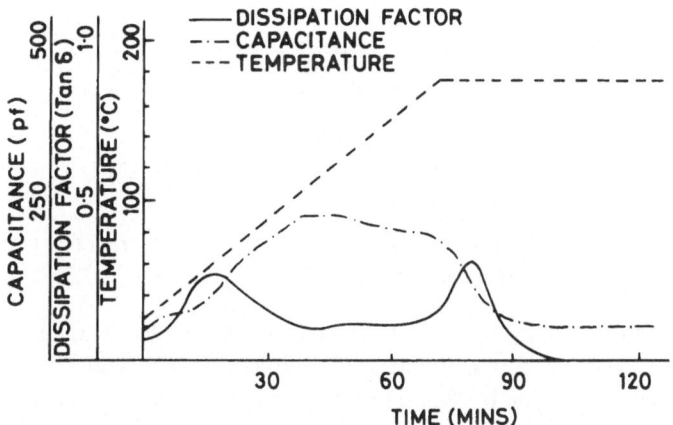

FIG. 2.23.   Dielectric analysis of 350 °F epoxy prepreg.

frequency and t is the time. Furthermore, we have $\bar{D} = \bar{E}\bar{F}$ where $\bar{E}$ is the complex dielectric constant and is equivalent to $(E' - iE'')$, the real and imaginary components.

The ratio $E''/E' = \tan\theta$ is called the loss factor and is related to the well-known G''/G' ratios for viscoelastic materials. The dielectric spectrometer is capable of giving a direct measurement of the dissipation or loss factor for epoxy resins when subjecting to an alternating electric field. Hence, changes in the flow characteristics are discernible by this method. The degrees of freedom of the dipoles are indicative of the viscous properties of a material. A typical dielectric analysis is shown in Fig. 2.23, where capacitance, dissipation factor, and temperature are plotted as a function of time.

The dissipation factor curve is related to the curing of the epoxy matrix. The left-hand peak of the curve corresponds to the softening and flow of the resin system. The right-hand peak is associated with the setting of the matrix. The portion between the peaks shows low dissipation since the epoxy resin displays a less viscous behaviour. This is the region where process changes, such as the application of pressure for consolidation of the laminates, are performed.

Examples of the use of dielectrometry in the study of composites have been reported in the literature. Sanjana and Rosenblatt[60] demonstrate the feasibility of using dielectric analysis as a means of monitoring and controlling the cure of epoxy composites in autoclave moulding. The mechanical properties are correlated with the variables on the dissipation factor profiles.

Another method related to dielectric changes when an epoxy cures is an

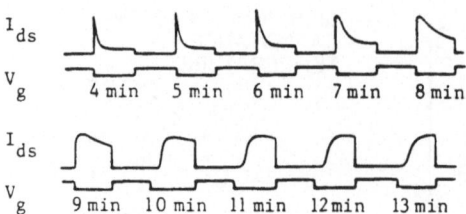

FIG. 2.24. CFT drain current waveforms (I_{ds}) during cure of a commercial
5-minute epoxy at room temperature.

electric monitoring technique[59] based on the charge-flow transistor, which
resembles a conventional metal-oxide-semiconductor-field-effect transistor, but with a portion of the metal gate replaced by the epoxy resin under
study. The dramatic change in the shape of the electrical signal during cure
can be related to corresponding changes in both the real and imaginary
parts of the dielectric constant and can be used for monitoring curing. A
typical signal output is shown in Fig. 2.24.

The instrumentation in dielectric analysis is simple, and since the peaks
in the curve, resulting from relaxation phenomena and associated with
softening and gelation of epoxy resin, are frequency dependent, they do not
define the point in time when softening or gelation occurs. Therefore, care
should be taken in interpreting the results.

2.14. CONCLUSIONS

Several methods are reviewed in this paper for characterising the epoxy
matrix in glassfibre-reinforced composites. TBA and dielectric analysis
yield the rheological changes during curing that are useful in composite
processing. DSC gives the extent of cure on the assumption that it is
proportional to the heat of reaction, but this assumption is invalid for
complicated crosslinking reactions. Infrared spectroscopy is very useful in
studying the mechanism of cure and extent of cure. FTIR is an ideal tool for
investigating the degradation behaviour of composites, and for controlling
the quality of epoxy matrices in fibre-reinforced composites. A combination of these techniques will become a powerful method in the study of
composites.

ACKNOWLEDGEMENT

We wish to express our gratitude to the US Army for their support of the
work described in this chapter, under contract No. DAAG29-80-C-0137.

REFERENCES

1. OLEESKY, S. S. and MOHR, J. G., *Handbook of Reinforced Plastics*, The Society of the Plastics Industry, Inc., New York, 1978, p. 71.
2. LEE, H. and NEVILLE, K., *Handbook of Epoxy Resins*, McGraw-Hill Book Company, New York, 1967, p. 13-1.
3. TATEOSIAN, J. H. and ROYER, J. R., Effect of mixing efficiency on epoxy properties, presented at the 15th ACS Middle Atlantic Regional Meeting, January 7, 1981, Washington, DC.
4. BELL, J. P., *J. of Appl. Polym. Sci.*, **27**, 3503 (1982).
5. CHIAO, T. T. and MOORE, R. L., A room-temperature-curable epoxy for advanced fiber composites, *Proc. of the 29th Ann. Tech. Conf. on Reinforced Plastics/ Compos. Inst.*, SPI, New York, 1974, Section 16-B.
6. KAPLAN, S. L., KATZAKIAN, A. and MITCH, E. L., Fast curing acid/epoxy-anhydride/epoxy-resins, *Proc. of the 30th Ann. Tech. Conf. on Reinforced Plastics/Compos. Inst.*, SPI, New York, 1975, Section 8-C.
7. BUCKLEY, L. and ROYLANCE, D., *Polym. Eng. Sci.*, **22**(3), 166 (1982).
8. DONNELLAN, T. and ROYLANCE, D., *Polym. Eng. Sci.*, **22**(13), 821 (1982).
9. SPITSBERGEN, J. C., LOENWRIGKEIT, P., BLUZSTEIN, C., SUGARMAN, J. and LAUZE, W. L., Improved epoxy resin from dihydroxy-diphenyl sulfone, *Proc. of the 26th Ann. Tech. Conf. on Reinforced Plastics/Compos. Inst.*, SPI, New York, 1971, Section 19-C.
10. KAPLAN, S. L., MITCH, E. L. and KATZAKIAN, A., Epoxy/imide resin systems, *Proc. of the 30th Ann. Tech. Conf. on Reinforced Plastics/Compos. Inst.*, SPI, New York, 1975, Section 19-A.
11. CATSIFF, E. H., DEE, H. B. and SELTZER, R., Hydantoin epoxy resins as matrix components, *Proc. of the 33rd Ann. Tech. Conf. on Reinforced Plastics/Compos. Inst.*, SPI, New York, 1978, Section 16-C.
12. BUSSO, C. J., Newey, H. A., BUCKMAN, T. D. and HOLLER, H. V., Chemical structure–mechanical properties studies on pure epoxy resin systems, *Proc. of the 25th Ann. Tech. Conf. on Reinforced Plastics/Compos. Inst.*, SPI, New York, 1970, Section 3-B.
13. MCGARRY, F. J. and SULTAN, A. S., R69-35, MIT, July 1, 1969.
14. SOLDATOS, A. C. and BURHANS, A. S., Cycloaliphatic epoxy resins with improved strength and impact coupled with high heat distortion temperature, *Proc. of the 25th Ann. Tech. Conf. on Reinforced Plastics/Compos. Inst.*, SPI, New York, 1970, Section 3-C.
15. SIEBERT, A. R., ROWE, E. H., RIEW, C. K. and LIPIEC, J. M., Toughness vs flexibility in epoxy resins (part A), *Proc. of the 28th Ann. Tech. Conf. on Reinforced Plastics/Compos. Inst.*, SPI, New York, 1973, Section 1-A.
16. SAMEJIMA, H., Formation and properties of elastomer modified epoxy resins, *Am. Chem. Soc. Div. Polym. Chem., Polymer Preprint*, **22**(2), 127 (1981).
17. FAVA, R. A., *Polymer*, **9**, 137 (1968).
18. HORIE, K., HIURA, H., SAWADA, M., MITA, I. and KAMBE, H., *J. Polym. Sci., part A-1*, **8**, 1357 (1970).
19. CIZMECIOGLU, M. and GUPTA, A., Curing kinetics of epoxy matrix resin by differential scanning calorimetry, *Proc. of the 37th Ann. Tech. Conf. on Reinforced Plastics/Compos. Inst.*, SPI, New York, 1982, Section 20-E.
20. KAMAL, M. R. and SOUROUR, S., *Polym. Eng. Sci.*, **13**, 59 (1973).

21. SOUROUR, S. and KAMAL, M. R., *Thermochim. Acta*, **14**, 41 (1976).
22. PRIME, R. B., *Polym. Eng. Sci.*, **13**, 365 (1973).
23. ACITELLI, M. A., PRIME, R. B. and SACHER, E., *Polymer*, **12**, 335 (1971).
24. CHOY, S. Y., *SPE Journal*, **26**, 51 (1970).
25. SMITH, I. T., *Polymer*, **2**, 95 (1961).
26. PAPPALARDO, L. T., *Soc. Plast. Eng. Tech. Papers*, **20**, 13 (1974).
27. DUTTA, T. A. and RYAN, M. E., *J. Appl. Polym. Sci.*, **24**, 635 (1979).
28. PARKER, B. G. and SMITH, C. H., *Modern Plastics*, December 1979, 58.
29. SCHNEIDER, N. S., SPROUSE, J. F., HAGNAUER, G. L. and GILLHAM, J. K., *Polym. Eng. Sci.*, **19**, 304 (1979).
30. ULUKHANOV, A. G., LAPITSKII, V. A., AKUTIN, M. S. and SKOKOVA, L. D., *Plast. Massy*, **9**, 59 (1981).
31. ASTM D2471-71, *Gel Time and Peak Exothermic Temperature of Reacting Thermosetting Resins*, 1970.
32. GILLHAM, J. K. and BENCI, J. A., *J. Polym. Sci., Symposium No. 46*, 279 (1974).
33. GILLHAM, J. K., *Polym. Eng. Sci.*, **16**, 353 (1976).
34. BABAYERSKY, P. G. and GILLHAM, J. K., *J. Appl. Polym. Sci.*, **17**, 2067 (1973).
35. SCHNEIDER, N. S. and GILLHAM, J. K., *Polymer Composites*, 1(2), 97 (1980).
36. SENICH, G. A., MACKNIGHT, W. J. and SCHNEIDER, N. S., *Polym. Eng. Sci.*, **19**, 313 (1979).
37. GOLDFARB, I. J. and LEE, C. Y. C., *Am. Chem. Soc. Div. Org. Coat. Plast. Chem.*, **41**, 386 (1979).
38. NEILSEN, L. E., *Polym. Eng. and Sci.*, **17**, 713 (1979).
39. TUNG, C. M. and DYNES, P. J., *J. Appl. Polym. Sci.*, **27**, 569 (1982).
40. REED, K. E., Dynamic mechanical analysis of fiber reinforced composites, *Proc. of the 34th Ann. Tech. Conf. on Reinforced Plastics/Compos. Inst.*, SPI, New York, 1979, Section 22-G.
41. GRIFFITHS, P. (ed.), *Transform Techniques in Chemistry*, Plenum Press, New York, 1978, 120.
42. COOLEY, J. W. and TUKEY, J. W., *Math. Comput.*, **19**, 297 (1965).
43. NORTON, R. and BEER, R., *J. Opt. Soc. Am.*, **66**. 3 (1976).
44. CHAMBERLAIN, J., *The Principles of Interferometric Spectroscopy*, Wiley, New York, 1979.
45. RABOLT, J. and BELLAR, R., *Appl. Spect.*, **35**, 132 (1981).
46. PEARCE, E. M., BULKIN, B. J. and LIN, S. C., *J. Polym. Sci., Polym. Chem. Edn*, **17**, 3121 (1979).
47. ANTOON, M. K., PhD dissertation, Case Western Reserve University, Cleveland, 1980, Appendix I.
48. PERKINSON, J. L., A method of obtaining a true infrared spectrum of the epoxy from a cured fibreglass–epoxy composite, *Govt. Rep. Announce (US)*, **71**(19), 88 (1971).
49. DANNENBERG, H., *SPE Trans.*, January 1963, 78–88.
50. ANTOON, M. K. and KOENIG, J. L., *J. Polym. Sci., Polym. Chem. Edn*, **19**, 549. (1981).
51. STEVENS, G. C., *J. Appl. Polym. Sci.*, **26**, 4279 (1981).
52. BYRNE, C. A., HAGAUER, G. L., SCHNEIDER, N. S. and LENZ, R. W., *Polym. Comp.*, 1(2), 71 (1980).

53. SPROUSE, J. F., HALPIN, B. M. and SACHER, R. E., Cure analysis of epoxy composites using FT-IR, *Gov. Rep. Announce Index* (*US*), **79**(17), 136 (1979).
54. ANTOON, M. K., STARKEY, K. M. and KOENIG, J. L., Applications of FT-IR to quality control of the epoxy matrix, *Composite Materials* (*Fifth Conference*), *Testing and Design*, ASTM, 1979, STP 674, 541.
55. ANTOON, M. K., ZEHNER, B. E. and KOENIG, J. L., *Polym. Comp.*, **2**(2), 81 (1981).
56. GEORGE, G. A., SACHER, R. E. and SPROUSE, J. F., *J. Appl. Polym. Sci.*, **21**, 2241 (1977).
57. ANTOON, M. K. and KOENIG, J. L., *J. Polym. Sci., Polym. Phys, Edn*, **19**, 197 (1981).
58. LU, C. S. and KOENIG, J. L., *Am. Chem. Soc. Div. Org. Coat. Plast. Chem.*, **32**(1), 112 (1972).
59. SENTURIA, S. D., SHEPPARD, N. F., POH, S. Y. and APPELMAN, H. R., *Polym. Eng. Sci.*, **21**, 112 (1981).
60. CHOTTINER, J., SANJANA, E. N., KODANI, M. R., LENGEL, K. W. and ROSENBLATT, G. B., *Polym. Comp.*, **13**(2), 59 (1982).
61. ELLERSTEIN, S. M., *Analytical Calorimetry*, eds R. S. Porter and J. F. Johnson, Plenum Press, New York, 1968, 279.
62. BORCHARDT, H. J. and DANIELS, F., *J. Amer. Chem. Soc.*, **79**, 41 (1957).
63. CRANE, L. W., DYNES, P. J. and KAELBLE, D. H., *J. Polym. Sci., Polym. Lett. Edn*, **11**, 533 (1973).
64. KISSINGER, H. E., *Anal. Chem.*, **29**, 1702 (1957).

Chapter 3

THERMOGRAPHY APPLIED TO REINFORCED PLASTICS

Kenneth L. Reifsnider and Edmund G. Henneke II

Materials Response Group,
Virginia Polytechnic Institute and State University,
Blacksburg, USA

SUMMARY

Thermography is an important and potentially extremely powerful technique for the non-destructive examination of materials, with special reference to the detection of flaws and other defects. This chapter explains the underlying principles, and the procedures currently available for applying them. Factors affecting the performance of the technique in practice are discussed, and examples are given which show the usefulness of thermography in the non-destructive examination of fibre-reinforced composite materials and structures.

3.1. BACKGROUND

Thermography is the general term given to the technique whereby contours of equal temperature—isotherms—are mapped over a surface. As a technique for nondestructive testing (NDT) and nondestructive evaluation (NDE), the application of thermography is based upon the assumption that defects, inhomogeneities, or other undesirable conditions of the test object will evidence themselves as local hot or cold spots in the isothermal mapping. To apply this technique one must, first, excite a thermal pattern in the test object to be studied; second, measure the areal temperature distribution on the surface of the examined test object; and, third, interpret the results according to well-established and well-understood physical principles. This chapter on thermography will describe the present state-of-the-art for performing these steps for the study of composite materials. The state-of-the-art is well advanced for performing steps one and two, but is still in the developmental stage for step three. While the physical principles of heat conduction and radiation are well-understood, the interpretation of these principles for the purpose of predicting the type and effect of damage or defects is an area of continuing development.

For convenience, thermography NDT techniques may be categorised as active or passive[1] when referring to the means by which heating is excited in the test object. Active methods refer to those for which heat is produced 'actively' in the specimen by transformation processes which occur while the test object is being subjected to normal operating, testing, or loading conditions. For example, an electrical component may overheat due to some abnormality when electrical current passes through it, or a structural material might develop hot spots under mechanical loading in regions where damage is occurring. On the other hand, passive methods refer to those for which the test object is treated as a path for heat from some external heat source to some external heat sink. In this case, local inhomogeneities or flaws cause local differences in thermal conductivities which evidence themselves in the thermal mapping. These categorisations will be dealt with in more detail in the section on the application to composites.

The means by which the temperatures are measured can also be conveniently categorised, in this case, as contact or noncontact methods.[2] Contact methods of temperature measurement refer to those by which a physical object is brought into contact with the surface of the test specimen to determine the temperature of the latter. Noncontact methods are those which rely upon noncontact observations of infrared radiation to determine the temperature of the radiating source. Both types will be discussed in the next section.

3.2. TEMPERATURE MEASUREMENT FOR THERMOGRAPHY

3.2.1. Contact Methods
As noted earlier, the second step in applying thermography is to measure the surface temperature of the object to be examined. (The first step, exciting the thermal pattern, will be discussed for various applications to

composite materials in the next section.) Perhaps the oldest and least expensive methods for doing this fall under the category of contact methods. Such methods require that actual contact be made between the surface of the examined object and the measuring device. There are a large variety of possible techniques for making contact surface temperature measurements, most of which rely upon some type of chemical reaction in an applied coating. Thermal coatings include paints, phosphors, papers, liquid crystals, and a number of other, special temperature-sensitive compounds. These methods have the distinct advantage of being relatively easy to apply and of requiring a low initial investment, particularly if only a small area of material is to be examined. On the other hand, these methods have the disadvantages of generally yielding only a qualitative indication of temperature and of affecting, themselves, the distribution of surface temperature. There are, however, some contact coatings such as liquid crystals, which can be calibrated for temperature measurement. A major disadvantage, common to all of the contact methods, is that one must know, *a priori*, the approximate value of the temperature that one wishes to measure.

To apply a contact thermographic method, a temperature-sensitive coating must be placed upon the surface of the examined test object. A change in temperature is evidenced by a change in colour or appearance of the coating due to a temperature-sensitive chemical or physical change. This change may be either reversible or permanent, depending upon the type of coating used. As mentioned in the previous paragraph, there are several types of coatings that have been used for thermographic NDT. Additional details may be found in the survey presented in ref. 2, but some of the highlights of their use are described in the subsequent paragraphs.

Temperature-sensitive paints have been developed for use between 40 and 1600 °C. These paints undergo a change at a temperature which, under optimum conditions, can be calibrated to within ± 5 °C. Some paints exist which will progress through several different colour transitions. Most often the colour transitions are observable under ordinary light, but certain paints exist which may require the use of ultraviolet.

Another coating, which normally requires ultraviolet light for its use, is the thermal phosphor type. These phosphors, as with many such compounds, emit visible light when excited by ultraviolet radiation. For a thermal phosphor, the intensity of the emitted visible light varies inversely with its temperature. The rate of decrease of light intensity with temperature is very rapid at a critical temperature, as much as 25 % per degree centigrade within a very limited temperature range around the critical point. Various phosphors exist for use between room temperature and 400 °C. There are several ways in which a phosphor may be applied to a surface: as a paint, a strippable coating, in a tape, or as a powder.

At least three different types of thermally sensitive papers exist for thermographic NDT: organic, plastic and infrared copy papers. Organic papers are coated with a meltable organic substance. Upon reaching the melting temperature, the organic substance melts and is absorbed by the paper, turning its colour from white to black. Plastic coating papers are constructed with a plastic coating containing a large density of air bubbles. These bubbles diffuse incident light giving the paper a pale colour. Upon reaching the critical temperature of the paper, the plastic melts, releasing the bubbles and revealing the black colour of the paper underneath. The infrared copy papers contain dye precursors which react to produce a colour when melted.

Reinitzer discovered liquid crystals in 1888, but they remained more or less unused until the 1950s. Liquid crystals refer to a special class of materials which at certain temperatures possess the properties of both a liquid and a crystal. These materials will flow as a viscous liquid while retaining the short range molecular structure of a crystalline solid. One form of liquid crystals, the so-called cholesteric, are dichroic. When unpolarised light is incident upon a dichroic material, it is decomposed into two components, one of which is circularly polarised clockwise while the other is circularly polarised counterclockwise. One of these components is transmitted and the other is reflected. The reflected component produces a colour when the material is illuminated by white light. As cholesteric liquid crystals are heated, the distances between molecular layers increases. This distance governs the wavelength of the reflected light. Eventually the liquid crystal melts, and with further increase in temperature, becomes more and more disordered until eventually it achieves the state of an isotropic liquid. One can select liquid crystals from a wide variety of possibilities such that the liquid crystal appears colourless at low temperatures, changes through a succession of colours as it melts and approaches isotropy, and again becomes colourless when the material is totally isotropic. It is possible to obtain many different liquid crystal compounds which exhibit colour changes at any temperature from $-20°$ to 250 °C. The various compounds have varying colour-change temperature ranges from 10 to 30 °C. The response time is relatively fast, 0·1 to 0·2 sec. The sensitivity for the detection of flaws by liquid crystals can be quite good.[3,4] Depending upon the thermal conductivity and the proximity of the flaw to the surface, flaw sizes as small as $\frac{1}{8}$ inch square have been detected. When liquid crystal paints or surface films are used to measure temperatures over long periods of time, special care must be taken to ensure that the compound remains stable, does not evaporate, and retains its calibration.

Several other temperature-sensitive compounds have found some application in thermographic NDT. Thermochromic order–disorder compounds are one class of these. Thermochromic order–disorder compounds change colour with temperature without breaking chemical bonds and hence are reusable. Variable surface-tension compounds have been developed with temperature-sensitive surface tension.[2] When sprayed upon a surface, these compounds are repelled from warmer areas and coalesce in the cool areas. Hence, these compounds are more sensitive to temperature differences than they are to the absolute value of the temperature. Finally, frost testing has been applied in situations where one wishes to find gross flaws or disbonds. To apply this method, one chills the surface of the examined object below the frost point so that a frosted surface is obtained. Differences in heat conductivities in the test object will then evidence themselves by the patterns formed as the frost melts.

3.2.2. Noncontact Methods

Noncontact methods for temperature measurement are based upon the phenomenon of infrared radiation. Sir Frederick W. Herschel discovered in 1800, in a very simple but elegant experiment, that there existed a portion of the electromagnetic spectrum, invisible to the human eye, with frequencies below that of the red portion of the visible spectrum (Fig. 3.1). This 'infrared' light caused higher temperature readings on thermometers than did visible light. Later workers found that, as a basic law of nature, all matter at temperatures greater than absolute zero spontaneously emits infrared, or thermal, radiation. Furthermore, the intensity and other properties of this radiation are predictable by a mathematical model, as

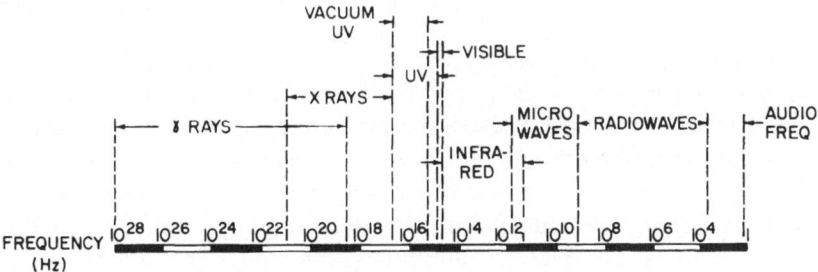

FIG. 3.1. The electromagnetic spectrum.

described later. The radiation is a result of the thermal agitation, or temperature, of the basic components of which all matter is composed: molecules, atoms and subatomic particles. The frequency range covered by infrared radiation is a result of the various energies possessed by the wide range of molecules, atoms, and subatomic particles and the various types of motions and quantum jumps available to these particles. Classically, the infrared spectrum was categorised into three subregions—near-, intermediate-, and far-infrared—depending upon the value of the infrared frequency in relation to the red spectrum. The initial reason for this classification was due to the different experimental techniques required to detect the radiation in these different bands. However, in general, one can discuss these three bands in relation to the mass of the source of the infrared:[5]

(i)   subatomic particles whose quantum jumps are responsible for the major part of near infrared radiation.
(ii) atomic particles whose thermal vibrations produce the major part of intermediate infrared radiation, and
(iii) molecules whose vibration and rotation motions produce far infrared radiation.

The basic properties of infrared radiation had been experimentally determined and were very well known, but no analytical model existed to account for them until Planck suggested his now quite famous model for black-body radiation exactly one hundred years after infrared radiation had been first discovered. Planck's major contribution to this field, and to modern physics in general, was his quantum hypothesis. Planck hypothesised that a harmonic oscillator, moving in one dimension, could not possess just any value of energy in a continuum of energies, but could only have a total energy value which would satisfy the relationship

$$E = nh\nu \qquad (1)$$

where E is the total energy of the oscillator, ν is the frequency of oscillation, n is an integer, and h is a universal constant (Planck's constant, $h = 6.625 \times 10^{-34}$ Wsec^2). Because of the discrete nature of energies available to a system of harmonic oscillators, such as the atoms or molecules composing a mass of matter, when one calculates the total energy contained therein, one must use a discrete summation of energies rather than an integration over a continuous function. The difference in mathematical properties of a discrete summation in comparison with a continuous integration, led Planck to derive his well-known distribution law for spectral radiation emittance:

$$W_\lambda = 2\pi(10^{-9})\frac{hc^2}{\lambda 5}\left(e^{\frac{hc}{\lambda kT}} - 1\right)^{-1} \tag{2}$$

where W_λ (W m^{-3} μm^{-1}) is the emitted intensity of radiation of wavelength λ, c is the speed of light ($2\cdot9973 \times 10^8$ m/sec), e is the Napierian or natural base of logarithms, k is Boltzman's constant ($1\cdot3804 \times 10^{-23}$ J/K), and T is the absolute temperature (K). Examples of the radiant emittance, W_λ, for different temperatures of a black body are shown in Fig. 3.2.

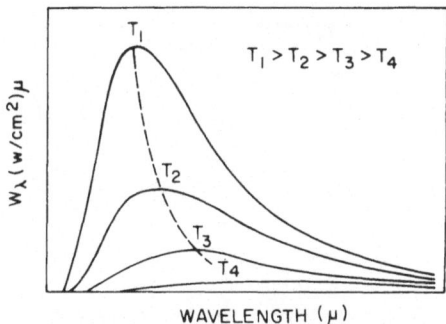

WAVELENGTH (μ)

FIG. 3.2. Radiant emittance for a black body at various temperatures.

Planck's Law applies only to black bodies, i.e., bodies which absorb all of the radiation incident upon them and emit the maximum possible amount of radiation at any given temperature. Such ideal bodies do not actually exist in nature but may be approximated in the laboratory by constructing a cavity inside a body and measuring the radiation emitted from a small hole in the surface,[5] Fig. 3.3.

All real bodies emit radiation which is some fraction of that given by Planck's radiation law. In order to account for this, a quantity known as the emissivity is defined and used as a multiplication factor on the right hand side of eqn (2). The emissivity is the ratio of the amount of spectral radiation emitted by a real body at temperature T to the amount emitted by a black body at the same temperature. Hence the emissivity is always a positive fraction less than one and is equal to one only for a black body. One may speak of a 'total' emissivity —the fraction of total energy emitted at a given temperature—or a 'spectral' emissivity—the fraction of energy emitted at a particular wavelength (and, of course, temperature). Figure 3.4 compares the spectral radiant emittance and the spectral emissivity of three

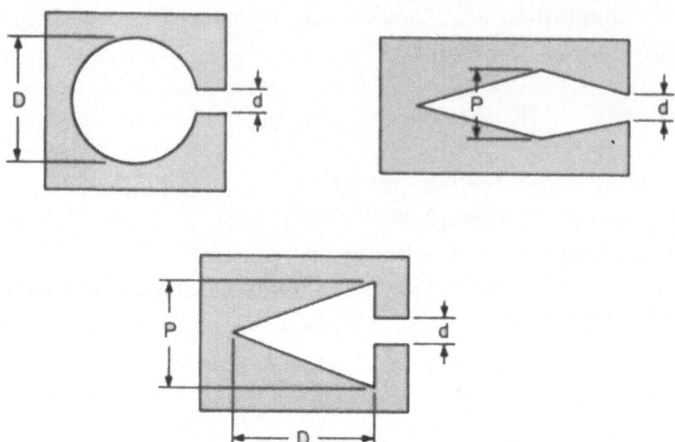

FIG. 3.3. Examples of cavities used to approximate ideal black body radiators.

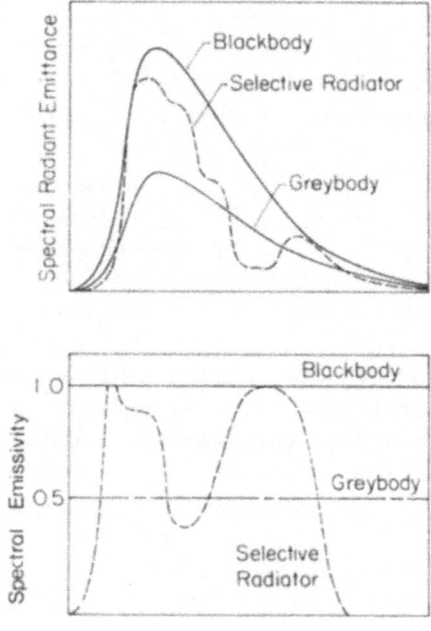

FIG. 3.4. Spectral radiant emittance for black body, grey body, and selective radiators.

kinds of bodies: black body, grey body, and selective radiator. A gray body is one which has the same emissivity at every wavelength. The total emissivity of a body might be quite low while the spectral emissivity for certain wavelengths may approach unity. Both types of emissivity may vary with temperature, physical state, surface finish, molecular surface layers, etc. The emissivity cannot be calculated from any basic physical model but must be determined experimentally for each body (and often for different points on the surface of the same body). It is strictly a surface characteristic for opaque materials. The emissivity ranges from zero for 'mirror-like' surfaces to nearly one for lamp black, zapon black, and such surfaces.[3]

Because of the variations in the emissivity of a surface, the determination of an absolute value of temperature by measurement of the intensity of emitted infrared radiation must be done very carefully. It is possible, for example, that a particular surface will have a sufficiently large reflection coefficient that the body heat emitted by an experimenter might reflect off the surface of the examined object and be interpreted as a high temperature for that object. One can reduce difficulties such as this, or surface variations of emissivity, when studying composite materials by painting the surface to be studied with a uniform coating of lamp black, for example. Other workers have used coatings such as powder spray deodorant.[6] While the difficulty involved with determining an absolute temperature value will still be present, this uniform surface coating will allow one to have some confidence that the temperature gradients observed on the surface are a result of material temperature differences and not spurious reflections from hotter objects in the surrounding environment. Practical methods for measuring temperatures using infrared radiation have been described in ref. 6.

Returning to the statement of Planck's Law, eqn (2), we might point out several other important physical characteristics of infrared radiation. Figure 3.2 presents a schematic representation of a family of spectral radiance curves for radiation from a black body at several different absolute temperatures. Such curves were determined experimentally long before Planck's mathematical law was found to predict their form quantitatively. One first notes that black body radiation is a smooth, continuous function of wavelength for solid bodies with a single maximum occurring at one value of wavelength. This observation led to the statement by Wien of his displacement law:

$$\lambda_m = \frac{b}{T} \qquad (3)$$

That is, the wavelength, in micrometres, at which the maximum intensity of radiation is emitted at a given temperature is inversely proportional to T, the absolute value of that temperature. The constant of proportionality, b, is known as the Wien displacement constant, and has a value of 2897 $\mu m/K$. Physically, the Wien law states that the higher the temperature, the more the radiation peak shifts towards shorter wavelengths and hence higher frequency and higher energy content (eqn. (1)). In addition to this shift λ_m, it is also apparent that higher temperature emittances become progressively richer in shorter wavelengths. The progression towards smaller wavelengths for increasing temperature is a phenomenon which most people have seen at one time or another. At moderate temperatures, the spectral emission peak is at wavelengths in the infrared and invisible to the human eye. As the temperature increases, the object will gradually take on a reddish hue, i.e. the wavelengths shorten to cover the red spectrum. At higher temperatures, the red brightens, changes into orange, yellow, and finally white when the emission spectrum extends to cover the green and blue bands. Perhaps the status of the earliest practitioners of thermography must be given to blacksmiths who quite early learned to gauge the proper tempering and heat treating temperatures of steel by judging the colour to which it was heated.

A second important observation which can be made from Fig. 3.2 is that there is a single emission curve for each temperature of the black body. Furthermore, no emission curve ever intersects another. In particular, each curve lies above all other curves corresponding to lower temperatures of the black body. This fact means that if proper techniques are developed for measuring the spectral emission curves, one can uniquely determine the absolute temperature of a black body. Recall, however, that because of emissivity variations for real emitting bodies, this absolute temperature determination offers many difficulties to the experimenter. The total radiant emittance, that is, the total radiant energy emitted over all wavelengths, is found by integrating under the spectral radiant emittance curve. This value is the Stefan–Boltzman law, given by.[7]

$$R = \frac{2\pi^5 k^4}{15h^3 c^2} T^4 = 5 \cdot 6687 \times 10^{-12} T^4 \tag{4}$$

where R is total radiant emittance, in units of W/cm^2, and the other parameters are as defined previously.

Two other considerations are of some importance to understanding noncontact thermographic NDE methods: the system geometric relationships and the fidelity of surface thermal patterns to subsurface flaws.

For an infrared system viewing any source, the received power at the system aperture is given by[3]

$$H = \frac{W_\lambda \omega}{\Omega} \qquad (5)$$

where W_λ is the spectral emittance, ω is the angular field of view of the viewing system (defined by the optical system and the detector) and Ω is the total solid angle about the source. Thus, it can be seen that the primary geometric parameter which governs the response of the infrared viewing system is the relationship between the angle subtended by the source compared to the angular field of view of the system. If the source is small compared with the field of view of the detector (a point source), the received radiation will vary with distance between the source and detector but not with angle about the source. Interestingly, if the source is large compared with the field of view, the received radiation varies with neither distance to the source nor angle about it. This latter fact is a result of Lambert's cosine law which states that the radiant energy emanating in a given direction from any point on a surface is a function of the cosine of the angle between the normal to the surface at that point and the given direction.[7] More specifically, the maximum radiation occurs along the normal direction to the surface and none occurs tangentially to it.

The significance of Lambert's law for infrared detection is that a detector viewing an emitting surface which is large compared with the detector field of view will always receive the same amount of energy no matter the angle between the detector's line of sight and the normal to the radiating surface. This statement is true for both planar and general, curved surfaces, although it is easier to see and understand for planar ones. As the angle between the line of sight and the surface normal increases, so does the area of the surface viewed by the detector. The increase in viewed area exactly matches the reduction in radiation as given by Lambert's cosine law. Hence the total energy incident on the detector is constant. This, of course, is valid only as long as the emitting surface completely fills the detector's field of view. A simple rule in utilising a detector system then is to make the source the only object in the field of view by appropriately controlling the field stop of the viewing system.

The second consideration that one must address when using noncontact thermography as an NDE technique is the fidelity of the observed surface thermal patterns to the interior inhomogeneities or flaws. The relationship between the surface isotherms and the interior thermal patterns is, of course, governed by the thermal conductivity of the material and the distance between the surface and an interior region of interest. For very thin materials, such as composite laminates, it has been shown that the fidelity of surface thermal patterns to interior flaws is quite good.[8] For bulk materials, thermography may be useful for qualitative indications of the presence of flaws but will not be nearly as good as other NDE methods for resolution of the flaw size and shape. For example, what might seem to be a rather sharp discontinuity may prove to have a weak thermal signature because of heat conduction through the material.[2] The thermal signature can be improved by making the cooling rate at the surface as large as possible. This might be done by making the surface as nonreflective as possible and by augmenting radiant cooling by forced air circulation over the test surface.

There are two basic types of infrared detectors: (i) photon-effect devices and (ii) thermal devices. The photon-effect devices are sensitive to the wavelength of the received radiation, while thermal devices respond only to the degree of heating caused by the incident radiation and are largely independent of wavelength. The performance of real-time thermography, such as with a video-thermographic system, requires that the entire field of view be scanned very rapidly so that the temperature of each field point can be measured and displayed many times each second. Such systems require the high sensitivity and very rapid response time of photon-effect devices.

Photon-effect devices utilise solid state materials which produce voltage, current, or resistance changes when irradiated by photons. These semiconductor materials are generally classified as photoemissive, photoconductive, or photovoltaic conductors. Photoemissive detectors, originally discovered by Hertz, are materials which emit electrons when irradiated by photons having wavelengths smaller than a critical value which depends upon the material. (Another way of looking at this, of course, is that the energy or frequency of the photon must exceed a respective critical value.) Photoemissive devices are primarily responsive through the visible wavelengths down to approximately one micrometre. Hence, such devices are not used extensively in infrared systems. Photoconductive devices are semiconductors whose conductivity changes when irradiated by photons. Photons of sufficient energy will cause bound electrons to jump into the conduction band and hence to become available as charge carriers. The response time of photoconductor materials is very fast, with times shorter than one microsecond having been reported. Photovoltaic cells are composed of p–n junction semiconductor materials. These cells produce a voltage when irradiated by photons. These devices also are characterised by a rapid response time. For maximum sensitivity, and to reduce extraneous thermal noise, it is usually necessary to cool the semiconductor material to low temperatures. Many commercially available thermographic detector systems require that liquid nitrogen be used for this purpose.

The rapid understanding and development of semiconductor materials after World War II also led to major changes in the detection of infrared radiation. The types of detectors discussed in the preceding paragraph have been incorporated into various commercially available infrared detection and viewing systems. One can obtain systems which scan the test surface in real-time as rapidly as 28 complete frames per second. The signal is conditioned and displayed upon television monitors in either grey scale or colour format. The grey scale display units display continuous temperature readings using shades of grey between black and white. The colour display units select a series of distinct colours (as many as ten) to display those regions of temperature which differ from one another by a temperature band equivalent to the total temperature range divided by the number of colours available ($\Delta T = 1/10$th the temperature range for displays having ten colours). With these devices, one can monitor simply and rapidly the temperature gradient profiles on the surface of the test objects for total temperature ranges between 1 and $1000\,°C$. Specific information and details on particular thermographic systems can be obtained from the manufacturers.

3.3. PROCEDURES

For the purpose of our present discussion, it will be assumed that all application procedures incorporate a thermographic detector of some type, such as a video-thermographic camera or other similar device. This part of the apparatus will be referred to simply as the 'detector'. It is further assumed that the heat pattern displayed or recorded by that device is interpreted by the user in a fashion which suits his particular engineering or scientific purposes. Hence, this section will be concerned primarily with the production of heat patterns which display suitable detail for nondestructive testing or evaluation purposes.

A schematic diagram of a basic thermographic nondestructive testing and evaluation testing system appears in Fig. 3.5. Regardless of how the system is used, the basic principles of operation remain essentially unchanged. Discontinuities in material properties or material geometry in

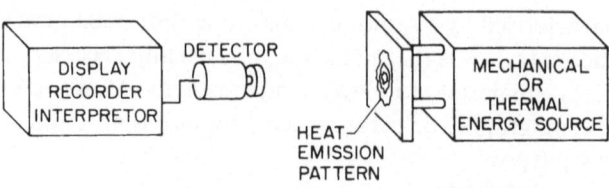

FIG. 3.5.   Schematic diagram of basic thermographic experimental configuration.

the specimen under examination are revealed by corresponding discontinuities in the heat emission pattern, which is produced by using the specimen as an emitter of radiant energy that is introduced into the specimen by an external mechanical or thermal excitation. The heat emission pattern so produced is detected by an infrared or thermographic camera (or other suitable device) and transmitted to the final system component which may display, record, and possibly interpret the heat pattern. Generally speaking, the detector and the display–recorder–intepreter are fixed for a given investigator since those components generally represent significant capital investments and sometimes require detailed integration and interfacing with other parts of the total engineering or investigative system.

On the other hand, the manner in which the heat emission pattern is produced frequently changes from specimen to specimen and from circumstance to circumstance. The choice of a specific method will generally depend on the type of flaw or defect to be detected, the material properties and geometry, the excitation devices which are available, and the nature of the thermographic detector that is available.

One of the basic choices to be made is whether the specimen is to be used as an active or passive heat emitter. While it is possible for the specimen to produce heat emission patterns by both processes simultaneously, the heat patterns emitted by most excitation devices usually fall almost completely into one of these two categories. As the terminology implies, passive heat emission occurs when the specimen is used simply as a conductor or conduit for energy transmission. The details of a heat pattern produced in that way are created by spatial discontinuities in that transmission process. Hence, for the most part, the detection of defects or flaws using this method depends on the degree to which the conductivity of the specimen is disturbed in local regions around these details to be revealed. Obviously, then, larger flaws which disturb the conductivity over larger areas are easier to find than smaller ones. However, the location of the flaw relative to an observable surface, the projected area of a flaw relative to the direction of

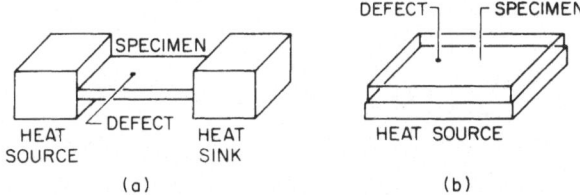

FIG. 3.6.   Schematic diagram of two arrangements for introducing heat into a specimen by conductivity.

principal heat flow, the conductivity of the material in the flaw (if it is not a void), and a number of other factors to be mentioned in Section 3.5, also affect the detectability of a defect using this method.

There are two commonly used schemes for introducing heat into a specimen which is to be examined by passive thermography. The first of these schemes is the introduction of heat into the specimen by conductivity, i.e. by placing a heat source in direct contact with the specimen at some point so that the specimen conducts heat away from the source, or by placing a heat sink in contact with a specimen at some point so that heat is conducted by the specimen to the sink. Of course, both a heat source and a heat sink can be used together at two different points to establish a fixed temperature gradient across which heat is conducted by the specimen. Figure 3.6 shows two arrangements whereby heat is introduced into the specimen by conductivity. In Fig. 3.6a the heat source and heat sink are arranged at either end of a specimen which has a small defect at the position indicated. In Fig. 3.6b, the heat source is in contact with the entire longitudinal dimension of the specimen and the other longitudinal surface is free to dissipate heat by convection. Generally speaking, the arrangement shown in Fig. 3.6b is likely to produce a more distinct image of the defect in the heat emission pattern observed by a detector which views the free longitudinal surface in both cases, if similar amounts of heat flow are involved in the two instances. In a simple sense, the defect represents a greater disturbance in the heat flow when the direction of flow is through the thickness of the plate than it does for the situation where the direction of the flow is along the longitudinal axis.

It is also possible to introduce heat into the specimen by radiation and convection. A flow of hot air or fluid can be directed at the surface of the specimen, for example. Two common arrangements are schematically depicted in Fig. 3.7. The common tendency is to direct the heat source at the surface opposite to the one viewed by the detector, as shown in Fig. 3.7a. However, it is also possible that an arrangement whereby the

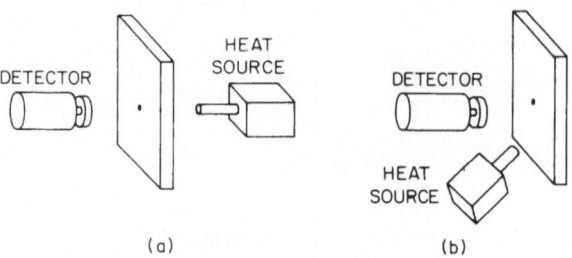

FIG. 3.7.   Two common arrangements for introducing heat into a specimen by
radiation and convection.

source of heat and the detector are on the same side of the specimen may be
preferable. While it is true that it is certainly easier to transfer energy into a
specimen by conductivity than by radiation or convection, the latter
transfer scheme does have a number of advantages. One of these
advantages is simplicity of operation since it is somewhat simpler to obtain
or construct a source of hot air (such as a common heat gun) or hot fluid
than to build a suitable device to transfer heat into a specimen by
conduction, especially if the specimen has an irregular shape or imperfect
surface. In fact, it is generally easier to obtain a uniform or reproducible
regular pattern of heat using radiative or convective energy transfer into a
specimen or component, a distinct advantage of such schemes.

Another basic matter of importance for passive thermography
procedures is the question of whether steady-state or transient conditions
should be used to develop the heat emission patterns. For a steady-state
experiment, the heat source, the specimen, and all of the heat loss
mechanisms are given sufficient time to reach thermal equilibrium before
the thermographic data are recorded. Once the steady-state has been
reached, the temperature distributions remain constant as a function of
time. Such a procedure has, in our view, the advantage of being more easily
reproduced and more amenable to both analytical representations and
quantitative interpretations and comparisons. Once the entire system has
reached thermal equilibrium, the number of operational or system-
dependent variables that enter into the interpretation of the data that are
recorded is minimised. Hence, the interpretation of the data is greatly
facilitated. Also it happens that the mathematical analysis necessary to
represent a steady-state heat transfer problem is generally simpler and
more easily applied to practical problems than the analysis which must be
used for transient heat transfer situations. On the other hand, transient
heating schemes have their advantages, too. One of the most obvious

advantages is brought to mind by realising that the largest temperature differences in a specimen occur at the first instant that a heat source is applied to the specimen. Following that first point in time, energy is transferred according to Fourier's law and distributed in such a way so as to balance the energy input against conduction, convection, and radiation losses. The temperature gradients are reduced by this process until thermal equilibrium (a steady-state condition) is reached. While a defect or flaw represents the same obstacle to the transfer of heat during the transient as well as during the steady-state situation, the thermographic detector depends on temperature differences for the resolution of data, and these temperature differences must be detected by the observation of a free surface. Hence, in some instances, transient heating schemes may reveal defects and other discontinuities that are not easily detected by steady-state methods. Some examples will be cited in Section 3.6, on applications.

The development of heat emission patterns by passive methods depends almost entirely on heat transfer characteristics for its resolution capabilities. To use the common electrical analogy between Fourier's and Ohm's law, the temperature difference details which reveal defects and flaws of interest are proportional to the conductive heat flow multiplied by the summation of thermal resistances in the region of the flaw. Hence, the effect of the flaw or defect can be thought of as a local variation in the thermal resistance which then is reflected in an observable temperature difference in that locality. In other words, the flaw acts like a local variation in the material property of conductivity for such passive measurements. Of course, in the engineering sense, the flaw is also a very important mechanical disturbance. That fact can be used to develop thermographic details when 'active' heat generation procedures are used. Three examples of such active procedures are indicated by the schematic diagrams in Fig. 3.8.

Figure 3.8a depicts the situation in which energy is generated by hysteresis when a specimen is deformed mechanically. It is a well-known fact that when most materials are cyclically deformed, even in the elastic range, some of the mechanical energy is dissipated by nonconservative micro-material deformation processes such as dislocation motion, impurity diffusion, and other complex local molecular or atomic activity. For polymeric materials, these processes bring about what is commonly called viscoelastic response. The amount of energy dissipated by such mechanisms may range from imperceptible amounts up to several percent of the input energy. Unfortunately, most of the attention that has been given to hysteresis has centred on crystalline solids and has been concerned

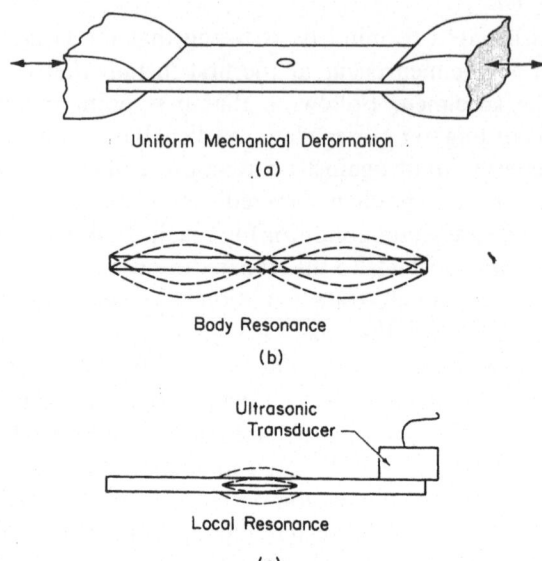

Uniform Mechanical Deformation

(a)

Body Resonance

(b)

Ultrasonic
Transducer

Local Resonance

(c)

FIG. 3.8.   Three types of excitation that produce 'active' heat generation.

almost entirely with a part of the field called 'anelasticity', which is commonly concerned with atomic diffusion events that are activated by stress, including such things as grain boundary motion, single or paired solution atom motion, and twin boundary activity. While these events may be prominent in the small strain range, the heat produced by them is not significant because of the low ratios of dissipated-to-input energy, a low level of input energy for small strains, and the low frequencies of oscillation at which these events are commonly excited. For composite materials with polymeric matrices, viscoelastic dissipative hysteresis frequently dominates the heat generation, but hysteresis energy heat patterns can be generated in metal matrix composites as well.

Two parameters are especially important in the generation of such heat patterns using mechanical excitation, as is indicated in Fig. 3.8a. The first of these is the stress or strain level at which the specimen is deformed. Depending upon the material used, the deformation process, the type of defect, and other variables, a stress or strain level greater than or equal to roughly one-third if the level required to fail the specimen is sufficient to produce observable heat patterns around defects of engineering interest in most cases. The second major parameter of importance is the cyclic frequency with which the mechanical load is applied. If other things are equal, the amount of power introduced into the specimen is directly proportional to the frequency of oscillation, so that the best frequency is the highest frequency available since temperature differences in the region of a defect will be proportional to the differential amount of energy generated by the defected and adjacent undefected regions. It should be mentioned, however, that energy dissipation mechanisms may be frequency- and temperature-dependent, so that the optimum frequency may not be the largest one easily obtainable with the test rig available. Trial and error and analysis is called for. Most heat patterns that are generated in composite materials using this active hysteresis energy emission scheme are created by cyclic frequencies between about 5 and 30 Hz, but higher frequencies may be useful. One of the advantages of the mechanical excitation scheme is the fact that the dissipation of the defect or defected region is in direct proportion to its mechanical importance, or in another sense, to its mechanical disturbance of the material response in that region. Hence, an element of interpretation is added to the information obtained in this way. It should also be mentioned that while this procedure can be used nondestructively, it is also possible to watch the growth and development of defects by simply observing the cyclic application of load levels and load histories which are sufficient to cause such growth. In that way, the chronology of damage development can be followed.

Figure 3.8b shows another example of an active heat generation procedure. If proper frequencies of mechanical excitation are chosen, resonant body vibrations of the specimen or component can be excited. Energy is dissipated, and a heat emission pattern is formed, in proportion to the stress distributions which are created by the resonance vibration mode. Defects and flaws, or damaged regions, are revealed not only by the manner in which they disturb the generation of the heat pattern, but also by the manner in which they disturb the vibration mode shape as revealed by the heat pattern.

Figure 3.8c indicated yet another method of mechanical excitation which has been dubbed 'vibrothermography' by the authors. That procedure involves the introduction of high frequency ultrasonic excitation into a specimen by affixing an ultrasonic transducer or shaker to the specimen at some point or points. The frequency of excitation is varied in such a way that local flawed regions are set into local resonance. In addition to the hysteresis energy dissipation in the region of local vibration, another very strong source of energy dissipation frequently develops due to the interaction of internal flaw surfaces when they interact with one another during local vibration motions. For example, in some cases it has been determined that the surfaces of a delamination rub together to produce a very strong dissipation source during this type of excitation. Vibrothermography has the advantageous capability of being able to resolve selectively different defects by the use of different frequencies of excitation, and the procedure has shown exceptional sensitivity to defects which have internal free surfaces which are in close proximity to one another.

While the commonly used procedures for thermographic detection of defects are outlined above, an enormous number of variations have been used and others are being developed. In fact, one of the most desirable aspects of thermography is the flexibility with which it can be applied to a variety of engineering situations. The reader and user is well advised to use his imagination and try his hand at the development of suitable procedures which may be based on the details mentioned above or on other innovative ideas.

3.4.   HEAT GENERATION BY REINFORCED PLASTICS

One of the most successful methods of using thermography to observe and investigate the behaviour of composite materials is based on the concept that under mechanical excitation these materials (and most others to various degrees) dissipate part of the mechanical energy as heat in an observable pattern that is directly and quantitatively related to the state of stress and state of the material. Such techniques are particularly useful for the observation and interpretation of damage development in these complex material systems. Examples will be provided in a later section. The success of this approach to the application of thermography to reinforced plastics depends on the heat generation characteristics of those materials. We will discuss briefly three major types of mechanism associated with heat generation: anelastic behaviour, viscoelastic effects, and 'mechanical' dissipation caused by such things as the rubbing of internal free surfaces.

If the response of a material to loading (or other applied environments) is entirely linear, i.e., the response is single values for a fixed input regardless of prior loading history, no energy loss will result and no corresponding heat will be produced. Hence, the dissipative mechanisms responsible for heat generation during loading, which form the basis of the vibrothermography technique are due to some departure from this strict linearity. In that context, it is interesting to note that most of the literature which deals with these mechanisms is based on analytical representations which introduce a damping or dissipative term into a linear theory by such means as assuming that the elastic stiffness has a real and an imaginary part, or that there is a phase lag between stress and strain, or some other scheme related to damped oscillator theory. As a consequence, it is commonly assumed that superposition of influences applies, that the specific dissipation of a material does not depend on the amplitude of excitation, and that other assumptions familiar to linear treatments hold.

Anelastic effects are nearly always present. One major reason for this is that the magnitude of temperature changes due to adiabatic straining at practical excitation frequencies is often significant, especially for the more conductive materials. The temperature change for adiabatic elastic deformation is given by,[9]

$$\Delta T = -\frac{eT\alpha Ke}{\rho c} \tag{6}$$

where α is the coefficient of thermal expansion, K is the bulk modulus, e is the volume dilatation, ρ is the mass density and c is the sepcific heat of the material being strained. For aluminium loaded over a range between the tensile and compressive yield strengths in a uniaxial test at 1 Hz the entire specimen may oscillate over a temperature range of $0.5\,°C$ or more, depending upon the type of aluminium, the conductive, convective and radiation losses and other experimental factors. Such large temperature oscillations which cycle in phase with the excitation are easily observed in many materials and have been observed for several metals and plastics in our laboratory. However, for homogeneous materials the dissipative loss associated with thermal currents induced by these temperature changes is generally rather small. In general, the losses are proportional to the difference between the adiabatic and isothermal elastic compliances of a given material, i.e.

$$S_{ij}\big|_{\text{adiabatic}} - S_{ij}\big|_{\text{isothermal}} = -\frac{\alpha_i \alpha_j}{c} T \tag{7}$$

where S_{ij} are the elastic compliance tensor components, α_i are the vector components of thermal expansion, T is the temperature and c is the specific heat of the material. The physical reason for the strength of the dissipative loss being controlled by eqn (4) is simply the fact that the difference between the adiabatic and isothermal behaviour controls the maximum area in the ideal hysteresis loop associated with what is commonly called the 'elastic after-effect', shown in Fig. 3.9. As an example of this situation,

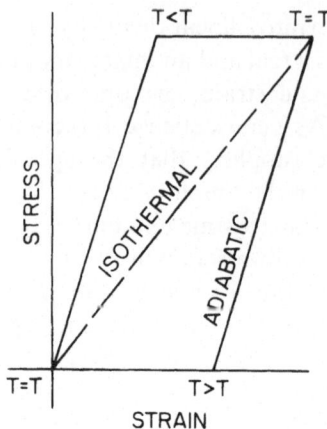

FIG. 3.9.   Schematic diagram of hysteresis loop for the elastic aftereffect.

for 'transverse thermal currents in a reed' (beam) of rectangular cross section Nowick gives the internal friction as,[10]

$$\tan \delta = \frac{E\alpha^2 T}{c}\left[\frac{\omega\tau}{1 + \omega^2\tau^2}\right] \tag{8}$$

where E is the elastic modulus; ω is the frequency of excitation; α, T and c are defined in eqn (6) and the relaxation time is

$$\tau = \frac{a^2}{\pi^2 D} \tag{9}$$

where a is the reed thickness and D is the thermal diffusivity. However, it should be emphasised that these expressions are for homogeneous materials. Virtually no basic characterisation of thermal current losses in inhomogeneous materials such as composites has been carried out. Values of relative damping given by eqn (8) usually fall in the range of 0·001 to 0·01; small by engineering standards.

Figure 3.10 shows an interpretation of this equation for some materials of interest for a range of region sizes of practical importance. Two types of practical information are illustrated by Fig. 3.10. If energy dissipation is due to thermal currents within the region of interest, then the frequencies of oscillation for a range of materials which will maximise the dissipative loss is given directly by the figure. It is important to notice the strong dependence on material properties in Fig. 3.10. (The ordinate is a logarithmic scale.) The material property that contributes the most to this dependence is the thermal conductivity. The density and specific heat rarely

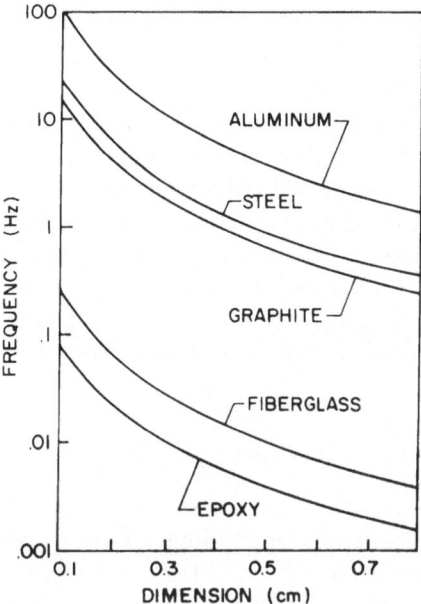

FIG. 3.10.   Frequency of oscillation for maximum dissipative loss due to thermal currents.

change by an order of magnitude, but the thermal conductivity may change by four (or more) orders of magnitude. This will be a recurring situation as we continue our characterisation: the thermal conductivity has a dominant influence over the nature of the heat patterns in comparison to all other physical variables such as surface film coefficients, emissivities or geometry effects, given a heat source of fixed intensity. It should also be noted that for regions of sufficient size to be readily observed as having a sharp difference in temperature, Fig. 3.10 indicates that excitations in the ranges below 1 Hz are required for materials which are poor conductors. If we consider frequencies which are significantly higher than those shown we can assume that the regions of interest have constant temperature distributions since conduction is not a major influence, i.e. this is essentially an adiabatic condition.

   In this regard, a few observations can be made concerning the evolution of heat by stress waves, either travelling or standing. The adiabatic condition is not quite satisfied for longitudinal waves since heat energy flows from the hot, compressed regions to the cooler, rarefied regions. This interchange of heat energy will, by necessity, remove energy from the mechanical wave, causing general heating of the body and eventual extinction of the stress wave. One might expect there to be preferential heating due to this effect in regions where the stresses are intensified due to material discontinuities. Upon closer examination, however, it is found that this effect is minimal at frequencies of practical interest. This heat loss will occur most readily at frequencies for which the wavelength becomes small enough that the compressed and rarefied regions are so close that heat exchange can most easily take place. In fact, for very high frequencies, one approaches an isothermal state where heat exchange takes place nearly instantaneously. In this region, the heat loss from the stress wave is a maximum. The frequency at which this transition from the adiabatic to the isothermal condition occurs can be estimated from[11]

$$f \gg \frac{1}{2\pi} \rho \frac{c_v v^2}{K} \tag{10}$$

where K is the bulk modulus, ρ is the mass density, c_v is the specific heat at constant volume and v is the longitudinal phase speed. For quartz, for example, the transition frequency is of the order of 10^{11} Hz. Hence in all practical cases, this transition will never be observed. The wave attenuation coefficient for this phenomenon in normal frequency ranges has been estimated to be[11]

$$A = \frac{2\pi^2 f^2}{\rho v^3} \left[\frac{K}{c_v} \frac{C_{11}^\sigma - C_{11}^\theta}{C_{11}^\sigma} \right] \tag{11}$$

where f is the frequency, and C_{11}^σ and C_{11}^θ are the isentropic and isothermal elastic moduli, respectively. For quartz, this value is of the order of 12×10^{-4} nepers/cm for $f = 100$ Hz. Hence, only at the very high frequencies normally used in ultrasonic testing does this heat generation effect become non-negligible.

For practical situations, most of the interest is in an intermediate frequency range, especially in view of the results of earlier work which show that heat images of defects in several materials can be observed at frequencies of excitation between about 30 Hz and the megahertz range.[8,12-14] For these frequencies, especially in the kilohertz range, the wavelength of travelling stress waves in most engineering materials is greater than the order of a centimetre. We can then assume that the stress or strain due to the external excitation is uniform over the regions of interest (1–10 mm). Situations for which the wavelength is of the order of the inhomogeneities or discontinuities produce what is often called a loss to propagating stress waves. One can refer, for example, to scattering losses incurred in polycrystalline materials when the wavelength is comparable to the grain size.[11] This, however, is only an apparent loss due to the scattering of wave energy in all directions, and hence away from the primary propagation direction. This loss does not directly result in a preferential transformation of mechanical to thermal energy at the scattering site, at least as far as presently accepted physical models predict.

While the energy dissipated by travelling waves is not large enough to create observable heat patterns under most practical circumstances, standing wave patterns are quite capable of heat generation at practical levels. Forced oscillation of resonant systems can produce energy levels sufficient to cause outright fracture, not only of structures, but of material coupons. This subject is addressed in detail by Mignogna and Green.[30] In general, the maximum temperature increases can be expected in the maximum strain positions, the nodal points in a one dimensional vibration, for example. Since the mechanics of body vibrations is a rather well developed field, heat patterns due to that type of excitation can be anticipated (or avoided).

Viscoelasticity is an 'older' and more familiar type of engineering behaviour than the anelasticity discussed above. It is also a more general classification and is frequently assumed to include anelasticity as a special case. Dissipative mechanisms such as grain boundary or phase boundary motion, eddy current effects, and various molecular and intermolecular motions (especially in long chain molecules) may contribute to viscoelastic dissipation or 'damping' as it is sometimes called. Only the last of these mechanisms appears to be of consequence for thermography since the other mechanisms dissipate such small amounts of energy that the consequent heat patterns cannot be discerned. Relative damping values of 0·1 to 1·5 can be associated with viscoelastic damping, enough to raise the temperature of some polymeric materials to the melting point with sufficient excitation. Viscoelastic dissipation is a common source of heat generation for polymeric matrix composite materials subjected to mechanical excitation.

Another closely related and very efficient heat production mechanism is plastic deformation, whereby slip occurs, creating permanent shape changes globally or locally. Most of the energy which is required to cause such plastic deformations is dissipated as heat. Such heat development is easily observed at the tip of a propagating crack even in excellent conductors such as aluminium.[15] This type of heat generation mechanism is unique in the sense that it is a very efficient heat producer and it requires 'permanent' deformation. It is possible for this mechanism to produce heat in a fixed location by reversed or alternating slip due to cyclic (reversed) applied loads, but such situations do not appear to be common. Large strain nonlinear deformations are, however, usually highly dissipative by their very nature, so that if such mechanisms are present they will usually produce very strong heat images.

Finally, it is also possible for internal adjacent surfaces such as those associated with internal cracks or delaminations to rub against each other and produce very strong local sources of heat and distinctly observable heat patterns. In composite materials where a variety of such flaws commonly occur in large numbers, this type of heat generation is very prominent, especially when the input excitation is such that a local resonance of the flaw surfaces is generated. Examples of this will be provided. The related techniques have the capability of locating and characterising individual flaws or defects.

In general, materials which have complicated microstructures and especially those which respond viscoelastically, usually dissipate energy efficiently and develop strong heat images. Since many of these materials, especially reinforced plastics, are also relatively poor conductors, they are ideal for the application of thermography as a nondestructive test and evaluation scheme. A special effort to characterise viscoelastic stress-induced heat generation has been made by McLaughlin and co-workers.[20,26-29] They began by characterising the heat generation of unidirectional composite materials during cyclic loading at various frequencies by 'calibrating' a Voigt model of the viscoelastic response. He then conducted stress analysis of each ply in a given laminate, assumed that the viscoelastic state of stress was effectively equal to the elastic one, and input those stresses into the characterisation of heat generation established for unidirectional material, to obtain a point-by-point distribution of heat generation for each ply in the laminate. Then a finite element scheme was used to solve the subsequent thermal transfer problem to predict the heat emission at the surface of such laminates. Among his important findings was the conclusion that viscoelastic heat generation in graphite epoxy and glass epoxy systems is not sufficient to delineate flaws such as internal delaminations, part-through holes, etc., using detection equipment currently available (which can detect temperature differences of about 0·2 °C) and using excitations up to 50 Hz with amplitudes below levels that would introduce further damage into the specimens. The experience of the present authors tends to confirm this finding. Stress-induced heat patterns are vividly evident, however, in material systems like boron/aluminium and boron/epoxy, when those materials are cycled at amplitudes up to about two-thirds of their ultimate stress and at frequencies between 10 and 40 Hz, even when no damage can be detected by other means, suggesting that viscoelastic losses in those cases do produce heat patterns of practical utility.[1,16,25] In certain special cases, such as fatigue of $[\pm 45]_s$ laminates, significant heat has been observed also in graphite epoxy due to the large interlaminar stresses.

3.5. FACTORS AFFECTING SUCCESS

There are a number of parameters which influence the capability of thermographic methods to detect flaws in materials. One of the most important of these is the conductivity of the materials to be observed. While the density and specific heat of various engineering materials rarely change by an order of magnitude, the thermal conductivity may change by four or more orders of magnitude. Hence, thermal conductivity has a dominant influence on this material interrogation scheme in many cases. It is certainly obvious that, since the image of defects to be detected is formed by temperature differences from point to point, materials which have a very high thermal conductivity are difficult to examine with thermographic techniques, since temperature differences are minimised by the high rate of heat energy transfer from point to point. There is a secondary consideration, however, which 'works the other way'. Thermographic detectors depend upon observations of a surface, so that any thermal information from the interior must be transmitted to the surface by conductivity. If the thermal conductivity of a material under observation is extremely low, it is possible to mask or hide thermal detail that is developed in the interior of the specimen. However, these two restrictions generally have the influence of requiring selections of different techniques and different information to be obtained rather than prohibiting the use of thermography. Thermography has been successfully applied to the detection of defects in materials as widely different in thermal properties as common metals and polymers.

A demonstration of the consequence of thermal conductivity differences is shown in Fig. 3.11, taken from ref. 16. This figure shows the apparent size of a thermal image (normalised by the actual size of the heat source) as a function of the sensitivity of the detector in units of percent of the thermal source strength over a distance of 0·1 of the ordinate value. For example, as shown on the figure, if the instrument used to observe a heat pattern can resolve one percent of the source strength, then the apparent size of the

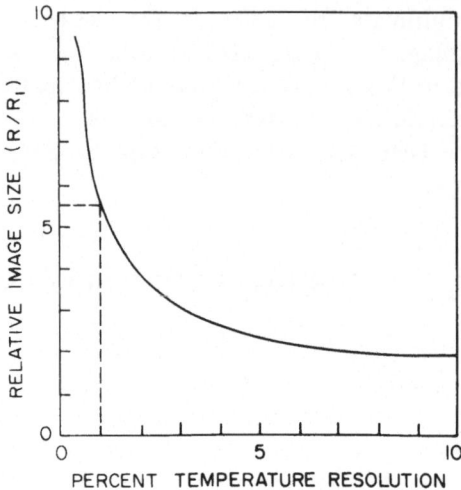

FIG. 3.11.   Relative heat source size as a function of detector thermal resolution.

pattern will be about 5 or 6 times the dimensions of the actual source, provided that the spatial resolution of the camera is not less than that apparent size. In fact, for many practical situations, this apparent size of five times the source size is a useful rule of thumb. The figure is also helpful in the interpretation of information from a heat image. For specific situations it is well to solve the heat transfer problem using the classical approaches discussed widely in the literature. It should be mentioned that heat transfer problems in composite materials, especially layered composite materials, are not as straightforward as those in homogeneous isotropic materials. Discussions of such problems can be found in refs. 16–19.

Surface reflectivity and emissivity are also parameters of importance as discussed previously. The reflection of radiant energy from the surface of a specimen under observation may interfere greatly with attempts to observe and record the heat emission pattern being developed by the specimen. For the thermal radiation band in the electromagnetic spectrum (defined as the portion of the spectrum between wavelengths of about 1×10^{-7} m and 1×10^{-4} m) reflectivity is influenced not only by the material properties but also by the surface condition. Hence, it is very difficult to classify different surfaces and corresponding quantitative reflectivities. It is best to determine by experiment in one's own laboratory the degree to which reflectivity will interfere with the observation of thermal data. The radiant energy from a lightbulb, a human body, or other heat source will generate glaring reflections from polished aluminium, for example, while a number of other surfaces, especially those painted with so-called 'flat' paints, show very little reflectivity.[20]

The emissivity of the surface is a measure of its ability to emit radiant thermal energy per unit time and unit area, usually normalised by the emissivity of an ideal 'black body' as defined in common physics texts. To that extent, for our purposes, the emissivity represents the ability of a surface to transmit the thermal information of interest that is developed in the interior of a body. Emissivity generally depends upon the material or substance, the surface condition (including roughness), and the temperature. Values of emissivity are widely tabulated and quoted (see ref. 21).

3.6. APPLICATIONS

The principal application of thermography for nondestructive testing and evaluation purposes appears to be for the detection of defects and flaws and the general monitoring of damage development and growth. In 1969 Schultz and Tsai[22] discussed the damping of composite materials and indicated that the energy loss due to damping was dependent upon the frequency of oscillation and the orientation of the orthotropic materials relative to the axis of mechanical excitation. In a subsequent report Segol and Tauchert[23] measured the heat generated during the torsional excitation of glass-fibre-reinforced plastic specimens at frequencies of 120, 210, and 300 cycles per minute. They found large temperature increases and observed discoloration and delamination in the glass epoxy composite materials with a consequent reduction in properties.

The use of thermography to detect flaws and defects, and to follow damage development, began in about 1974.[24,25] It was found that thermography is especially useful for following the complex development of damage in the vicinity of notches such as holes in composite laminates.[14] Boron–aluminium and boron–epoxy materials were used in those early studies. The heat patterns developed quite early during fatigue in the vicinity of the hole, at locations where ultimately significant microstructural damage occurred. That is, thermography served in these cases as a sensor of major stress concentrations which led during fatigue to areas of major damage development. This damage included matrix cracking, ply delaminations, and longitudinal splitting in the outer $0°$ layers.

A good demonstration of using a transient external heat source can be found in the work of McLaughlin and co-workers.[20,26–29] They found

TABLE 3.1

PREDICTED AND OBSERVED FREQUENCIES OF FLAW HEATING

(after Russell[31])

Flaw size	Frequencies analytically predicted (simply supported–clamped) (kHz)	Experimentally observed peak frequencies (kHz)	Temperature difference from far field (°AGA)[a]	Frequency variation which causes detectable drop in temperature from peak (kHz)	Minimum detectable difference in temperature (°AGA)[a]
0·134 × 0·126 in (0·34 × 0·32 cm)	23·6–24·8	23·9	1	0·1	0·5
0·201 × 0·193 in (0·51 × 0·49 cm)	10·37–10·57 17·53–17·92 24·72–24·98	10·2 18·35 26·1 13·0	14 2·5 0·5 1·5	0·04 0·01 0·035 0·03	2 0·5 0·5

a °AGA refers to degrees measured by the AGA thermovision camera. These are close to, but not precisely °K; the difference depending on the room temperature, infrared detecting material used, etc.

that in certain cases, such as for the detection of subsurface flaws, the best passive method may be a transient heat transfer scheme. McLaughlin demonstrated this by developing a technique whereby a surface heater is applied to a $(0, \pm 45, 90)_s$ graphite epoxy laminate having a part-through hole, for short time periods of the order of 60 s or less. (Surface exposure was about $0 \cdot 36$ W/cm^2.) Then the heat source was removed and the cooling patterns observed for similar times, during which the flaws showed up as hot spots due to the disturbance in free convection cooling which they caused. It appears that transient methods may indeed have some important utility, especially for near-surface disturbance.

The detection of delamination-type defects in composite laminates using active excitation schemes has been extensively studied. Russell[31] has found that heat patterns developed by low amplitude mechanical excitation in the range of 10–30 kHz are very sensitive to the excitation frequency. Frequency changes of $\pm 5\%$ or less are sufficient to cause a heat pattern developed by a damaged region to completely disappear. More precisely, as one sweeps through this frequency range, heat patterns will develop only around certain frequency values. A centre frequency can easily be identified by maximum heating levels. As one departs from the centre frequency, the hot spots which develop on the specimen surface over the damaged regions disappear, and for frequency variations of $\pm 5\%$ or less around the centre frequencies, the heat pattern of the specimen is uniform. Russell has found that the centre frequencies at which heating develops are controlled by mechanical system resonances of either a global or local nature. That is, in some cases, the hot spots develop when two conditions are met:

(i) a mechanical resonance of the shaker–specimen (global) system is attained, and

(ii) a flaw exists in the neighbourhood of one of the nodes of the particular mechanical resonance mode.

In other cases, Russell has found that the heating occurs when a local mechanical resonance occurs, that is a mechanical resonance of the local surfaces composing the flaw. For known flaws (manufactured square delaminations in the specimens), Russell has predicted the frequencies at which local resonance occurs and has found close correlation with the established heat patterns, see Table 3.1.

We have mentioned that thermography is especially sensitive to debonds and delaminations. This appears to be true when either mechanical excitation or convective radiative incident heat is used.[8,29] In fact, resonant vibration (of the body vibration or local vibration type) has been

FIG. 3.12.   Thermogram of glass epoxy gate rotor seal showing hot spots over locally damaged regions.

FIG. 3.13.   Photograph of gate rotor seal shown in Fig. 3.12.

found to be an especially sensitive method for the excitation of heat patterns generated by internal defects which have free surfaces such as debonds, delaminations, or shear cracks.[8,12,32,33] Pye and Adams[12,32] have analysed and predicted heat patterns developed by mechanical resonances in specimens having shear cracks. They observed, as already mentioned, that thermal conductivity plays a major role in the observability of damage. They predicted minimum detectable crack sizes of 28 mm in carbon-fibre-reinforced plastics and 14 mm in glass-fibre-reinforced plastics by the thermographic technique. Mignogna and Green have reported that resonant vibrations created by high-power ultrasound can also be used to detect the nature of internal boundaries.[30] Wilson and Charles[33] used a radiant, passive technique to detect adhesive-bond-line flaws in glass epoxy, graphite epoxy, and high density moulding compound. An example of damage detection by an active thermographic technique is given in Fig. 3.12. This is a thermogram of a gate rotor seal, shown in Fig. 3.13, which had been tested in service for several thousand hours. It is interesting to note that hot spots occur at locally damaged regions even in a geometrically complex component.

Since thermography works on the basis of optical field techniques, it can be applied to the quick observation of large surfaces. Because of that characteristic, it is especially suited to the quick scanning of engineering structures for the detection of defects and the identification of regions that should be examined more carefully. Also, if video thermographic equipment is used, the time resolution of the system permits the detection and recording of some of the dynamic aspects of damage development, and in some instances can be a major advantage in the sense that systems that are being dynamically tested or used for some engineering purposes do not have to be interrupted for observation.

Another application that should be mentioned is based on a technique called Thermal Pulse Video Thermography developed by the UK Atomic Energy Authority's Harwell Research Laboratory.[34] Pulse thermography involves exposing the test material to bursts of radiant energy from an intense source. The resulting transient conductivity patterns are monitored for short periods (sometimes for periods as brief as one second) by an infrared video camera which outputs a record of the test to a videotape recorder. That record can subsequently be analysed, enhanced, and processed. The method is a non-contact technique which requires access to only one side of a component. It appears to have possible applications in quality control, post-installation and in-service inspection routines. At the time of writing, the technique has been shown to be sensitive to a number of defects in composite sheets, including delaminations, inclusions, variations in thickness, and disbonds of selected materials. Research on the method continues, primarily at the Nondestructive Testing Centre at Harwell.

There is a commercial device and related technique that is based on the principle of thermoelastic temperature changes due to rapid deformation. Under adiabatic conditions, the temperature change in a component due to an increment, S, of increase in the sum of the principal stresses can be written as

$$\Delta T = -K_m T \tag{12}$$

where T is the average temperature of the component and K_m is the thermoelastic constant

$$K_m = \frac{\alpha_x}{\rho C_p} \tag{13}$$

where α_x is the coefficient of linear expansion, ρ is the mass density and C_p is the coefficient of specific heat at constant pressure (cf eqn (6)). Hence if some type of dynamic loading is applied to a structure, a corresponding dynamic heat pattern will develop in proportion to the distribution of stress in the component. A video camera system developed by the Sira Institute, Ltd, is designed to perform 'Stress Pattern Analysis by Thermal Emission' (a so-called 'SPATE' system[35]). The system records temperature patterns during cyclic excitation (with sensitivities of the order of 0·001 °C), and processes the data to provide thermoelastic stress profiles (scans or contour plots) which can be interpreted in terms of strain or stress. Since the thermoelastic change in temperature is proportional to the coefficient of thermal expansion of a material, the stress resolution of the system is material dependent. Commonly stress changes of the order of 1 N/mm² in steel or about 0·4 N/mm² in aluminium can be resolved. Calibration routines are generally required for each material to be examined. The scheme is a field technique which can be used to measure stress distributions quite rapidly and conveniently over large areas in components having very complex shapes. The stress contours can be interpreted in terms of quasi-static load levels or in terms of dynamic (including fatigue) load levels. The spatial and time resolution of such a system depends on the specific capabilities of the equipment to be used, but generally falls within ranges common to high-quality closed circuit video systems.

An example of the use of thermography to study stress fields is given in Fig. 3.14. In this instance, a proposed streamline specimen designed by Oplinger of the Army Materials and Mechanics Research Center is compared with a ASTM D638 standard dogbone specimen of the same material and under the same loading conditions. The thermograms show

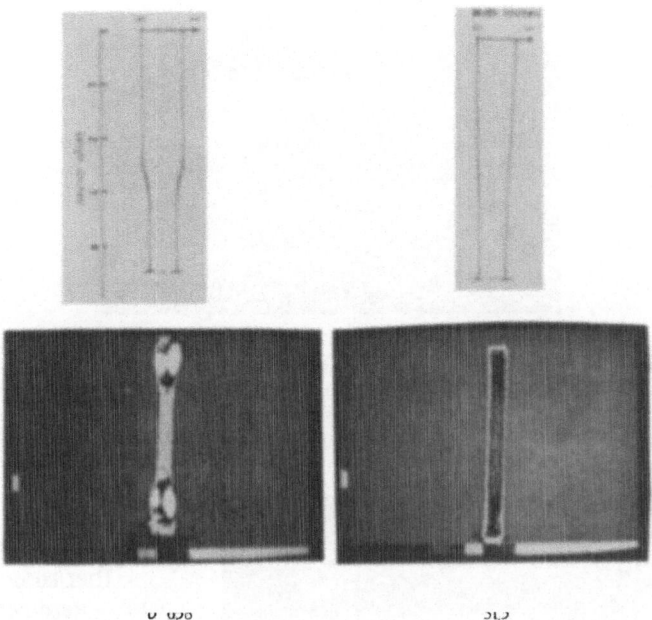

FIG. 3.14. Infrared thermograms of ASTM D638 and proposed streamline SL3 specimens during fatigue loading at 80% UTS, 5 Hz, $R = 0.1$. Thermograms were taken with 10 °C full-scale after 200 cycles (after ref. 36).

that the dogbone specimen develops significant stress concentrations in the regions of the shoulders. On the other hand, the stresses in the streamline specimen are more uniform.

Other interesting uses of thermography for investigating composite material behaviour have also been reported.[36] A thermogram taken by an active technique (mechanical vibration of the specimen) is shown in Fig. 3.15. It is easily seen that a glass epoxy specimen which had been partially soaked in water for six days became much warmer over that length soaked in water. Thus vibrothermography appears to be a good method for detecting areas of moisture absorption in composites. Proper processing control of the curing of composite laminates has been enhanced through the use of thermography. Figures 3.16 and 3.17 show thermograms obtained during the cure of a 200 ply (50 mm thick) S-2 glass/SP250 epoxy laminate. If cured too rapidly, and uncontrolled exotherm can occur in this material and result in thermal degradation and oxidation. At 110 minutes into the cure cycle, the top side of the laminate was significantly hotter than the middle and bottom side, Fig. 3.16. Platen heaters were then adjusted to

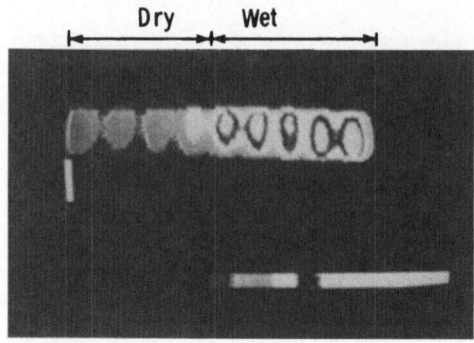

FIG. 3.15.   Thermogram of S-glass/SP250 laminate at 10·4 kHz mechanical excitation taken at 20 °C full-scale. Laminate had been partially soaked in water at 80 °C for 6 days (after ref. 36).

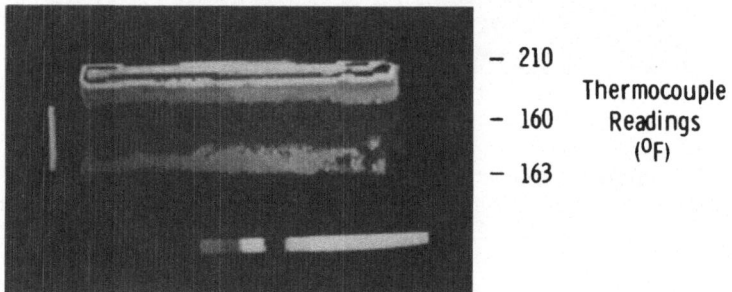

FIG. 3.16.   Thermogram of 200 ply S-glass/SP250 laminate during press cure after 110 minutes, 50 °C full-scale (after ref. 36).

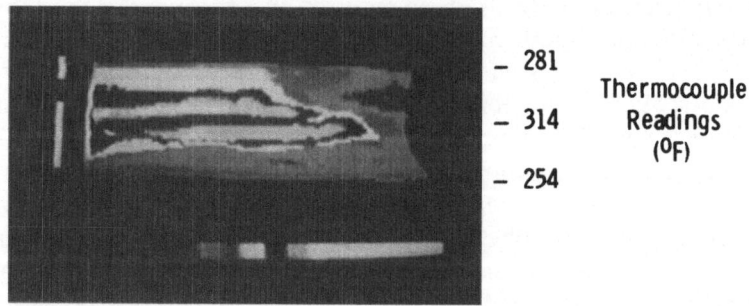

FIG. 3.17.   Thermogram of 200 ply S-glass/SP250 laminate during press cure after 160 minutes, 100 °C full-scale (after ref. 36).

Clamp

Strip heater

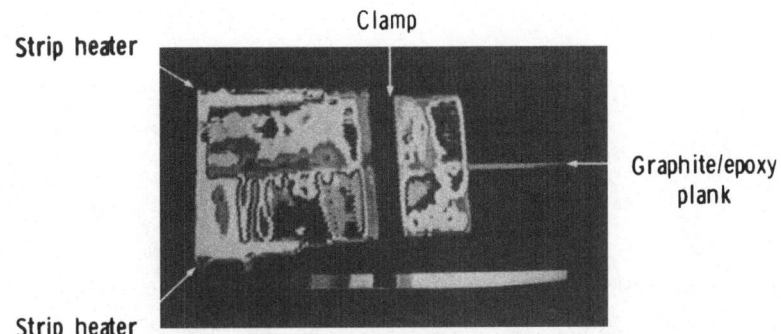

Graphite/epoxy plank

Strip heater

FIG. 3.18.   Thermogram of graphite epoxy pultrusion process, 100 °C full-scale (after ref. 36).

reduce this uneven heating. At 160 minutes into the cure cycle, an exotherm developed in the centre of the laminate, Fig. 3.17. Normal temperature monitoring procedures measure the temperature only at the top and bottom platens and would have missed this. Thermography has been used, in addition, to monitor the fabrication of pultruded graphite epoxy planks, Fig. 3.18. The pultrusion at these plants requires very careful temperature control so that the epoxy resin is almost completely cured as it leaves the die.

We close our survey of some selected applications by mentioning a method that is still under development. We have discussed the emission of heat by various composite (and conventional) materials during mechanical excitation. Such heat emission can be used in a special way to study damage development in composite materials during dynamic and cyclic loading. Since, as we have seen, heat is emitted in proportion to stress distributions, by strain energy release and by internal dissipation around defects—especially by flaw surface rubbing—the heat patterns associated with damage events not only indicate the shape and extent of damage but also indicates the severity and nature of damage. If our understanding of such patterns was complete, this information could be the basis of a direct interpretation of the mechanics of damage development. An example of a sequence of damage related heat patterns is shown in Fig. 3.19 for cyclic loading of a $[0, \pm 45]_s$ boron/epoxy laminate with a centre hole from the initiation of damage (A) to incipient fracture (I). In composite materials, damage development is generally very complex, involving great varieties and combinations of damage modes and events. Without a method of characterising the collective effect of these patterns of defects, very little useful information has been gained by only locating them. It is precisely

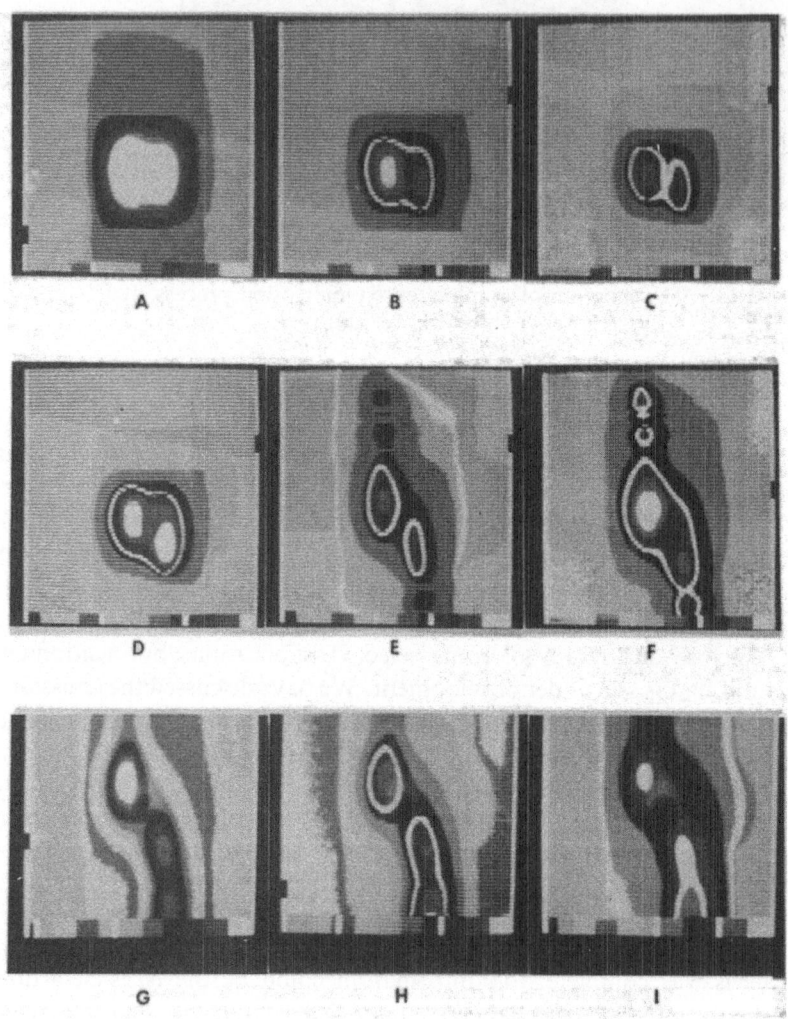

FIG. 3.19. Thermographs showing heat patterns associated with damage development in a $[0, \pm 45]_s$ boron/epoxy specimen with centre hole under cyclic loading from initiation (A) to incipient fracture (I).

FIG. 3.20.   Thermographic pattern of damage development in a helicopter blade
during full-scale fatigue testing.

this potential capability to characterise such complex damage development
in a way that is proportional to the net effect of all events, is directly related
to the collective mechanical situation (which produces the heat pattern),
and which provides an analogue interpretation of complex damage states,
that is particularly unique and exciting. The authors are aware of no other
technique that has such a capability (cf. refs. 16 and 37).

This is especially true in the context of other practical capabilities of thermography. Since video-thermography has a time resolution of the video system involved, dynamic service or test conditions can be monitored without interrupting the loading sequence or disturbing the load history. Thermography is generally a noncontact measurement, and is an optical scheme. Hence, high spatial resolutions can be obtained using microscopes and large areas can be viewed with proper lenses. Figure 3.20 demonstrates the utility of some of these features. That figure shows a thermogram recorded during the full scale fatigue testing of a helicopter rotor blade (about 1 m wide by 8 m long in a Sikorsky resonance test frame. The 'hot spot' in that figure developed near the root end of the blade near the rotor hub joint where a box section is flared into the blade skin and blade structure. The region indicated eventually became the failure location of the blade after several million additional cycles of loading. This early indication of damage development in a complex structure made from a complex material (glass/epoxy in this case), in a region where the mechanical state of stress and damage were changing as a function of time, was not reproduced by any other NDT technique attempted, including acoustic emission and ultrasonic attenuation examinations.

It is clear that thermography is presently a useful and valuable examination and investigation technique for monitoring and studying composite materials. It is also clear that much work must yet be done to complete our understanding of how heat is produced, how heat patterns are associated with internal structure and events, and how the engineering strength, stiffness and life of composite materials can be associated with and anticipated by the interpretation of thermographic data.

REFERENCES

1. HENNEKE, E. G. II, REIFSNIDER, K. L. and STINCHCOMB, W. W., Thermography—an NDI method for damage detection, *J. of Metals*, **31**, 11–15 (1979).
2. ENGELHARDT, R. E. and HEWGLEY, W. A., Thermal and infrared testing, *Nondestructive Testing—A Survey*, NASA SP-5113, US Government Printing Office, pp. 119–40, 1973.
3. WILLIAMS, J. H., JR., FELENCHAK, B. R. and NAGEM, R. J., Quantitative geometric characterization of two-dimensional flaws via liquid crystals thermography, *Materials Evaluation*, **41**, 190–201 (1983).
4. WILLIAMS, J. H., JR. and NAGEM, R. J., A liquid crystals kit for structural integrity assessment of fibreglass watercraft, *Materials Evaluation*, **41**, 202–10 (1983).

5. VANZETTI, R., *Practical Applications of Infrared Techniques*, John Wiley & Sons, New York, 1972.
6. ORLOVE, G. L., Practical thermal measurement techniques, *Thermal Infrared Sensing Diagnostics (Thermosense V)*, Proceedings of the Int. Soc. for Optical Engng, Vol. 371, Oct. 25–27, 1982, pp. 72–81.
7. BRAMSON, M. A., *Infrared Radiation, A Handbook for Applications*, translated by R. B. Rodman, Plenum Press, New York, 1968.
8. HENNEKE, E. G., II and JONES, T. S., Detection of damage in composite materials by vibrothermography, *Nondestructive Evaluation and Flaw Criticality for Composite Materials*, STP 696, American Society for Testing and Materials, Philadelphia, 1979.
9. BIOT, M. A., Thermoelasticity and irreversible thermodynamics, *J. Applied Physics*, **27**(3), 240–53 (1956).
10. NOWICK, A. S., Internal friction in metals, *Prog. in Metal Physics*, **4**, 1–70 (1953).
11. MASON, W. P., *Piezoelectric Crystals and Their Application to Ultrasonics*, D. Van Nostrand Co., Inc., New York, 1950.
12. PYE, C. J. and ADAMS, R. D., Detection of damage in fibre reinforced plastics using thermal fields generated during resonant vibration, *NDT International*, June 1981, 111–18.
13. REIFSNIDER, K. L. and STINCHCOMB, W. W., Investigation of dynamic heat emission patterns in mechanical and chemical systems, *Proc. of 2nd Biennial Infrared Information Exchange*, AGA Corp., St. Louis, MO, 1974, pp. 45–58.
14. HENNEKE, E. G., II, REIFSNEIDER, K. L. and STINCHCOMB, W. W., Thermography—an NDI method for damage detection, *J. of Metals*, **31**, 1115 (1979).
15. HSIEH, JAMES C., Temperature distribution around a propagating crack, Doctoral Dissertation, College of Engineering, Virginia Polytechnic Institute and State University, April 1977.
16. REIFSNIDER, K. L., HENNEKE, E. G. and STINCHCOMB, W. W., The mechanics of vibrothermography, *Mechanics of Nondestructive Testing*, Ed., W. W. Stinchcomb, Plenum Press, New York, 1980.
17. JONES, T. S., Thermographic detection of damaged regions in fiber reinforced composite materials, Thesis for Master of Science, Engineering Science and Mechanics Dept., School of Engineering, Virginia Polytechnic Institute and State University, April 1977.
18. MAGEAL, A., BACHE, T. C. and HEGEMEIR, G. A., A continuum model for diffusion in laminated composite media, *J. of Heat Transfer*, **98**, 133–5 (1976).
19. ESTES, R. C. and MULHOLLAND, G. P., Diffusion through orthotopic rectangular laminated composites, *Fibre Science and Technology*, **7**, 257–9 (1974).
20. MCLAUGHLIN, P. V., MCASSEY, E. V. and DEITRICH, R. C., Non-destructive examination of fibre composite structures by thermal field techniques, *NDT International*, April 1980, 58–62.
21. SUEC, J., *Heat Transfer*, Simon and Schuster, New York, 1980.
22. SCHULTZ, A. B. and TSAI, S. W., Measurements of complex dynamic moduli for laminated fiber-reinforced composites, *J. of Composite Materials*, **3**, 434–43 (1969).

23. SEGOL, G. and TAUCHERT, T. R., *An Experimental Study of the Heat Generated During Cyclic Deformations of Glass–Epoxy Composites*, Technical Report No. 21, Dept. of the Army, US Army Research Office, Durham, Contract DAHC04-69-C-0062, December 1979.

24. REIFSNIDER, K. L. and STINCHCOMB, W. W., Investigation of dynamic heat emission patterns on mechanical and chemical systems, *Proc. of 2nd Biennial Infrared Information Exchange*, AGA Corp., St. Louis, MO, 1974, pp. 45–8.

25. REIFSNIDER, K. L. and WILLIAMS, R. S., Determination of fatigue-related heat emission in composite materials, *Experimental Mechanics*, **14**, 45–8 (1974).

26. MCLAUGHLIN, P. V., MCASSEY, P. V., JR. and DEITRICH, R. C., *Aerostructure Nondestructive Evaluation by Thermal Field Techniques*, Report No. NAEC-92-131, Naval Air Systems Command, March 1979.

27. MCLAUGHLIN, P. V., MCASSEY, E. V., EMANY, V. R., KOERT, D. N. and SPITZER, J. M., *Aerostructure NDE by Thermal Field Detection; Phase I—Fundamental Information and Basic Technique Development*, Report No. NAEC-92-157, Naval Air Systems Command, May 1982.

28. MCLAUGHLIN, P. V., JR., MCASSEY, E. V., JR. and KOERT, D. N., *NDT of Composites by Thermography*, Technical Report AFWAL-TR-81-4080, Air Force Wright Aeronautical Laboratory, September 1981.

29. MCASSEY, E. V., JR., MCLAUGHLIN, P. V., JR., KOERT, D. N. and DEITRICH, R. C., Thermographic NDT of composites using externally applied thermal fields, *Proc. 35th Conf. on Reinforced Plastics*, SPI, pp. 26A.1-8, February 1980.

30. MIGNOGNA, R. B. and GREEN, R. E., JR., Nondestructive evaluation of the effects of dynamic stress produced by high-power ultrasound in materials, *The Mechanics of Nondestructive Testing*, Plenum Press, New York, 1980.

31. RUSSELL, S. S., An investigation of the excitation frequency dependent behavior of fiber reinforced epoxy composites during vibrothermographic inspection, PhD Dissertation, Dept. of Engineering Science and Mechanics, Virginia Polytechnic Institute and State University, November 1982.

32. PYE, C. J. and ADAMS, R. D., Heat emission from damaged composite materials and its use in nondestructive testing, *J. Phys. D.: Appl. Phys.*, **14**, 927–41 (1981).

33. WILSON, D. W. and CHARLES, J. A., Thermographic detection of adhesive-bond and interlaminar flaws in composites, *Experimental Mechanics*, **21**, 276–80 (1981).

34. HAYDON, E., UK program will advance thermography, *Industrial Research and Development*, June 1983. 58.

35. OLIVER, D. E. and WEBBER, J. M. B., Absolute calibration of the SPATE technique for non-contracting stress measurement, *Proc. V Intl. Congress on Experimental Mechanics*, Montreal, 10–15 June 1984 (SESA), 539–46.

36. HENNEKE, E. G., II, REIFSNIDER, K. L., SHUFORD, R. J., HINTON, Y. L. and MARKERT, B. R., Thermography—applications to the manufacture and nondestructive characterization of composites, *Thermal Infrared Sensing Diagnostics (Thermosense V)*, SPIE Vol. 371, 1982, pp. 98–104.

37. REIFSNIDER, K. L. Automated time-resolved data acquisition during static and dynamic loading of composite materials, *Proc. of Sixth Int. Conf. on Experimental Stress Analysis*, Munich, FRG, September 18–22, 1978 (VDI-Berichte Nr. 313, 1978) pp. 699–703.

Chapter 4

COMPRESSIVE PROPERTIES OF RESINS AND COMPOSITES

M. R. PIGGOTT

*Department of Chemical Engineering and Applied Chemistry,
University of Toronto, Ontario, Canada*

SUMMARY

*In this chapter the compressive properties of resins used in high performance
composites are described, as a prelude to a detailed look at methods for
testing aligned fibre composites and at the results of these tests. Data on
aligned fibre composites, various laminates and hybrids are provided,
including fatigue data, and results of biaxial tests. Finally, failure processes
are discussed, and it is shown that, though much progress has been made
recently, there is much work still to be done in obtaining data, and
investigating failure mechanisms.*

4.1. INTRODUCTION

Compressive properties of resins have not received much attention: for
example, in the case of polyesters and epoxies the Plastics Encyclopedia[1]
lists compressive strengths but not moduli. The compressive properties of
composites, on the other hand, have received a great deal of attention since
the 1960s, and compressive strength is now widely used as a measure of the
quality of high performance (aligned fibre) composites, such as laminates.

The understanding of what factors affect the compressive strength of
fibre composites has been slow to develop. It is instructive to note the
similarities between this development and that of our understanding of the
tensile strength of the more traditional materials such as metals and

131

ceramics. First came an approximate theory for the perfect material. This produced results which were much too high. The theory was therefore refined, but still produced results which were much too high. Finally, the theory was abandoned for use with ordinary materials. Instead, theories which took account of imperfections were developed.

The ideal material considered by the theoreticians was perfectly homogeneous down to the atomic level, and the theoretical strength was determined only by the strength of the chemical bonds. We come close to realising this ideal strength with single crystals or whiskers about 1 μm in diameter. More practical materials have their strengths determined by imperfections, i.e. grain boundaries and dislocations in the case of ductile metals, and microcracks in the case of ceramics, such as glass.

With fibre composites, the ideal structure had perfectly straight fibres, which were surrounded by, and perfectly bonded to, an ideally elastic matrix. Composite strength was then determined by elastic buckling of the fibres, and for one mode of failure the composite strength,[2] σ_{1u}, was given by

$$\sigma_{1u} = E_m/2(1 + v_m)(1 - V_f) \tag{1}$$

In eqn (1), E_m and v_m are the Young's modulus and Poisson's ratio of the matrix, and V_f is the fibre volume fraction. (The other mode of failure gave a result that was not very different from eqn (1).)

This equation had two problems:

1. it did not give the correct form for the variation of compressive strength with fibre volume fraction, and
2. it predicted values for σ_{1u} that were much too great.

There have been many attempts to refine the theory,[3] but they have not solved the basic problems. It has even led to experiments with model composites,[4] but these have only indirectly helped our understanding of the compressive strength of practical composites. Unfortunately this theory is treated as substantially correct in almost all texts on fibre composites, including several recent ones.

Equation (1) implies that the fibres have infinite strength, since the composite strength goes off to infinity as V_f approaches one. If the fibres are not perfectly strong we can thus expect eqn (1) to break down. It was first shown in 1970 that this was indeed the case.[5] In 1975 the effect of adhesion breakdown was demonstrated,[6] and in 1980 the effect of matrix yielding was thoroughly explored.[7] It was noted in this work that the composite modulus was also influenced by matrix yielding. This added strong support to a fairly recent suggestion[8] that fibre straightness was an important factor, and was quickly verified.[9] These observations paved the way for an explanation of the strength and modulus of aligned fibre composites based on the imperfections.[10]

In this chapter we will first describe the compressive properties of the resins used for high performance composites, and also filled resins, including chopped-glass-fibre-filled resins. The rest of the chapter is devoted to high performance composites: how their compressive strengths and moduli are measured, and values so obtained. Failure processes are described, and factors affecting compressive properties are discussed.

4.2. COMPRESSIVE PROPERTIES OF RESINS

Strength and modulus of resins and filled resins are normally measured by standard methods. Thus the ASTM D695 test method for compression strength uses, as standard specimen, a cylinder having an aspect (length/diameter) ratio of two. Other shapes of specimen are permitted, but thin specimens ($<3\cdot2$ mm thick) must have lateral support. Testing is carried out at $1\cdot3$ mm/min.

Polyester casting resins are generally weaker (90–110 MPa)[1] than epoxies (110–180 MPa),[1] though some individual epoxies can be less strong than some polyesters. The state of cure of the resin has an important effect, as shown for an isophthalic polyester resin in Fig. 4.1.[7] The lowest curve is for the resin when it has not been post-cured. Partial curing (1 h at 80 °C,

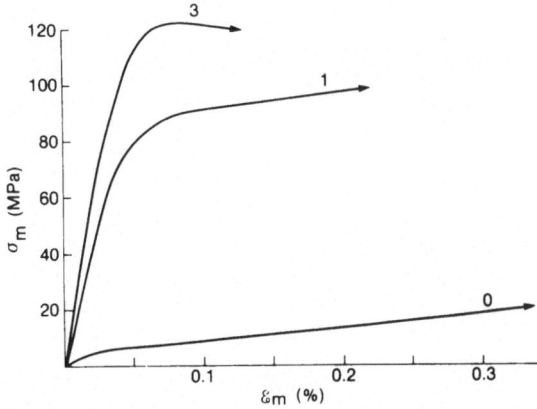

FIG. 4.1.   Stress–strain curves for a polyester resin.[7] The figures on the curves indicate the number of hours of post-curing at 80 °C.

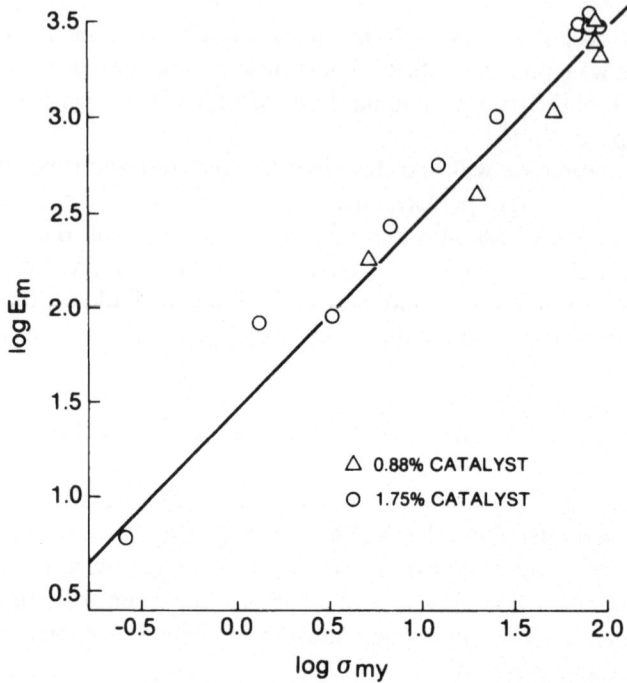

FIG. 4.2.   Logarithmic plot indicating proportional relation between modulus and
yield strength of a polyester resin at various stages of cure.[7]

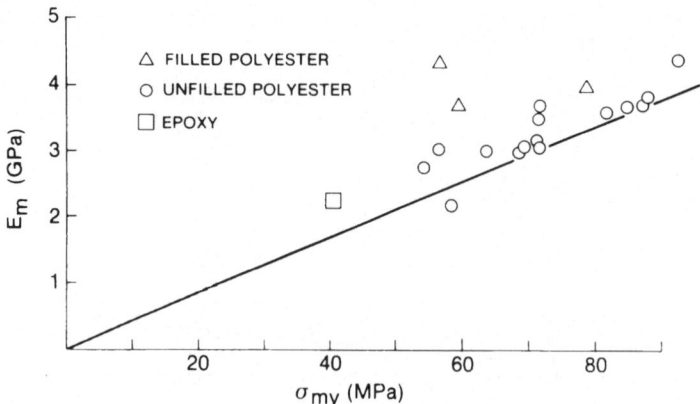

FIG. 4.3.   Relation between modulus and yield strength for various polyester
resins, and an epoxy resin.[7] Line is $E_m = 43\sigma_{my}$.

middle curve) increases both yield stress (σ_{my}) and strain (ε_{my}), and increases the level of the stress at all strains. Full post-curing of this resin (3 h at 80 °C) increases the yield stress and strain a little more.

All these polymers have easily identified yield stresses which terminate the initial linear region of the stress–strain curve. The fully cured resins have an apparent compression failure, while the partially cured resins appear to work-harden continuously; the stress–strain curve for the fully cured resin has a peak, followed by a decline in stress for further increases in strain. These specimens were subjected to very large strains, about a halving in length. When put on one side after testing, they recovered their original dimensions in about 30 min, and when re-tested, followed the same stress–strain curve, within a few percent. This type of recovery has been looked at in detail with thermoplastics, see for example, Benjamin et al.[11]

For the partially cured resins, the modulus (E_m) is about 43 times the yield stress. Figure 4.2 shows a log–log plot of modulus vs yield stress. It can be seen that the relationship is not significantly affected by using a lower quantity of catalyst. The relationship is also obeyed, approximately, by a wide range of fully cured polyesters containing different acids and glycols, Fig. 4.3, though fire-retardant fillers slightly increase the modulus above $43\sigma_{my}$. A Ciba-Geigy epoxy resin (Araldite VY 219) also has E_m slightly greater than $43\sigma_{my}$.

The apparent compression strength (σ_{mu}), when it exists, is also approximately proportional to σ_{my}, as shown in Fig. 4.4 for the fully cured resins. In Fig. 4.4 the regression line gives $\sigma_{mu} \simeq 1\cdot7\,\sigma_{my}$. Both these relationships have been observed with epoxy resins softened by the addition of dimethyl formamide, and when copolymerised with spiro ortho carbonates to reduce the shrinkage stress.[12]

Compression strength and modulus are not very rate sensitive, at least in

TABLE 4.1

EFFECT OF TESTING SPEED ON STRENGTH AND MODULUS OF POLYESTER RESIN, AND A GLASS-FIBRE PULTRUSION MADE WITH THE SAME RESIN[7]

Speed ($mm\,min^{-1}$)	Resin		Pultrusion	
	Yield strength (MPa)	Modulus (GPa)	Strength (GPa)	Modulus (GPa)
0·05	74	3·3	0·45	19·3
0·5	94	3·3	0·47	19·5
5·0	96	3·6	0·53	20·2
50·0	117	3·7	0·52	26·2

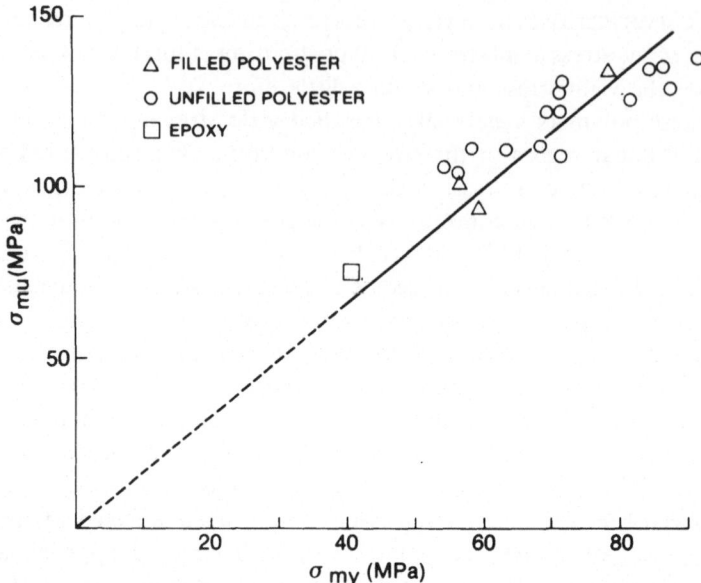

FIG. 4.4. Relation between apparent ultimate compression strength and yield strength for various polyester resins, and an epoxy resin.[7] Line is $\sigma_{mu} = 1{\cdot}7\sigma_{my}$.

the range 3×10^{-5} sec^{-1} to $0{\cdot}03$ sec^{-1}. Table 4.1 shows the results of tests on 25 mm long cylinders of a polyester resin.[7] It can be seen that the strength increases by 58% and the modulus by 12% for a thousand-fold increase in speed. The ratio modulus/yield strength declines from 46 to 32.

A CRC handbook[13] containing a compilation of data on resins used as matrices for fibre composites, and on the composites themselves, includes compression data on a wide range of resins.

Silica and aluminium particulate fillers increase the compression strength of epoxy casting resins from 110–180 MPa to 110–250 MPa; similarly, chopped glass-fibres increase the strength of rigid polyesters from 90–110 MPa to 110–210 MPa.[1]

Finally, it is interesting to note that, with thermoplastics at least, there is a correlation between glass transition temperature, T_g, and compression yield strength,[14] σ_{my}:

$$\sigma_{my} \simeq A T_g^2$$

where A is a constant. The relationship also describes the yield stress of a wide range of other materials, including ceramics.

4.3. COMPRESSIVE TESTING ON FIBRE COMPOSITES

4.3.1. Introduction
Special techniques are needed for compressive tests on aligned fibre composites because of their high degree of anisotropy. Both strength and modulus anisotropies have important effects.

1. Strength anisotropies: aligned fibre composites are relatively weak when tested at right angles to the fibre direction, this strength being controlled by the resin (or the interface) rather than the fibres. Compression testing in the fibre direction produces transverse stresses, and although these are not very great, they cause the composite to split instead of failing in compression, unless steps are taken to confine the ends of the specimen.

2. Modulus anisotropies: elastic buckling can occur very easily with fibre composites, especially aligned fibre composites tested along the fibre direction. For example, with pultruded rods, elastic buckling occurred when the aspect ratio was about seven.[7] With such rods, Euler buckling should occur when the aspect ratio is about 11. This difference is probably due to shear stresses, arising from a slight misalignment, initiating shear instability. Such instability easily occurs because of the very low shear modulus of these materials. (With isotropic materials the shear modulus is about 38 % of the Young's modulus; with aligned fibre composites it can be as little as 5 % of the Young's modulus.[15])

Two conditions must thus be met by any compression test method:

(a) adequate end confinement to prevent splitting, and
(b) very short gauge length, or adequate lateral support over the gauge length, to prevent buckling.

In addition, the load must be accurately aligned along the specimen axis, so that bending moments are not applied to the specimen.

A great number of different methods of compression testing have been developed. In this section we will describe examples of the methods currently in use. The subject is well reviewed in a recent text.[16]

4.3.2. Simple Method for Pultrusions
Pultrusions are often characterised by cross-sections which have substantial thicknesses as well as widths. A method suitable for routine testing of this type of pultrusion is illustrated in Fig. 4.5.[7] The specimen

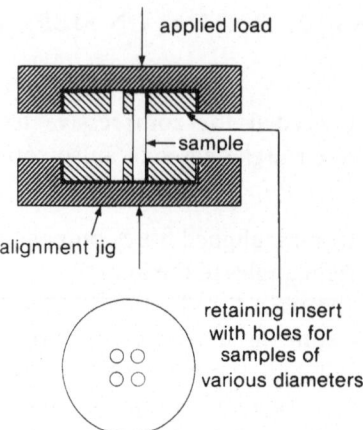

FIG. 4.5.   Simple method for compression testing of circular section pultrusions. It may also be used for other sections, by using different cut-outs in the end plates.

aspect ratio should not exceed about six. The specimen should fit snugly into the holes in the end plates, and should have ends which are polished and normal to the specimen axis. (This is easily accomplished with a simple jig.)

With 6 mm diameter rods this method gave results which did not vary significantly with gauge length, Fig. 4.6, until elastic buckling occurred at a gauge length of about 45 mm. Some differences in failure mechanism were noted, however, as gauge length was increased, Fig. 4.7.

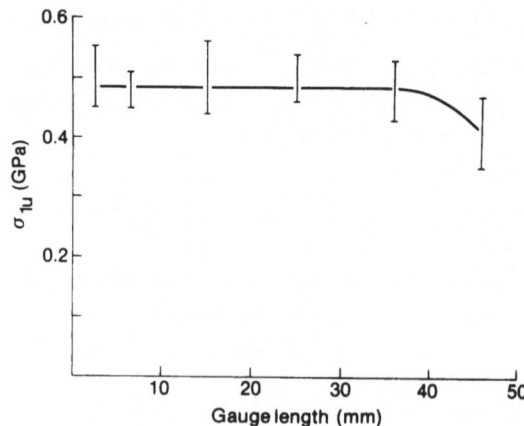

FIG. 4.6.   Effect of gauge length on strength of 6 mm diameter glass–polyester pultrusions.

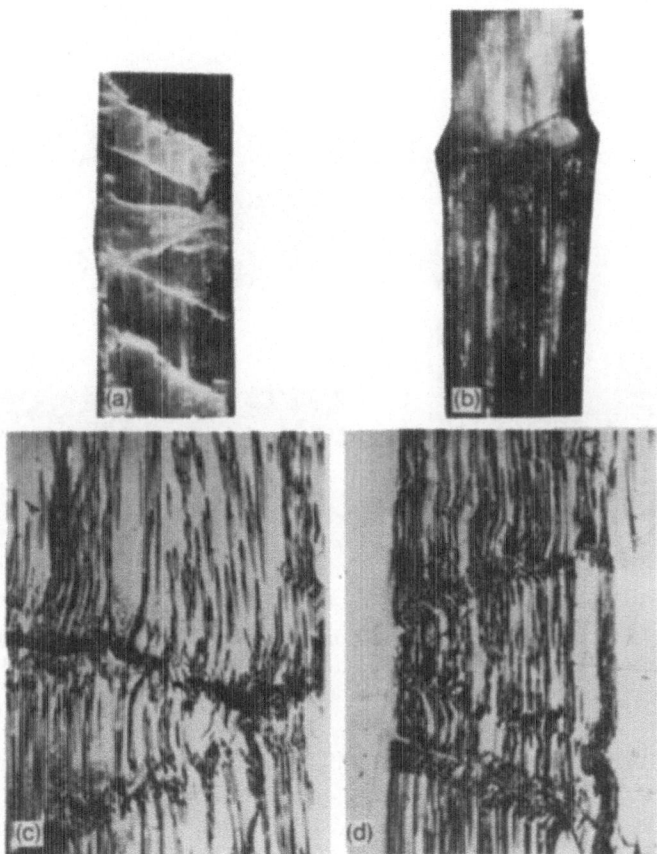

FIG. 4.7. Effect of gauge length on failure mode. It changes from multiple cracking in the shortest specimen (a and c) to kinking in the longest (b and d). Glass–polyester pultrusions with $V_f = 0.31$.[7] In (c) and (d) the square size corresponds to 0.1 mm. The rods were 6 mm in diameter.

4.3.3. Sandwich Beam

In this method,[17] the specimen is fully supported, so that elastic buckling and end brooming are prevented. The method has been widely used in the past for testing laminates, and is easily adapted for biaxial and fatigue testing.

The sandwich beam is made with a metal skin on one face (aluminium, titanium or steel, for example) or a reinforced plastic skin. On the other face is the laminate to be tested, and in the centre is a honeycomb, normally

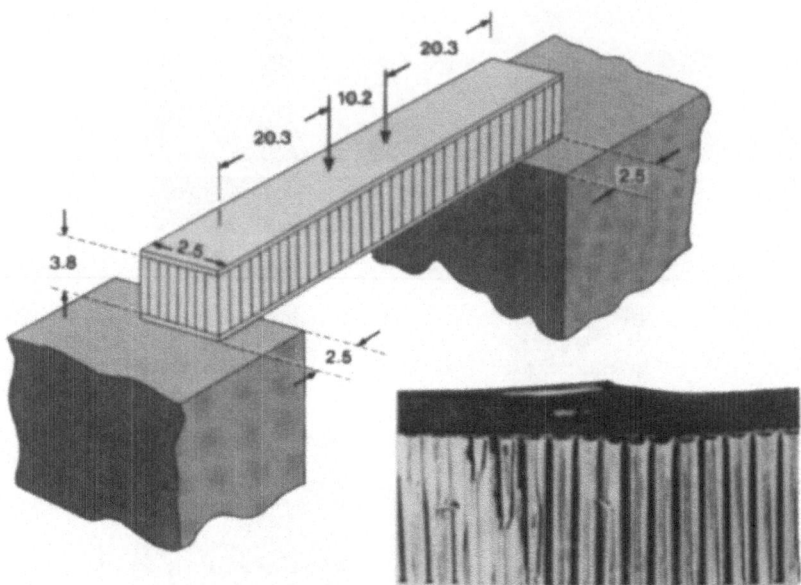

FIG. 4.8.   Sandwich beam method for compression testing. The beam is put in four point bending. At the bottom right is shown the type of failure obtained with an aligned carbon fibre laminate. Dimensions are in mm.

aluminium. The two skins are usually bonded in one operation, and great care is required to ensure uniformly good adhesion. The diagram in Fig. 4.8 shows the test specimen; it is tested in four-point bending, as indicated schematically in the figure. The loads are applied through pads, so that they are distributed over an area of about 4 cm^2.

Figure 4.8 also shows a typical failure observed with a unidirectional carbon–epoxy laminate. The line of the fibre fracture is not, in general at right angles to the edge, and the honeycomb is severely distorted underneath the failed region.

This method produces reliable results, but uses rather large specimens which are extremely expensive to make. It is not suitable for making Poisson's ratio measurements. It has recently been adapted so that the sandwich is reusable.[18]

4.3.4. Unsupported Specimen

This method is exemplified by the Celanese method which has been adopted by ASTM (D3410). It uses a short gauge length (12·7 mm). To ensure that

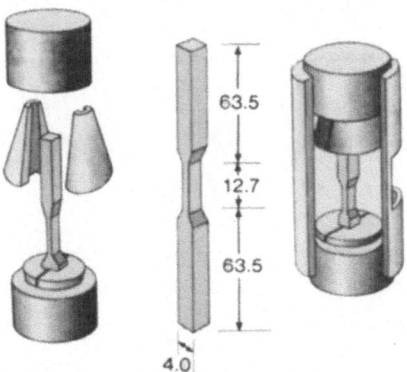

FIG. 4.9. Celanese specimen, and gripping and compression fixture (ASTM D3410). Dimensions are in mm.

alignment is maintained, both ends of the gripping mechanism are inside a tube, Fig. 4.9. The grips are conical, and the forces are transferred by friction when the tapered sleeve is loaded in compression. The specimen has end tabs of steel or fibreglass, which are bevelled to reduce stress concentrations.

The ITTRI method is similar,[19] will accept the same specimen, but also fits specimens wider than 6·35 mm. It avoids some of the problems associated with the Celanese method, and it is easier to insert the specimen, as it uses steel pins for alignment control, rather than a tube.

A comparison[19] between ITTRI results and those obtained with the sandwich beam showed good agreement between the ITTRI method and the sandwich beam method for unidirectional laminates and $0/90/\pm45$ degree laminates made with various carbon–epoxy prepregs. The ITTRI method gave results for the strength of aligned-boron-fibre composites which were about 50–75 % of those obtained by the sandwich beam, however, and moduli were about 10 % lower. Thus the method is not recommended for laminates made with aligned very high modulus fibres.

There are a number of other methods which do not use lateral supports for the specimen. They all require specimens which have been end tabbed with great precision, so that the faces are very close to being perfectly parallel.

4.3.5. Partially Supported Specimen

These methods vary only in the shape of the specimen supports. The specimens themselves are usually quite long, typically 90 mm, and have end

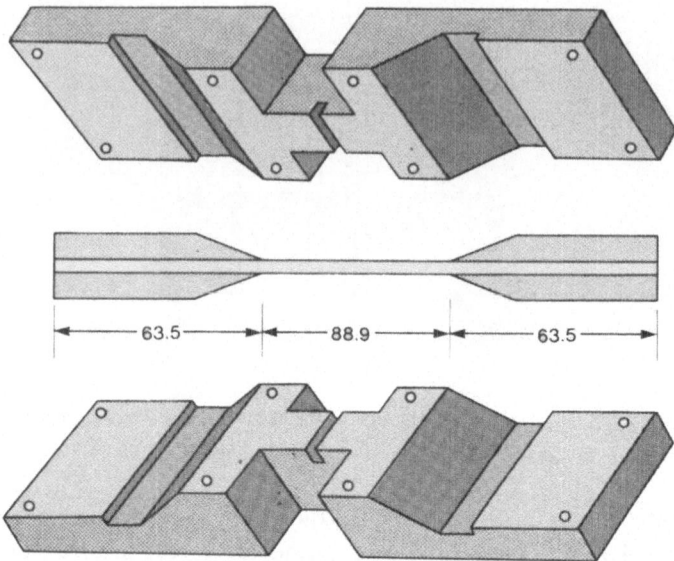

FIG. 4.10.   SWR1 specimen supports.[20] Test specimen is standard tensile
specimen with end tabs (ASTM D638). Dimensions are in mm.

tabs. The South West Research Institute specimen[20] and fixture are shown
in Fig. 4.10.

These methods have the advantage that they can take standard ASTM
tensile specimens. They give results for multidirectional laminates that
agree with those obtained with the Celanese method, but give lower results,
in the fibre direction, for unidirectional laminates.

4.4. COMPRESSIVE PROPERTIES OF FIBRE COMPOSITES

4.4.1. Aligned-fibre Composites

Manufacturer's data for fibre properties and prepreg properties are freely
available for tension, but data for compression are not so common. Table
4.2, [6,7,9,19,21,22] lists the compressive strengths of a number of different
aligned fibre composites. It will be observed that boron gives the highest
values and Kevlar the lowest. Different carbons give different values;
generally the higher modulus–lower tensile strength varieties give lower
values than the higher tensile strength ones. Glass-fibre composites are
usually about as strong as those made with high strength carbons.
Pultrusions are generally weaker than laminates.

TABLE 4.2
COMPRESSION STRENGTHS AND MODULI OF ALIGNED-FIBRE-REINFORCED EPOXY AND
POLYESTER RESINS

Fibre	Method of manufacture	Resin	V_f	Strength (GPa)	Modulus (GPa)	Reference
Glass	Pultrusion	Polyester	0·5	0·9	35	9
Glass	Filament winding	Epoxy	0·7	1·4	56	21
Stiff carbon	Hand layup	Epoxy	0·6	0·7	—	6
Stiff carbon	Laminate	Epoxy	~0·6[a]	0·44	211	19
Strong carbon	Hand layup	Epoxy	0·6	1·3	—	6
Strong carbon	Laminate	Epoxy	~0·6[a]	1·70	159	19
Boron	Laminate	Epoxy	~0·6[a]	2·5	217	19
Kevlar	Pultrusion	Polyester	0·31	0·17	21	7
Kevlar	Laminate	Epoxy	0·6	0·28	—	22

[a] No volume fraction given. This method usually gives $V_f \simeq 0.6$.
Note: Additional data for boron and carbon are given by Lynch,[13] and prepreg tape
manufacturers usually have data available (e.g. 3M's S glass prepregs should make laminates
with $V_f = 0.55$, $\sigma_{1u} = 1.01$ GPa, and $E_1 = 48.0$).

Also listed in Table 4.2 are moduli. These are often somewhat less than
tensile (Young's) moduli. Kevlar, however, displays a much lower modulus.
The poor compressive properties of Kevlar composites are often mitigated
by the use of hybrids with carbon fibres.

Testing speed has very little effect on strength and modulus, at least in the
case of glass[7] at moderate speeds, Table 4.1. The slight upward trend as
speed is increased mirrors that for the unreinforced polyester. At extremely
high speeds, however, there is a large increase in strength.[23]

4.4.2. Laminates
With laminates, an almost infinite variety of configurations is possible. In
addition, hybrids, with layers made with different fibres, are commonly
used. In this section we will consider a few examples of homofibre
composites. (See Section 4.4.4 for hybrids.)

Typical data obtained for carbon-fibre–epoxy laminates are given in
Table 4.3.[24,25] These laminates are symmetrical about their mid-planes,
and the fibre directions are indicated in the column labelled structure. In
the cases of the ±45 and 0/90 laminates, there were equal numbers of layers
in each orientation. For the ±45/0/90 laminates there were 5 layers in the
45° directions, 16 in the 0° and 4 in the 90° directions.

It is noteworthy that the ±45 laminate is nearly 10% weaker than the
aligned-fibre laminate tested at right angles to the fibres. Also important is

TABLE 4.3

Structure	Strength (MPa)	Modulus (GPa)	Reference
0	1 400	110	24
90	259	11·7	24
±45	236	16·6	24
±45/0/90	724	71·7	24
0/90	603[a]	64·3	25

[a] Result is strongly affected by stacking sequence and specimen thickness. A 2 mm thick laminate with 0/90/90/0 sequence gave 603, but with 90/0/0/90 sequence gave 516. For 1 mm thick laminates the results were 598 and 365 MPa respectively.
Note: Additional data for boron and carbon laminates are given by Lynch.[13]

the effect of stacking sequence, as indicated in the footnote to Table 4.3. The stacking sequence effect (and possibly, also, the weakness of the ±45 laminates) is considered to be the result of the stresses that arise in laminates near edges and discontinuities.[26] However, it is not possible, at present, to predict these effects from established theory.[25]

For stresses at an angle to the fibres in aligned-fibre laminates the Tsai–Hill theory may be used to estimate the strength. Let the tensile and compressive strengths in a laminate be X and X' in the fibre direction, Y and Y' at right angles to the fibre direction (the compressive strengths are indicated by the primes). Let S denote the shear strength in the plane of the laminate, assumed to be independent of the shear direction. We can construct the compression strength–angle relationship from a knowledge of X', Y' and S, thus

$$\sigma_{cu\phi} = \frac{\cos^4 \phi}{(X')^2} + \left(\frac{1}{S^2} - \frac{1}{(X')^2} \right) \cos^2 \phi \sin^2 \phi + \frac{\sin^4 \phi}{(Y')^2} \qquad (2)$$

where $\sigma_{cu\phi}$ is the strength in a direction ϕ to the fibre direction. (Note: the maximum stress theory is sometimes used instead, see any standard text, e.g. Piggott,[15] Whitney et al.,[16] or Jones.[27]) Figure 4.11 shows a plot obtained with hand laid-up aligned carbon–epoxy.[6] It may be seen that the theory (eqn (2)) tends to underestimate the strengths slightly.

The compressive strength of a laminate is very sensitive to the interlaminar bonding. If this is poor, buckling can easily occur, and

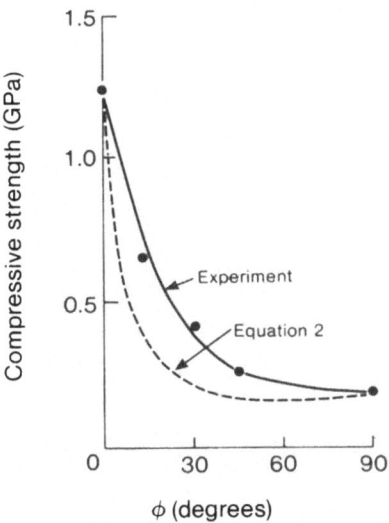

FIG. 4.11. Effect of testing aligned fibre composite at an angle, ϕ to the fibres.[6] The dashed line is the theoretical result from eqn (2).

strength is much reduced. This is demonstrated in Fig. 4.12,[28] where an artifically poor bond was introduced by placing a teflon pad between layers in a laminate, which was then tested in compression. The compression strength decreases monotonically with increasing size of debonded area. In Fig. 4.12 the pad occupies the full width of the specimen. When the pad occupies a fraction of the width, a mixed failure is obtained, delamination occurring over the pad, and fibre fracture elsewhere.

FIG. 4.12. Effect of poor bonding between laminae on compression failure.

TABLE 4.4

EFFECT OF HEAT AND MOISTURE ON THE COMPRESSIVE STRENGTH AND MODULUS OF
CARBON FIBRE LAMINATES[24]

	Moisture content (% weight)	Temperature (°C)	Layup			
			0	90	±45	±45/0/90
Strengths	<0·4	23	1·40	0·259	0·236	0·724
(GPa)	1·1	23	1·24	0·199	0·211	0·661
	<0·4	103	1·18	0·186	0·144	0·615
	1·1	103	1·13	0·141	0·143	0·523
Moduli	<0·4	23	110	11·7	16·6	71·7
(GPa)	1·1	23	111	8·8	16·6	69·6
	<0·4	103	116	11·0	14·7	72·4
	1·1	103	115	14·5	13·1	71·7

The compressive strength is also affected by heat and moisture, and the effects are different for different layups,[24] Table 4.4. With aligned-fibre composites, the properties in the fibre direction are not much affected, while both strength and modulus at right angles to the fibre direction are drastically reduced by moisture at 23 °C, though strength only is significantly reduced by heat, whether wet or dry. The effects for the other laminates are intermediate between these extremes.

4.4.3. Biaxial Compression

Biaxial properties are usually given in the form of failure envelopes, which normally include shear failure as well. The failure envelope is derived from a tensor polynomial failure criterion. For example, when the lamina is symmetric with respect to shear direction (this is usually assumed to be true) the quadratic polynomial criterion is expressed as follows:

$$F_1\sigma_1 + F_2\sigma_2 + F_{11}\sigma_1^2 + F_{22}\sigma_2^2 + F_{66}\tau_{12}^2 + 2F_{12}\sigma_1\sigma_2 = 1 \qquad (3)$$

The σ terms are stresses applied to the lamina in the fibre direction (the 1 direction) and normal to it, in the plane of the lamina (the 2 direction). τ_{12} is the in-plane shear stress.

The F terms are the principal strength tensors for the lamina, and it can be quite easily shown that[29] $F_1 = 1/X'$, $F_{11} = 1/XX'$, $F_2 = 1/Y - 1/Y'$, $F_2 = 1/YY'$ and $F_{66} = 1/S^2$.

Thus, in principle, the failure envelope, which is actually a surface in three dimensions, can be determined from the five strength parameters X,

X', Y, Y' and S, together with an extra experiment to determine F_{12}, using multiple loading.[30]

Tennyson et al.[31] have recently shown that the quadratic polynomial is not adequate to determine the failure envelope in the all-compression quadrant. Instead, a cubic polynomial should be used. This introduces the following cubic terms into the polynomial given in eqn (3): $3F_{112}\sigma_1^2\sigma_2 + 3F_{122}\sigma_1\sigma_2^2 + 3F_{166}\sigma_1\tau_{12}^2 + 3F_{266}\sigma_2\tau_{12}^2$.

When the applied shear stresses are zero, as in biaxial compression, the full equation is

$$F_1\sigma_1 + F_2\sigma_2 + F_{11}\sigma_1^2 + F_{22}\sigma_2^2 + 2F_{12}\sigma_1\sigma_2 + 3F_{112}\sigma_1\sigma_2^2 + 3F_{122}\sigma_1\sigma_2^2 = 1$$

(4)

Two extra experiments are needed to determine the interaction terms F_{112} and F_{122}. The stresses used in the experiments must be chosen judiciously in order for this approximate analysis to be successful. One of them should be in the all-compression quadrant. Such an experiment has been carried out, using the sandwich beam method[32] and Fig. 4.13 shows the failure envelope derived from glass–epoxy unidirectional laminates. The quadratic polynomial predicts values which are far too large in the compression-quadrant.

4.4.4. Hybrids

A composite made by combining more than one type of fibre in a resin matrix is called a hybrid composite. This method of manufacture is used to improve the toughness of carbon-fibre composites (glass or Kevlar are used) and to increase the compression strength and compression modulus of Kevlar composites (carbon is usually used for this).

There are two main types of hybrid:

1. intermingled hybrids, in which the different fibre types are intimately mixed, and
2. laminar hybrids, in which laminates are made with different layers containing different fibre types.

Hybrids do not usually follow the simple mixture rule indicated by the equation:

$$P_h = V_{fa}P_{fa} + V_{fb}P_{fb} + V_m P_m$$

(5)

Here P is the property (Young's modulus, strength, toughness, etc.) and V is the volume fraction of the components. The subscripts are: h for hybrid composite, fa for one fibre type, fb for the other fibre type, and m for the

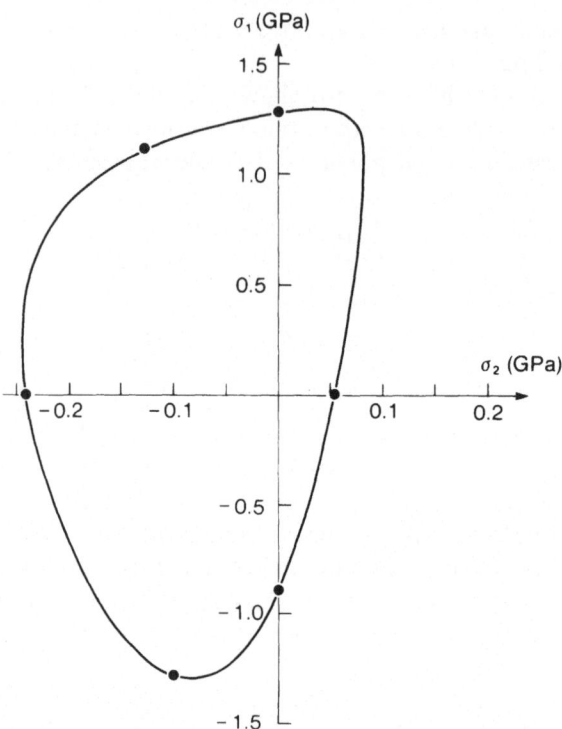

FIG. 4.13. Failure envelope for biaxial compression of unidirectional glass–epoxy.[32] Experimental points are marked.

matrix. (Hybrids are seldom made with more than two different fibre types.)

When eqn (5) is not obeyed, there is said to be a 'hybrid effect'. This can either be positive or negative, i.e. synergistic or cohibitive. In the case of compression strengths and moduli the hybrid effects are usually cohibitive, i.e. strengths and moduli are less than given by eqn (5) when V_{fa} and $V_{fb} > 0$.

A systematic study[33] has been made of compression strengths and moduli of pultruded hybrids, all having the same total fibre content (30 %), the same dimensions, etc. A polyester resin matrix was used.

When Kevlar was used, the strengths obeyed eqn (5), but the moduli fell below the mixture rule values, Figs 4.14 and 4.15. The Kevlar fibres appear to behave like an elastic–perfectly-plastic metal. Thus, when surrounded by other fibres, and so prevented from buckling, they yield, and then support a constant load while the load on the other member of the hybrid is increased. The stress–strain curves show continually increasing ductility as the Kevlar

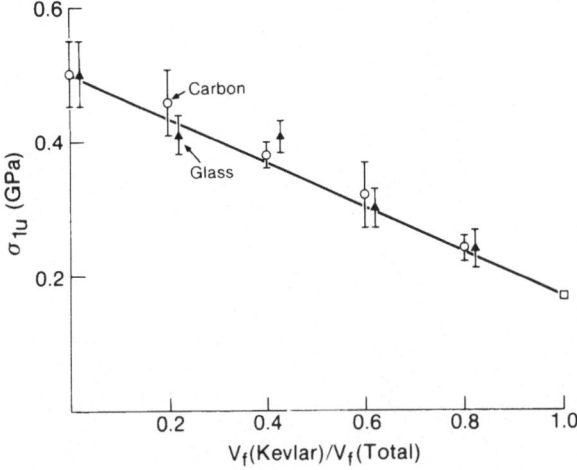

FIG. 4.14. Compressive strengths of Kevlar hybrids with glass and strong carbon.[33]

content is increased, Fig. 4.16. The origin of the modulus hybrid effect remains obscure.

When both components are brittle the effects show no consistent pattern. Figure 4.17 shows the compressive strengths of hybrids made with high modulus carbon combined with glass, and with high tensile strength carbon. The glass appears to give a negative hybrid effect, while the high strength carbon gives a positive one. In the case of the moduli, the glass

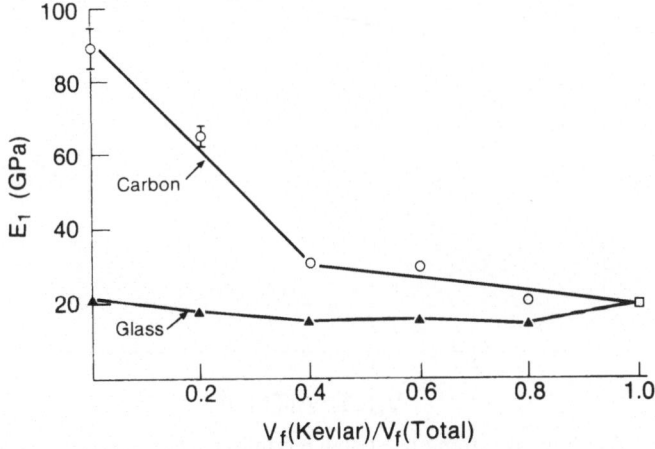

FIG. 4.15. Compressive moduli of Kevlar hybrids with glass and strong carbon.[33]

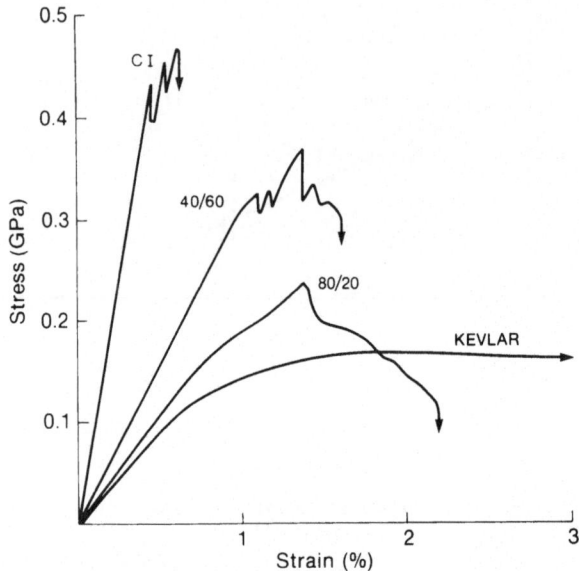

FIG. 4.16.   Stress–strain curves for Kevlar–carbon hybrids.[33]

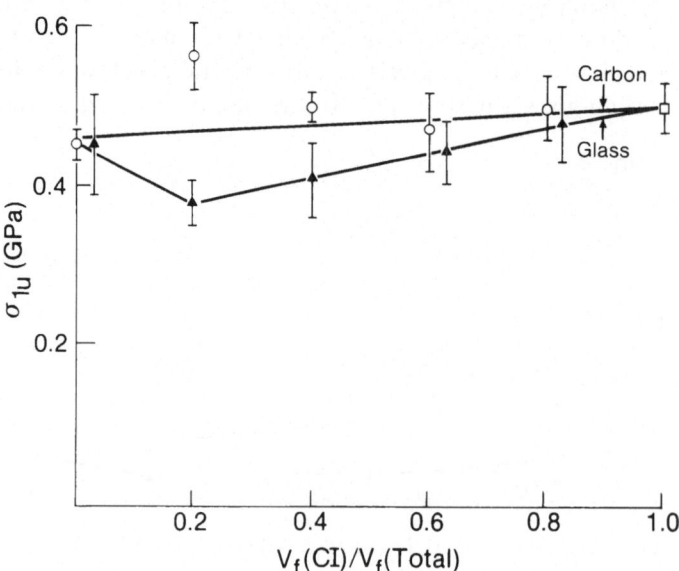

FIG. 4.17.   Compressive strengths of high modulus carbon (CI) hybrids with glass
and strong carbon.[33]

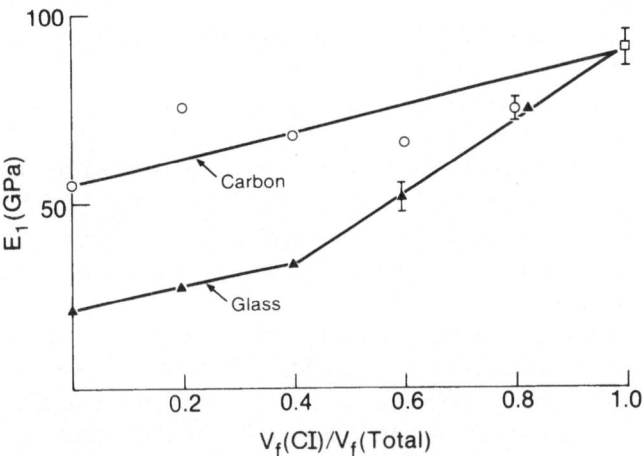

FIG. 4.18.   Compressive moduli of high modulus carbon (CI) hybrids with glass and strong carbon.[33]

gives a negative effect while the high strength carbon gives both positive and negative effects, Fig. 4.18.

4.4.5. Fatigue

Glass- and boron-fibre composites have better resistance to compression fatigue than to tension fatigue. In the case of boron this is due to the composite being stronger in compression than in tension. With glass, on

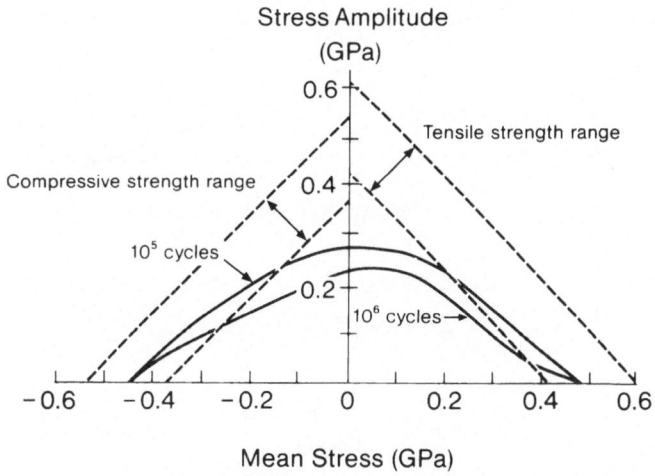

FIG. 4.19.   Fatigue failure envelope for carbon–epoxy $0/\pm45$ laminates.[35]

the other hand, the effect is due to water promoting microcrack growth in the fibres.[34]

Kevlar and carbon-fibre composites fare worse in compression fatigue than tension fatigue, and this appears to be related to their generally poorer performance in compression. This effect is demonstrated by the fatigue failure envelopes for carbon-fibre laminates shown in Fig. 4.19.[35] It can be seen that for failure at 10^5 and 10^6 cycles the mean stress and stress amplitude are smaller on the left (compression) side of the diagram than on the right hand side.

Grimes included compression-fatigue–moisture effects in his compression study.[24] In the case of the aligned-fibre laminates, tested in the fibre direction, moisture appeared not to have much significant effect. These laminates had an endurance limit of about 61 % of their static compressive strengths, at 23 °C, irrespective of moisture content. In the case of the ±45 laminates the endurance limits were 49 % and 45 % of the static strengths

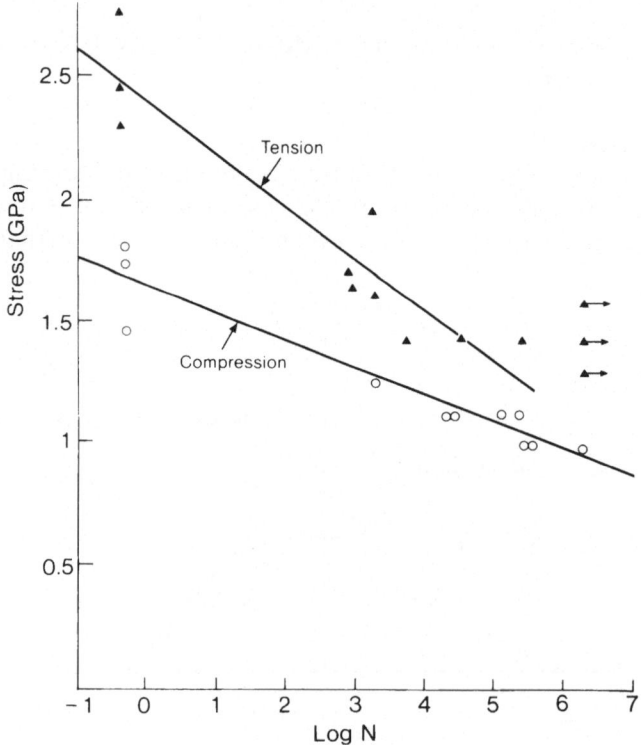

FIG. 4.20. S–N curves for unidirectional-carbon-fibre laminates with $R = 20$, stressed along the fibre direction.[36]

when dry ($<0.5\%$ moisture) and wet ($1.1 \pm 0.2\%$ moisture) respectively. The $\pm45/0/90$ laminates were not significantly affected by moisture, and had endurance limits of about 29% of their static strengths.

$S-N$ curves for aligned fibre composites in compression fatigue are usually convex upwards; Grimes'[24] results typically show this effect. This may be due to an underestimation of the static compressive strength of the composite. Recently Tratt[36] has obtained linear, or if anything concave upwards, $S-N$ curves, Fig. 4.20. A feature of Tratt's work is the high mean static compression strength of the laminates. Note the large difference between tension and compression. The situation is reversed when the composites are tested normal to the fibre direction: the static compressive strength (300 Mpa) is about five times the tensile strength and both tensile and compressive strengths decline to about half their static values after 10^5 cycles. The fatigue resistance of notched composites is also less good in compression than in tension.[37] It is, in addition, very dependent on stacking sequence in laminate layup.

4.5.   FAILURE PROCESSES

Polyester and epoxy resins do not appear to fail under compression. Instead they yield, work-harden to a degree, and then soften, if the strain is increased beyond the point where the stress is a maximum, as mentioned earlier (see Fig. 4.1). If left to stand for a short while, they recover their original dimensions and follow the same stress–strain curve as they did when first tested.

Fibre composites, by contrast, do fail in compression. Aligned-fibre composites tested in the fibre direction usually fail by a complex buckling process which leads to fibre breakage. When the fibres are not aligned, or the stress is not in the fibre direction the composite may break into two or more pieces.

A fibre composite has three entities that can fail: the fibres, the matrix, and the fibre–matrix interface. The failure of one can initiate failure of the others, and the actual process that takes place in any particular case is determined by the stress required to activate each individual mechanism. The mechanism activated by the lowest stress will normally govern composite failure.

4.5.1. Fibre Failure
When the fibres are relatively weak in compression, fibre failure governs

composite failure. (It is not necessary for the fibres to be weaker than the matrix for this type of failure to occur.) This is the case with Kevlar, which gives composites with quite low compressive strengths and moduli; Kevlar pultrusions fail when the fibre's stress has reached about 0·28 GPa;[7] compare this with Kevlar's tensile strength of 3·5 GPa.

The fibre failure controlled case has been carefully examined[38] using steel 'fibres' which can be work-hardened to give different fibre strengths. It was found that resins reinforced with aligned steel fibres obeyed the modified rule of mixtures expression:

$$\sigma_{1u} = V_f \sigma_{fu} + V_m E_m \sigma_{fu}/E_f \tag{6}$$

where E_f and σ_{fu} are the fibre Young's modulus and compression strength respectively. This expression was obeyed over the range tested: V_f from 0·15 to 0·34 and σ_{fu} from 1·3 to 2·2 GPa.

With the very straight fibres used in these experiments, matrix support appeared to prevent the fibres buckling when they yielded, giving σ_{fu} values about 80 % above the fibre yield stresses. Failure occurred explosively with the harder steels, and the specimens broke into small fragments. With the softer steels failure occurred by slow buckling.

In the case of glass-fibre-reinforced plastics, compressive failure is almost certainly not governed by fibre compressive failure. Very short lengths of production E glass-fibres have tensile strengths of 4 GPa or more.[39] The compressive strength of these fibres must be at least as high as this, so that pultrusions with $V_f = 0·6$ should have compressive strengths of at least 2·4 GPa, if fibre failure controlled composite failure. Most aligned-glass-fibre composites have lower compressive strengths than this.

In the case of carbon, the fibre compressive strength has been estimated to be 1·3 GPa for high modulus (450 GPa) fibres and 2·5 GPa for lower modulus (300 GPa) fibres.[40] (These values were obtained from tests in which single fibres were embedded in epoxy blocks and compressed until the fibres could be seen to fail.) These values are somewhat too high to account for the strength values reported for carbon–epoxies (Table 4.2). In addition, the decrease in compression strength observed by Ewins and Potter,[41] as testing temperature is raised from −80 to 20 °C, does not support a fibre compressive failure mechanism, in this temperature range at least. (Above 100 °C they observed a drastic decrease in strength, and the failure mode changed from a 'shear' mode to a fibre buckling mode.) Further evidence against fibre failure governing compressive failure comes from experiments in which hydrostatic pressure was superposed on uniaxial compression.[42] The application of 300 MPa of hydrostatic pressure increased the strength of carbon–epoxy pultrusions from 1·50 to 2·05 GPa.

With boron composites also, fibre compressive strength does not govern composite compressive strength. Instead, it has been shown that fibre divagation plays an important role.[8]

4.5.2. Matrix Yielding

The effect of the matrix can only be understood when it is appreciated that fibre composites cannot be made with perfectly straight fibres. Even laminates made with boron fibres, which have relatively large diameters (100 μm) and high Young's moduli (420 GPa) have been shown to contain fibres whose axes follow wavy lines, with amplitudes of 50 μm and wavelengths of 20 mm.[8] Some, at least, of this lack of straightness could be engendered by matrix shrinkage along the fibre direction during cure.

Composites carefully made with fibres that are as straight as possible are relatively strong[43] while those with fibres that are deliberately kinked are relatively weak, in compression, and the strength loss depends on the degree of fibre curvature.[9]

The effect of this fibre divagation can be understood[10] if we let the fibre axes assume sinusoidal shapes, as shown in Fig. 4.21. (Any fibre axis

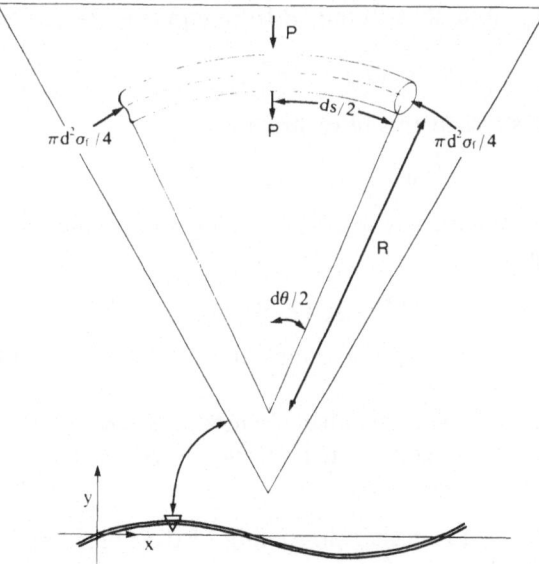

FIG. 4.21.   Curved fibre, with stresses acting on fibre antinode shown in inset.[10]

trajectory can be reduced to sine waves using Fourier methods.) We assume, for the moment, that the fibres adhere perfectly to the matrix.

The curve can be conveniently expressed using the dimensionless parameters x, y, a and λ.

$$y = a \sin (2\pi x/\lambda) \tag{7}$$

where xd and yd are distances, as indicated in Fig. 4.21, d being the fibre diameter (or fibre bundle diameter, if many fibres act in concert). ad and λd are the amplitude and wavelength of the sine wave.

The radius of curvature of the fibre has a minimum value, R, given by

$$R = d\lambda^2/4\pi^2 a \tag{8}$$

at $x = \pi/4$, $3\pi/4$, etc., and the fibre here presses strongly against the matrix, with pressure P, Fig. 4.21. When P reaches the matrix yield stress, σ_{my}, the matrix will be pushed aside by the fibre, the radius R will decrease, causing the pressure to increase, further reducing R and increasing P, since

$$P = \pi d\sigma_f/8R \tag{9}$$

(σ_f is the compressive stress in the fibre). The maximum fibre compressive stress, σ_{fmax}, at the onset of this unstable state is obtained by replacing P in eqn (9) with σ_{my}, and substituting R from eqn (8). Hence

$$\sigma_{fmax} = 2\lambda^2 \sigma_{my}/a\pi^3 \tag{10}$$

The composite stress at this fibre stress is

$$\sigma_{1u} = \sigma_{fmax}(V_f + V_m E_m/E_f) \tag{11}$$

assuming that the matrix is still elastic. This is normally the case.[7] Using eqn (10) now gives

$$\sigma_{1u} = (2\lambda^2 \sigma_{my}/a\pi^3)(V_f + V_m E_m/E_f) \tag{12}$$

This is the failure stress of the composite for the fibre-buckling–matrix-yielding mode of failure.

The fibre curvature will also affect the modulus, since P will cause elastic deformation of the matrix when $P < \sigma_{my}$. When this happens, the composite has two compliances:

1. $1/E_f$, due to the elastic shortening of the fibres, and
2. $1/E_{f1}$, resulting from the increase in a and decrease in λ due to the action P.

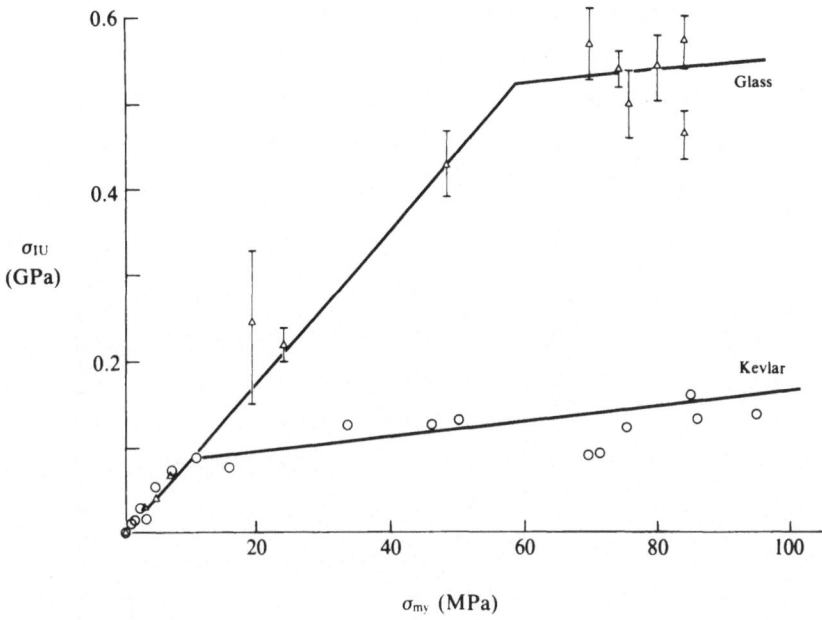

FIG. 4.22. Composite strength vs matrix yield strength for Kevlar and glass pultrusions.[10]

Using Swift's analysis,[44] modified for perfect adhesion between fibres and matrix[10] we have

$$E_{f1} = \lambda^4 E_m / \pi^5 a^3 \qquad (13)$$

and the composite modulus is

$$E_1 = V_f / (1/E_f + 1/E_{f1}) + V_m E_m \qquad (14)$$

These results (eqns (12)–(14)) can be used to explain the moduli and strengths of pultrusions made with soft (partially cured) matrices.[7,10] Thus, Fig. 4.22 shows that at low matrix yield stresses

$$\sigma_{1u} = 9\sigma_{my} \qquad (15)$$

using both Kevlar and glass fibres. (With Kevlar, the relation breaks down when $\sigma_{my} \simeq 10$ MPa, while with glass it continues till $\sigma_{my} \simeq 60$ Mpa.) Thus, under the appropriate conditions

$$(V_f + V_m E_m / E_f) 2\lambda^2 / a\pi^3 = 9 \qquad (16)$$

M. R. PIGGOTT

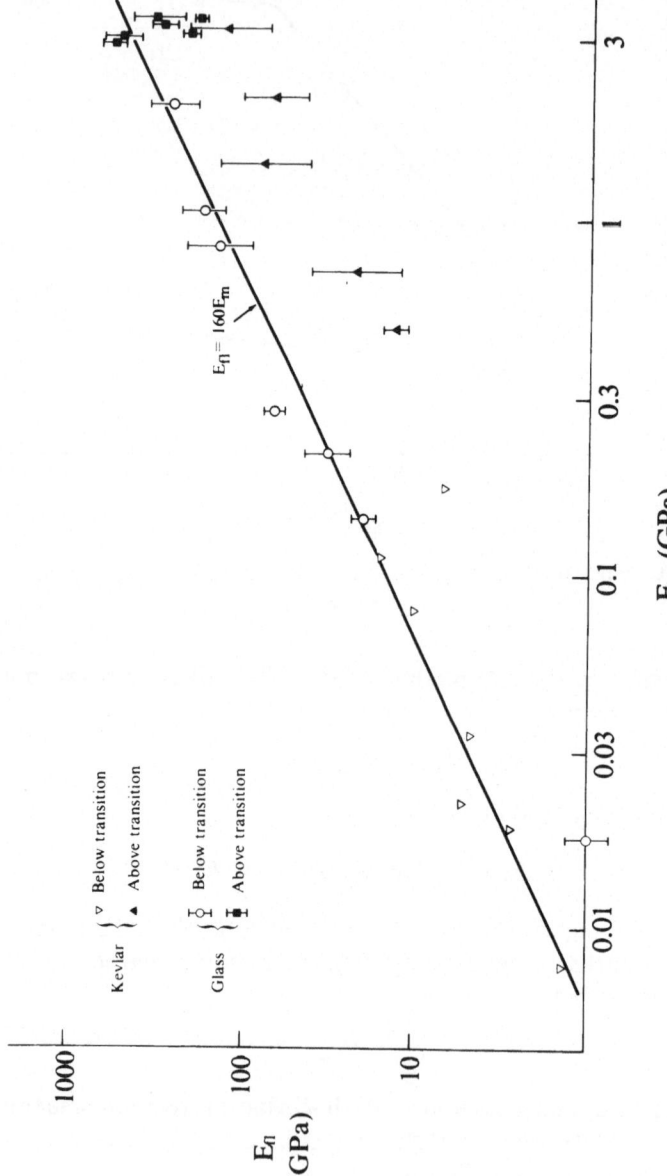

FIG. 4.23.   Fibre buckling stiffness, E_{f1}, vs matrix Young's modulus.[10]

The moduli of these composites obey eqn (14) with $E_{f1} \simeq 160E_m$, again up to the same limits, Fig. 4.23. Thus, using eqn (13) we have

$$\lambda^4/\pi^5 a^3 = 160 \qquad (17)$$

Equations (16) and (17) enable us to estimate that $a \simeq 4$ and $\lambda \simeq 43$. Then from eqn (8), $R \simeq 11d$. These results are all relative to fibre or fibre bundle diameter. We make them absolute by making use of the results of experiments in which the fibres were deliberately kinked.[9] This gives $R \simeq 5$ mm and $d \simeq 0.45$ mm,[10] indicating that some 2100 fibres act together, since individual fibre diameters are typically $10\,\mu$m.

Additional support for the fibre-buckling–matrix-yielding model comes from the experiments in which hydrostatic pressure was used, as well as uniaxial compression.[42] It was observed that the compressive strength increased with pressure, approximately in step with the increase in matrix yield stress brought about by the pressure. Also, the drastic decrease in strength above 100 °C observed by Ewins and Potter[41] could well be a manifestation of this mode of failure being activated by the matrix softening at 100 °C.

4.5.3. Interface Failure and Matrix Tensile Failure

When the interface is weak, the pressure P, which is negative, i.e. equivalent to a tensile stress on the inside of the curve in Fig. 4.21, can cause separation between the fibre and matrix. This can be followed by matrix splitting, as shown in Fig. 4.24. Here, σ_{mtu} is the matrix tensile strength, and σ_a is the adhesion strength.

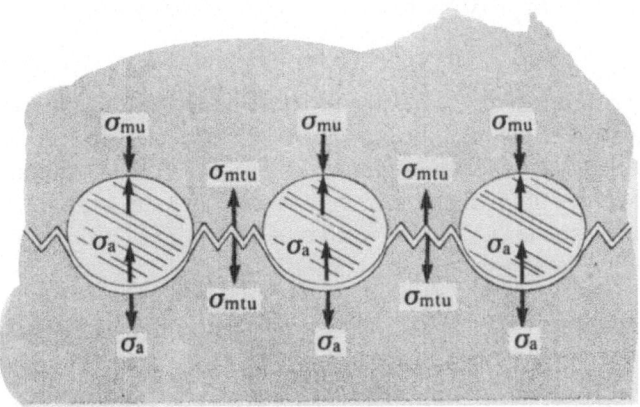

FIG. 4.24.   Interface and matrix failure, showing stresses involved.[10]

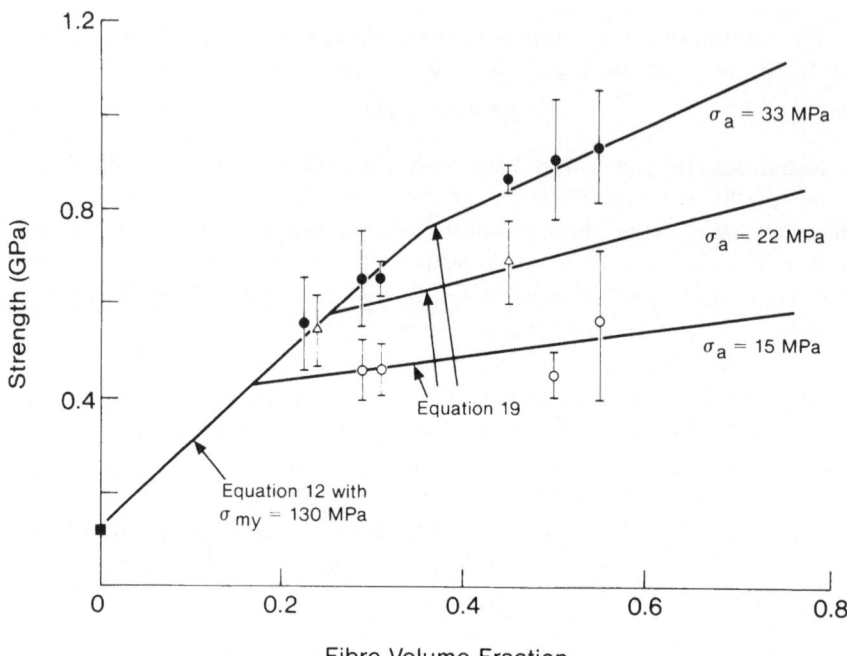

FIG. 4.25.   Fitting poor adhesion results[9] for glass pultrusions by use of eqns (12) and (19). The solid circles are as-received fibres (sizing intact), triangles are solvent soaked fibres, and the open circles are fibres which have had their sizing removed by heat.

When we take into account the relative areas over which these stresses act, we get an approximate equilibrium of forces when

$$P = \pi\sigma_a + [\sqrt{(P_f/V_f)} - 2]\sigma_{mtu} \tag{18}$$

(We have emphasised the effect of σ_a by supposing it acts over an area equivalent to the whole fibre surface area), P_f is a factor representing the fibre packing arrangement, and is equal to $2\pi\sqrt{3}$ for hexagonal packing.

At the onset of failure, P is given by eqn (9) with $\sigma_f = \sigma_{fmax}$, as before, so that using eqns (11) and (18), with eqn (9), we obtain, for the composite compressive strength:

$$\sigma_{1u} = (4R/\pi d)(\pi\sigma_a + \{\sqrt{(P_f/V_f)} - 2\}\sigma_{mtu})(V_f + V_m E_m/E_f) \tag{19}$$

This equation may be used to account for the anomalous volume fraction effects observed with poorly adhering fibres.[6,9] For low volume fractions, where σ_{1u} increases linearly with V_f, we use eqn (12). At higher volume fractions, where the linear relation between σ_{1u} and V_f no longer holds, we use eqn (19). In doing this we are assuming that the failure process activated in any given instance is the one that works at the lowest stress.

A good fit to the experimental results can be obtained, as shown in Fig. 4.25, by using eqn (12) with $\sigma_{my} = 130$ MPa for the linear region near the origin, and eqn (19) with $\sigma_{mtu} = 33$ MPa together with the various values of σ_a indicated on the figure for the rest of the results. In addition, we use the value of 11 for R/d derived earlier (Section 4.5.2). Note that the adhesion is apparently essentially perfect for the fibres with sizing intact, i.e. $\sigma_a = \sigma_{mtu}$.

6. CONCLUSIONS

Compression strength is now being given a great deal of attention, but data on strengths and moduli are far from plentiful, and relatively little work has been done on compression fatigue with glass and Kevlar composites. Data on the effects of biaxial stresses are extremely rare. However, reliable methods of testing appear to be available.

A number of failure mechanisms have been identified recently, and progress is now being made on understanding the separate roles played by the fibres, the matrix and the interface in the failure process. Although a consensus has not been reached on the appropriate failure mechanism in any particular instance the way is now open for rapid progress.

ACKNOWLEDGEMENTS

I wish to acknowledge with thanks the assistance of Dr J. Hansen of University of Toronto, Institute for Aerospace Studies, in the provision of pictures, and recent data from his group's studies. Thanks also go to J. Clifford, for invaluable help in typing and organising this manuscript.

REFERENCES

1. Modern plastics encyclopedia, *Modern Plastics*, **56**(10A), 503 (1979).
2. ROSEN, B. W., *Fibre Composite Materials*, ed. S. H. Bush, American Society for Metals, Cleveland, Ohio, 1965, p. 37.
3. ORRINGER, O., AFOSR TR 71, 1971, p. 3098.
4. GRESZCZUK, L. B., AFML TR, 1972, p. 231.
5. DE FERRAN, E. M. and HARRIS, B., *J. Comp. Mater.*, **4**, 62 (1970).

6. HANCOX, N. L., *J. Mater. Sci.*, **10**, 234 (1975).
7. PIGGOTT, M. R. and HARRIS, B., *J. Mater. Sci.*, **15**, 2523 (1980).
8. DAVIS, J. G., ASTM STP 580, 1975, 364.
9. MARTINEZ, G. M., PIGGOTT, M. R., BAINBRIDGE, D. M. R. and HARRIS, B., *J.Mater. Sci.*, **16**, 2381 (1981).
10. PIGGOTT, M. R., *J. Mater. Sci.*, **16**, 2837 (1981).
11. BENJAMIN, T. A., CHANG, J. C. and LI, L., *J. Mater. Sci.*, **16**, 889 (1981).
12. LAM, P. W. K., PhD Thesis, Chemical Engineering and Applied Chemistry, University of Toronto, 1983.
13. LYNCH, C. T. (ed.), *Handbook of Materials Science*, Vol. II, CRC Press, Cleveland, Ohio, 1975, 403.
14. NORTHOLT, M. G., *J. Mater. Sci.*, **16**, 2025 (1981).
15. PIGGOTT, M. R., *Loading Bearing Fibre Composites*, Pergamon, Oxford, 1980, 169.
16. WHITNEY, J. M., DANIEL, I. M. and PIPES, R. B., *Experimental Mechanics of Fiber Reinforced Composite Materials*, Society for Experimental Stress Analysis, Brookfield Center, CT, 1982, 175.
17. LANTZ, R. B., *J. Comp. Mater.*, **3**, 642 (1969).
18. GRUBER, M. B., OVERBEEKE, J. C. and CHOU, T. W., *J. Comp. Mater.*, **16**, 162 (1982).
19. HOFER, K. E. and RAO, P. N., *J. Testing and Evaluation*, **5**, 278 (1977).
20. GRIMES, G. C., FRANCIS, P. H., COMMERFORD, G. E. and WOFFE, G. K., AFML TR 72 40, 1972.
21. CLEMENTS, L. L. and MOORE, R. L., *Composites*, **9**, 93 (1978).
22. KULKARNI, S. V., RICE, J. S. and ROSEN, B. W., *Composites*, **11**, 217 (1975).
23. AMIJIMA, S. and FUJII, T., Proc. ICCM3, 1980, 399.
24. GRIMES, G. C., ASTM STP 734, American Society for Testing and Materials, Philadelphia, 1981, 281.
25. DITCHER, A. K. and WEBBER, J. P. H., *J. Comp. Mater.*, **16**, 228 (1982).
26. PAGANO, N. H. and PIPES, R. B., *J. Comp. Mater.*, **5**, 50 (1971).
27. JONES, R. M., *Mechanics of Composite Materials*, McGraw-Hill, New York, 1975, Chap. 4.
28. MABSON, G. E., HANSEN, J. S., TENNYSON, R. C. and WHARRAM, G. E., Report for DNS/DSS (Canada) Contract No. 05SB 97708-1-0043, 1982.
29. TSAI, S. W. and WU, E. M., *J. Comp. Mater.*, **5**, 58 (1971).
30. TENNYSON, R. C., MACDONALD, D. and NANYARO, A. P., *J. Comp. Mater.*, **12**, 63 (1978).
31. TENNYSON, R. C., NANYARO, A. P. and WHARRAM, G. E., *J. Comp. Mater.*, Suppl., **14**, 28 (1980).
32. ELLIOTT, W. G., MASc Thesis, University of Toronto, Institute for Aerospace Studies, 1983.
33. PIGGOTT, M. R. and HARRIS, B., *J. Mater. Sci.*, **16**, 687 (1981).
34. GURNEY, C. and PEARSON, S., *Proc. Roy. Soc. (London)*, **A192** 537 (1978).
35. BEVAN, L. G., *Composites*, **8**, 227 (1972).
36. TRATT, M., MASc Thesis, University of Toronto, Institute for Aerospace Studies, 1983.
37. ROSENFELD, M. S. and HUANG, S. L., *J. Aircraft*, **15**, 264 (1978).
38. PIGGOTT, M. R. and WILDE, P., *J. Mater. Sci.*, **15**, 2811 (1980).

39. CHUA, P. S., MASc Thesis, Chemical Engineering and Applied Chemistry, University of Toronto, 1982.
40. HAWTHORNE, H. M. and TEGHTSOONIAN, E., *J. Mater. Sci.*, **10**, 41 (1975).
41. EWINS, P. D. and POTTER, R. T., Phil. Trans. Roy. Soc. (London), **A294**, 507 (1980).
42. PARRY, T. V. and WRONSKI, A. S., *J. Mater. Sci.*, **17**, 3656 (1982).
43. CHAPLIN, C. R., *J. Mater. Sci.*, **12**, 347 (1977).
44. SWIFT, D. G., *J. Phys. D.*, **8**, 223 (1975).

Chapter 5

NOVEL CURING AGENTS FOR EPOXY RESINS

Daniel A. Scola

United Technologies Research Center,
East Hartford, Connecticut, USA

SUMMARY

A review of the literature for curing agents and accelerators for epoxy resins covering the years 1966 to 1982 reveals that many new compounds and compositions have been disclosed in this time period. Most of the innovations are related to novel diamines, polyamines and amine–epoxy adducts. Novel approaches to introduce latency in epoxy resin compositions using aminimides, organic and metallic salts and complexes have been disclosed. Novel approaches in epoxy cure, utilising cationic complex salts and organometallic compounds directed toward fast moulding processes, have also been developed since 1966. However, most innovations are related to curing agents and compositions which affect the cure and properties of films, coatings and adhesives. Very few curing agents have been developed for advanced composite applications. Several novel curing agents and curing agent compositions have been commercialised, but these do not nearly approach the number reported in the literature.

5.1. INTRODUCTION

This review covers the years 1966 to 1982. For an excellent, comprehensive treatment of epoxy resins and curing agents the reader should consult the *Handbook of Epoxy Resins.*[1] Progress in curing agent developments for epoxy resins was reviewed by Suzuki and Inoue[2] in 1972 and by Kaman[3] in

1974. Curing agents and curing agent compositions have been reviewed by DiStasio[4] in a series *Epoxy Resin Technology Developments since 1979.* Some of the highlights of these reviews will be discussed in the present review. Several novel aliphatic, alicyclic and aliphatic amines containing heterocyclic atoms were reviewed, as well as novel amine/epoxy adducts, dianhydrides, and acidic, basic and inorganic accelerators for cure of diglycidyl ether of Bisphenol A (DGEBA) and other epoxy resins.

The technical and patent literature reveals that many compounds and compound mixtures exhibit the capacity to cure or accelerate the cure of epoxy resins. However, very few of these compounds or formulations are commercially available. The curing agents and accelerators will be discussed in the following order:

Novel commercial curing agents
Aliphatic and alicyclic amines and modifications
Amine–epoxy and thioalcohol–epoxy adducts as curing agents
Aromatic amines and modifications
Dianhydride curing agents
Latent curing agents
Complex and salt curing agents
Miscellaneous curing agents and accelerators

5.2. NOVEL COMMERCIAL CURING AGENTS

Since 1966 several new curing agents have been introduced commercially. Those that have been introduced usually contain two amine components, such as eutectics of various amines or adducts of amines and epoxies. Some of the commercial aromatic amines contain a blend of 4,4'-methylene-dianiline (4,4'-MDA) with polymethylene dianilines (poly 4,4'-MDA)[5] or 4,4'-MDA and poly 4,4'-MDA with *m*-phenylenediamine (*m*-PDA)[5] or 4,4'-MDA with 3,3'-diethyl 4,4'-MDA.[6] These curing agents are not very different from curing agents introduced before 1966, such as eutectics of *m*- and *p*-phenylenediamines or the single component curing agents, 4,4'-MDA or 4,4'-diaminodiphenylsulphone (4,4'-DDS). The major difference is that these tend to be slower curing because of the presence of polyaromatic polyamine species and hence have a longer pot-life by about 3–4 hours at room temperature, a distinct advantage for some applications. The polyaromatic amines do present toxicity problems. Hence, this deficiency, which is common in aliphatic amines, has not been eliminated.

The lower vapour pressures of these polyaromatic amines have reduced the toxic effects of breathing amine vapours.

Polyamines containing flexible oxyethylene and/or oxyisopropylene units in the backbone of the chain, are examples of aliphatic type amine curing agents introduced within the last 15 years.[7,8] The advantages of these materials over typical polyaliphatic amines is the longer pot-life, making these materials applicable to filament winding applications. The disadvantage of these materials, as with most amines, is the distinct toxicity by ingestion or by skin contact.

An unusual commercial curing agent is a dispersion of a solid polyamide in a liquid epichlorohydrin–Bisphenol epoxy resin.[9] This material can be formulated with epoxies to develop one part systems with at least six months stability at room temperature. The formulated epoxies can be cured in 10 minutes between 100 °C to 177 °C for adhesive and coating applications. The latent properties are derived from the room temperature stability of a complex material (organometallic salt, aminimide, sulphonium salt) which is thermally decomposed to reactive species which initiate cure. The advantages of these materials are: (1) their obvious long time room temperature storage ability and (2) their ability to develop an adhesive bond with moderate strength ($\sim 1500\,$psi or $\sim 10\,$MPa) within minutes. The disadvantage of these systems is that they are not amenable to high temperature performance (100–150 °C) and in bulk they exhibit a highly exothermic curing reaction, limiting them to adhesives and coating applications.

Modified polymeric phenolic materials for reactivity with epoxies above 110 °C, having long term shelf stability with powder epoxy resins, are also commercially available.[10] These materials can be utilised in epoxies as transfer moulding compounds, for cure between 110 and 200 °C for short press cycles. The advantages of these materials are their latency in solid resin powder mixtures, the short press cycles and the low mould shrinkage. Their disadvantages are the limited shelf-life (1 day) in liquid epoxy resins and the relatively poor thermo-oxidative stability of the cured materials relative to 4,4'-DDS cured epoxies.

2,2'-Diethyl-4,4'-methylenedianiline (**1**) (2,2'-diEt-4,4'-MDA) is an

2,2'-diethyl-4,4'-methylenedianiline (ref. 11)

example of a commercial curing agent[11] introduced within the last ten years, which is available as a pure compound. This aromatic amine is a low viscosity liquid (20–25 CP at 23 °C), soluble in epoxy resins and in most polar and non-polar solvents. DGEBA (EEW 190) cured with this diamine gives an HDT of 129 °C, compared with 157 °C for 4,4'-methylenedianiline (4,4''-MDA) and 140 °C for 20% MDA and 80% diethyl MDA. The mechanical, electrical and chemical resistance properties of 2,2'-diethyl-4,4'-MDA cured epoxy are similar to the 4,4'-MDA cured systems. The advantage of this curing agent is that it exhibits greater latent properties over 4,4'-methylenedianiline (4,4'-MDA), because of the presence of the ethyl groups, and resins cured with 2,2'-diEt-4,4'-MDA have properties equivalent to 4,4'-MDA cured resins. A disadvantage of this material is the higher cost relative to 4,4'-MDA. 3,3'-Diaminodiphenylsulphone (3,3'-DDS) has also been introduced within the last 10 years.[11] This material is more reactive at elevated temperature, more soluble, and lower melting than 4,4'-diaminodiphenylsulphone (4,4'-DDS). It can be used in place of, or combined with, 4,4'-DDS. Because of the higher reactivity of this material compared with 4,4'-DDS, catalysts such as BF_3-MEA are not essential. The higher reactivity of 3,3'-DDS over 4,4'-DDS suggests that systems containing this material will have a shorter shelf-life than systems with 4,4'-DDS.

   m-Xylylenediamine (MXDA) (2) and 1,3-bis(aminomethyl)cyclohexane (1,3-BAC) (3), are also commercially available as epoxy curing agents.[12]

2
m-Xylylenediamine (MXDA)
(ref. 12)

3
1,3-bis(Aminomethyl)cyclohexane (1,3-BAC)
(ref. 12)

The advantage of these amines over traditional aliphatic diamines and polyamines is that they are less hazardous to handle, provide improved latency characteristics, and provide cured epoxy systems with better solvent and chemical resistance than the aliphatic cured epoxies. These materials are effectively used as curing agents when modified as adducts, such as an epoxy adduct, phenol–formaldehyde adduct, Michael addition or Mannich addition adduct, or fatty acid dimer polyamide adduct. These materials exhibit similar disadvantages to the aliphatic diamines, such as absorption of carbon dioxide from the air, and they represent a potential skin and vapour hazard. These diamines are more costly than the typical aliphatic diamines.

1,2-Diaminocyclohexane (**4**) is a commercially available,[13] colourless liquid diamine which, because of the proximity of the two amine groups,

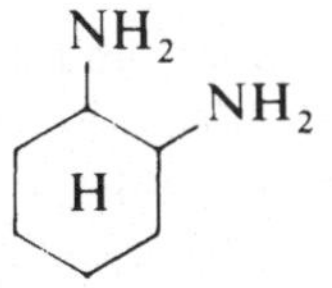

4

1,2-Diaminocyclohexane (ref. 13)

exhibits some latent properties relative to other aliphatic and cycloaliphatic diamines. For example, the pot-life of 1,2-diaminocyclohexane/DGEBA (EEW $\sim$ 200) epoxy mixture at room temperature was found to be 93 min, while for isophorone diamine, a less hindered amine, the pot-life with the same epoxy resin was found to be 50 min. A comparison of the heat distortion temperatures of DGEBA (EEW $\sim$ 200) cured with several types of cycloaliphatic diamines is shown in Table 5.1. 1,2-Diaminocyclohexane yields epoxy resins with HDTs equivalent to those of other cycloaliphatic diamines. The tensile strength and elongation of 1,2-diaminocyclohexane cured DGEBA (EEW $\sim$ 200) was reported to be 56·7 MPa (8·22 ksi) and 2·08 %, while for the isophorone diamine cured resin, the values reported were 37·2 MPa and 1·60 %. This cycloaliphatic diamine, like straight chain

TABLE 5.1

HEAT DISTORTION TEMPERATURE OF CYCLOALIPHATIC/DGEBA (EEW $\sim$ 200) CURED RESINS[a] [13]

Cycloaliphatic diamine	PHR[b]	HDT (°C)
1,2-diaminocyclohexane	17·0	155·3
3-aminomethyl-3,5,5-trimethylcyclohexylamine (Isophorone diamine)	23·0	146·5
Menthane diamine	25·0	144·6
Bis(4-aminocyclohexyl)methane	28·0	154·9
Bis(2-methyl-3-aminocyclohexyl)methane	35·2	157·8

[a] Cure cycle 80 °C/2 h + 150 °C/2 h.
[b] Parts per hundred of epoxy resin.

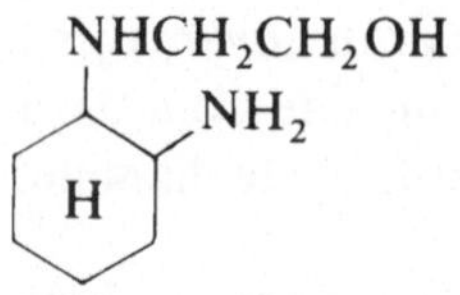

DANIEL A. SCOLA

NHCH$_2$CH$_2$OH

5

1,2-Diaminocyclohexane–ethyleneoxide adduct (ref. 13)

aliphatic diamines, has some distinct handling and chemical disadvantages, namely fuming in air, carbonate formation in air and yellow discoloration.

The ethyleneoxide adduct (**5**) of 1,2-diaminocyclohexane is a commercial product[13] which has none of the disadvantages of 1,2-diaminocyclohexane, but can be used in most of the same applications as the 1,2-diamino compound.

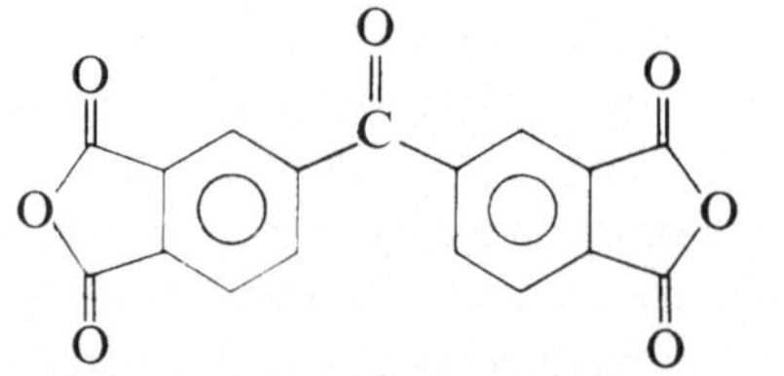

6

3,3′,4,4′-Benzophenonetetracarboxylic acid dianhydride (BTDA) (ref. 14)

The dianhydride of 3,3′,4,4′-benzophenonetetracarboxylic acid (**6**), is well established as a curing agent for epoxy resins and as an intermediate for polyimides and other polymers.[14] This dianhydride is a high melting solid and this precludes to some extent its use as a curing agent. The literature does not reveal the availability of modified, e.g. more soluble forms of this dianhydride, such as the carbinol.

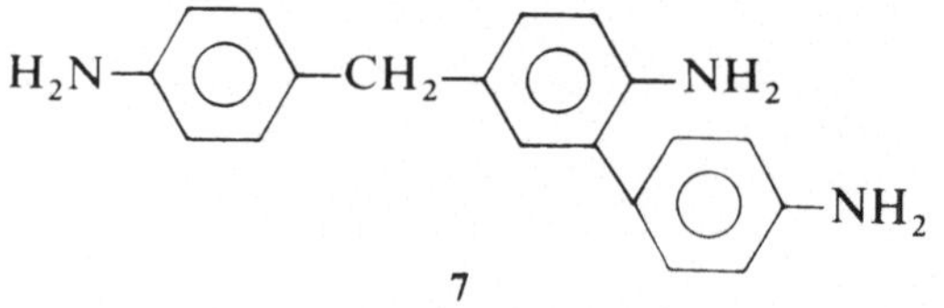

7

2,4-bis(p-Aminobenzyl) aniline (ref. 15)

A commercial aniline derivative, 2,4-bis(p-aminobenzyl) aniline (**7**), has recently been introduced as an intermediate for epoxy, urethane and isocyanate applications.[15] This material is most likely to be used in blends with aromatic diamines.

Several derivatives of imidazole have been introduced as curing agents

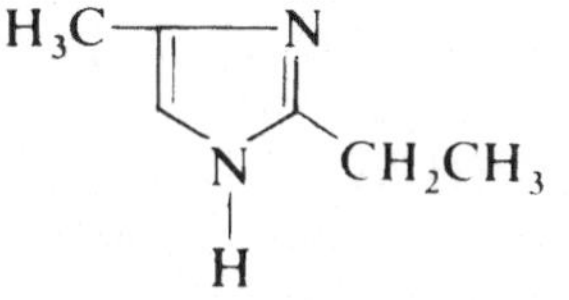

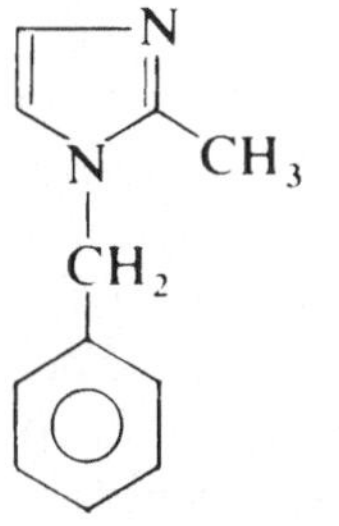

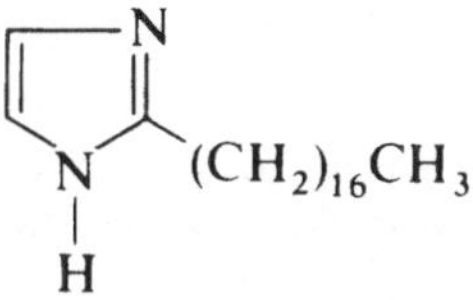

8   9   10

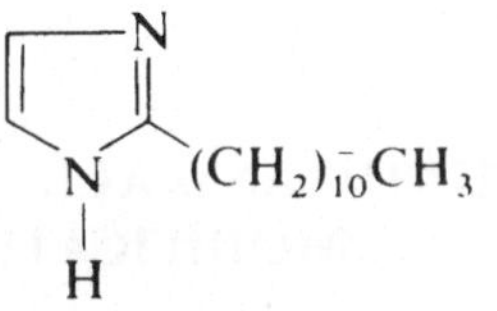

11

Imidazole derivatives (ref. 16)

over the last several years.[16] Examples **8**, **9**, **10** and **11** of several commercially available derivatives are shown.

These compounds are noted for their latent action with epoxy resin, and are therefore used in adhesive formulations where shelf-life is important. The imidazole derivative, polyaminoimidazoline, is also a commercial product[17] for cure of epoxy resins.

Long chain aliphatic dimer amine N,N'-bis(3-aminopropyl) dimer diamine (**12**) containing on an average 42 carbon atoms in the chain is commercially available as a curing agent[18] for epoxy resins. This is a mobile liquid at ambient temperature, and is soluble in methanol, benzene, chloroform and petroleum hydrocarbons. A similar fatty acid diamine (dimeryl diamine) **13** containing 36 carbon atoms in the backbone is also commercially available.[19]

$$H_2NCH_2CH_2CH_2\!-\!NH\!\!\left(\!CH_2\!\right)_{\!n}\!NHCH_2CH_2CH_2NH_2$$

12

N,N'-bis(3-Aminopropyl) dimer diamine (ref. 18)

$$H_2N\!\left(\!CH_2\!\right)_{\!36}\!NH_2$$

13

Dimeryl diamine (ref. 19)

Modified aromatic amines for cure down to $-5\,°C$ or under wet conditions and even under water are commercially available.[20] These aromatic amines are derivatives of phenol, and are probably prepared by aminomethylation of phenol or phenol-formaldehyde resins.

4-Methylhexahydrophthalic anhydride (MHHPA) is a liquid, low viscosity (65 cP at 20 °C) curing agent which is now commercially available.[20] This material is hygroscopic, and hydrolyses readily to the diacid in the presence of moisture. The diacid is insoluble in MHHDA. Therefore it must be stored in containers which protect it from moisture. MHHPA is a potential replacement for nadic methyl anhydride (NMA), because it appears to give cured resin specimens with higher T_g values than NMA cured epoxies.

5.3. ALIPHATIC AND ALICYCLIC AMINES AND MODIFICATIONS

Primary aliphatic diamines and polyamines have been used as curing agents for epoxy resins for over 30 years. The changes in molecular structure which have occurred over the last 16 years are those which change the reactivity of the amine with the epoxy and/or change the chemical resistance, environmental resistance and mechanical performance of the resin as a coating or adhesive or as a matrix in fibre-resin composites. In this time period, several aliphatic and alicyclic amines have been invented. The need for latency in resin systems intended for filament winding applications, and for improved toughness, has led to the development of polyether amine curing agents.[7,8] Variations of the polyether amine structures are indicated by structures 14, 15 and 16.

For each structure, the values of x, y and z can vary, thereby yielding curing agents with varying molecular weights. Thus the properties of epoxy resins can be tailored to fit a particular application.

$$H_2N-CH-CH_2 \{ O-CH-CH_2 \}_x (O-CH_2-CH_2)_y (O-CH_2-CH)_z NH_2$$

with CH_3 groups on the first, second and last CH units.

14

Polyoxyethylenepropylenediamine (ref. 7)

The available polyoxyethylenediamines (14) can be described by the values for x, y and z shown in Table 5.2. The structural features of these compounds suggest that they may be regarded as essentially poly(oxyethylene) diamines, because of the low number of oxypropylene units in the

TABLE 5.2
COMPOSITION OF POLYOXYETHYLENEPROPYLENE DIAMINES[7]

Polyoxyethylenepropylenediamine	Approximate value of		Molecular weight
	x	$y+z$	
1a ED-600	13·5	3·5	600
1b ED-900	20·5	3·5	900
1c ED-2001	45·5	3·5	2 001

molecule. Some properties of commercial batches of the polyetheramines are listed in Table 5.3. Properties of diglycidyl ether of Bisphenol A (DGEBA) cured with the polyetherdiamines of molecular weights 600 and 900 are listed in Table 5.4. Commercially available polyoxypropylenediamines (**15**) of molecular weights 230, 400 and 2000[20] and the

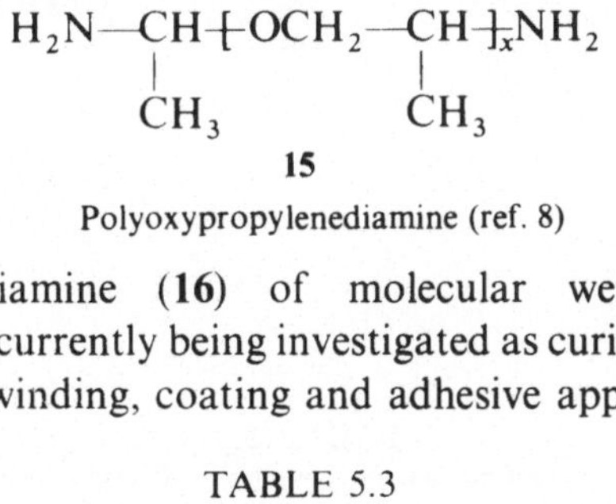

$$H_2N-CH+OCH_2-CH+_xNH_2$$
$$CH_3 CH_3$$

15

Polyoxypropylenediamine (ref. 8)

polyoxypropylenetriamine (**16**) of molecular weight 403,[8] where $x + y + z = 5·3$, are currently being investigated as curing agents for epoxy resins for filament winding, coating and adhesive applications.

TABLE 5.3
ANALYSES OF TYPICAL BATCHES OF POLYOXYETHYLENEDIAMINE COMPOUNDS[7]

	ED-600	ED-900	ED-2001
Total amine, wt %	97·5	95·3	91·7
Primary amine, wt. %	95·9	93·9	90·8
Total acetylatable, meq/g	3·27	2·15	0·96
Colour, Pt–Co	40	40	50 (Est.)
Water, wt. %	0·1	0·09	0·14
Specific gravity, g/cm³	1·045 4, 20 °C/20 °C	—	—
	1·029 3, 50 °C/20 °C	1·055 6, 50 °C/20 °C	1·080 8, 50 °C/20 °C
	0·990 5, 99 °C/20 °C	—	1·041 8, 99 °C/20 °C
Flash point, °C (°F)	257 (495)	268 (515)	260 (500)
Pour point, °C (°F)	−6·7 (20)	—	—
Freezing point, °C (°F)	—	−5·6 (22)	6·1 (43)
Viscosity, cSt	72, 20 °C	119, 25 °C	—
	20, 50 °C	41·4, 50 °C	133·7, 50 °C
	6·7, 99 °C (210 °F)	11·9, 99 °C (210 °F)	35·1, 99 °C (210 °F)

TABLE 5.4

PROPERTIES OF POLYOXYETHYLENEDIAMINE CURED EPOXY CASTINGS[7]

	1	2
Composition, pbw		
DGEBA (EEW ~ 185)	86·5	136
Nonylphenol	43	—
polyoxyethylenediamine ED-600[a]	59·4	—
polyoxyethylenediamine ED-900[a]	—	140
Accelerator 298[a]	6·0	9·2
Cure Schedule		
Room temperature (ca. 25 °C) for 2 weeks.		
Both systems were tack-free after 24 hours.		
Properties of cured systems		
Tensile strength, MPa (ksi)	3·32 (481)	2·87 (410)
Ultimate elongation, %	68	67
Hardness, shore A	67	64
Tear strength, kg/cm	10·1	8·09
Tensile set, %	Unknown	0

[a] Product of Jefferson Chemical.

TABLE 5.5

TYPICAL ANALYSES AND PHYSICAL PROPERTIES OF POLYOXYPROPYLENE AMINES[8]

	Polyoxypropylene amines			
	D-230	D-400	D-2000	T-403
Total acetylatables, meq/g	8·75	5·17	1·01	6·75
Total amines, meq/g	8·45	4·99	0·96	6·45
Primary amines, meq/g	8·30	4·93	0·95	6·16
Water, wt.%	0·10	0·13	0·10	0·08
Colour, Pt–Co	30	50	100	10
Viscosity, cSt, 25 °C	8·7	22	—	76·5
cSt, 20 °C	14·4	30	344	97
Specific gravity, 20/20 °C g/cm^3	0·948 0	0·970 2	0·996 4	0·981 2
Refractive index, 20 °C g/cm^3	1·446 6	1·480 1	—	1·460 6
Flash point, COC, °C (°F)	124 (256)	175 (347)	238 (460)	193 (380)
pH, 1% aqueous solution	11·3	11·3	10·1	11·2
pH, 5% aqueous solution	11·7	11·6	10·5	11·6
Vapour pressure, mm Hg/°C	1/101, 10/133	1/165, 10/193	—	1/181, 5/207

$$CH_3CH_2-\underset{\underset{\displaystyle CH_2\text{+}O-CH_2-CH\text{+}_zNH_2}{\overset{\displaystyle CH_2\text{+}OCH_2-CH\text{+}_xNH_2}{|}}}{\overset{\displaystyle |}{C}}-CH_2\text{+}O-CH_2-\underset{\displaystyle CH_3}{\overset{\displaystyle |}{CH}}\text{+}_yNH_2$$

16

Polyoxypropylenetriamine (ref. 8)

Typical analytical and physical property data for these polyoxy-propyleneamines are shown in Table 5.5. Properties of polyoxypropylene-amine-cured DGEBA resins are illustrated in Table 5.6. Of significance in these resin formulations is the low viscosity at 25 °C for the catalysed resin mixtures. Chiao and Moore[21] investigated the polyoxypropylenetriamine **16** (where $x + y + z = 5\cdot3$) as a room-temperature-cured epoxy resin. These workers determined some physical properties of the polyoxypropylenetri-amine-cured DGEBA epoxy resin (Table 5.7), and also investigated the effect of cure cycle at a fixed resin/hardener ratio (Table 5.8) on mechanical properties. The effect of resin/hardener ratio on mechanical properties, at a fixed cure cycle was also determined (Table 5.9). The good handlability of this resin system as a filament winding resin is indicated by its low viscosity at room temperature ($\sim 1000\,cP$) and at 50 °C ($100\,cP$) (Fig. 5.1). The effect of cure vs time and temperature for a fixed ratio of epoxy/hardener of 100/36 and degree of cure vs time and composition at 100 °C are illustrated in Figs 5.2 and 5.3 respectively. The highest degree of cure is obtained at 90 °C in 8 h, while the stoichiometric ratio 100:42 yields essentially 100 % cure in 2 h. The epoxy system is only 40 % cured at an epoxy/hardener ratio of 100/36 at 30 °C, and 85 % cured at 100 °C at the same ratio.

Chiao and co-workers[22] further investigated the polyethertriamine as a moderate-temperature-curable filament-winding epoxy resin. In this study, the diglycidyl ether of Bisphenol F (DGEBF) was used as the epoxy resin. The temperature–viscosity and time–viscosity curves (Figs 5.4 and 5.5) of this epoxy/polyethertriamine system (100/49) indicate a working life at least 4–5 h at 20 °C, and a very low viscosity at room temperature and particularly at 60 °C. Table 5.10 summarises the process characteristics of this epoxy system, while Tables 5.11 and 5.12 list the physical property data of the DGEBF/polyethertriamine epoxy system at various ratios of epoxy to amine. It is apparent from the moisture absorption, heat distortion

TABLE 5.6

PROPERTIES OF POLYOXYPROPYLENE AMINE/DGEBA RESIN SYSTEMS AND CLEAR CASTINGS[8]

	1	2	3	4	5	6	7	8
Composition, pbw								
DGEBA (EEW ~ 190)	100		100	100	100	100	100	
DGEBA (EEW ~ 185)		100						100
Polyethertriamine D-230	35	35	44					
Polyethertriamine D-400				57	57	60	77	
Polyethertriamine F-403								35
Catalysed resin viscosity, 25°C, cP	800	—		500	500		350	
Pot-life								
400 g, 25°C, h	4·5			9–20	9–20		—	—
500 ml, 25°C, h	—	2·5				5·5	—	3
Gel time								
150°C, gel plate, min	2·3			3·9	3·9		4·5	—
415 ml, 25°C, h[b]	—	7				24	—	—
Cure cycle, h/°C	2/100	16/85	2/100	168/23	Gel/55 2/100	16/85	2/100	16/85
HDT, °C	68	75	53		31	49		63
Izod impact strength, 23°C, kg-cm/cm notch	5·7		5·6		5·4		6·5	—
Tensile strength, MPa, −40°C					8·49		88·9	—
23°C	51	73	41	34	17	55·9	21	78·6
Tensile elongation, %, −40°C					3·7		4·7	—
23°C	7·1	7	8·6	16	58		90	—
Flexural strength, MPa[c]		117				87·6		128
Flexural modulus, GPa[c]		3·1				2·8		3·4
Tensile shear strength, 23°C, MPa	13·1						23	—
Hardness, Shore D		85				82		88
24-h water boil, % wt gain	3·0		4·1		2·3		3·5[d]	—
3-h acetone boil, % wt gain	8·5		17		29		46[d]	—
Dielectric constant, 23°C, 1kC	3·8			3·7			—	—
Dissipation factor, 23°C, 1kC	0·015			0·013			—	—

[a] Epoxy Resin Formulators (ERF) Method 13-64-A. [b] ERF 2-61-D. [c] ASTM D 790-66. [d] 7-week soak at 23°C.

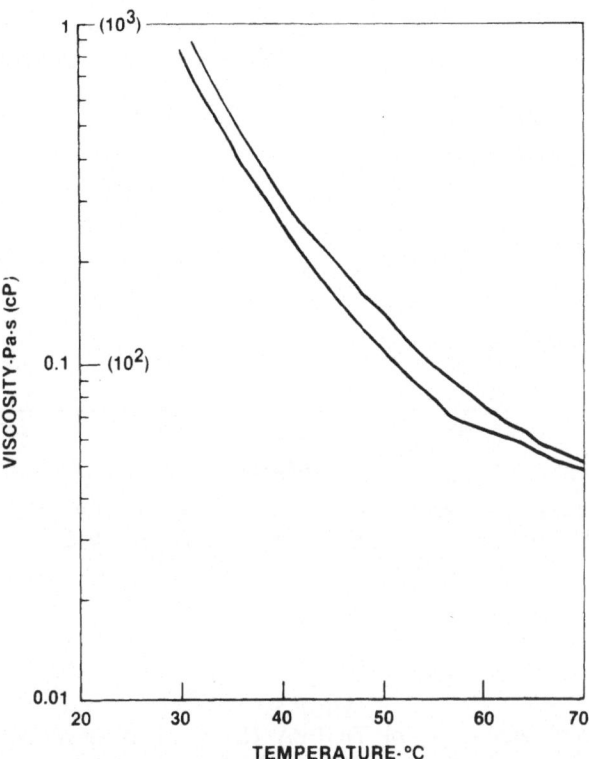

FIG. 5.1. Viscosity versus temperature of DGEBA (EEW 190)/polyethertriamine F-403 (100:36) epoxy system.[21]

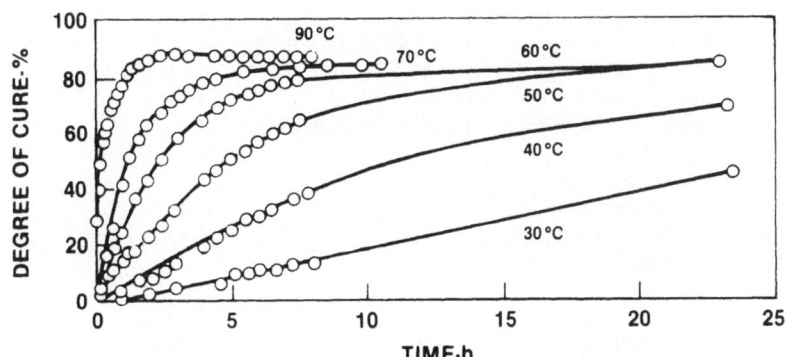

FIG. 5.2. Degree of cure versus time and temperature of DGEBA (EEW 190)/polyethertriamine T-403 (100:36).[21]

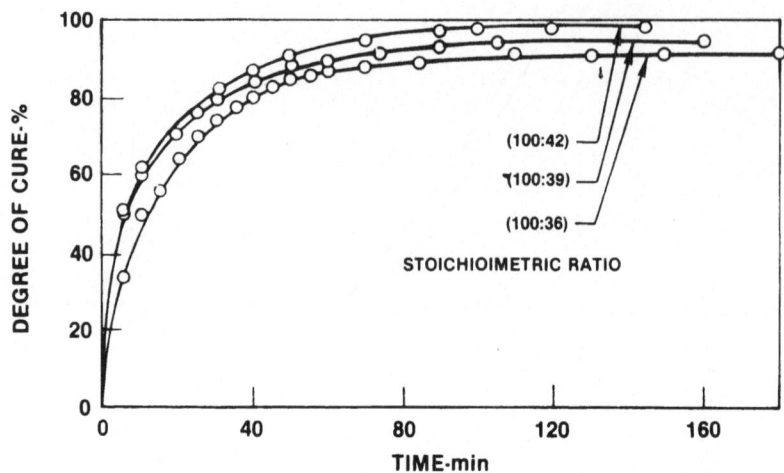

FIG. 5.3.   Degree of cure versus time and curing agent ratio for DGEBA (EEW
190)/polyethertriamine T-403 epoxy system at 100 °C.[21]

TABLE 5.7

GENERAL   PROPERTIES   OF   DGEBA   (EEW ~ 175)/POLYETHER-
TRIAMINE T-403 (100:36)[21] [a]

Heat distortion, °C	
(ASTM D-648-56) (revised 1961)	
at 455 kPa (66 psi)	65·5
at 1 820 kPa (264 psi)	62·5
Density, g/cm^3 at 20·7 °C	1·162
Water absorption, % gain (ASTM D-570-63)	0·75
Specific heat, J/g/°C (ASTM E-351-61)	1·75
Heat capacity, J/°C (ASTM C-351-61)	6·12
Impact strength Izod, kg-cm/cm of notch	
(ft-lb/in)	1·12 (0·206)
Shrinkage, total %	
after gelation	4·4
after cure (no change)	4·4
Thermal conductivity, w/m K	
at 298·52 K	0·133 5
at 317·15 K	0·173 8
at 336·13 K	0·210 2

[a] Cure cycle: 60 °C/24 h + 77 °C/24 h.

TABLE 5.8

TENSILE PROPERTIES OF DGEBA (EEW ~ 175)/POLYETHERTRIAMINE T-403 CURED AT VARIOUS CONDITIONS (RESIN/HARDENER FIXED RATIO, 100:36)[21]

Cure cycle (time/°C)	No. of specimens	At maximum				At rupture				Secant modulus to 0·01 strain	
		Strength MPa (psi)	CV%[a]	Elongation %	CV%[a]	Strength MPa (psi)	CV%[a]	Elongation %	CV%[a]	MPa (ksi)	CV%[a]
8 d/23	8	52·6 (7629)	4·2	1·7	5·6	52·6 (7629)	4·2	1·7	5·6	3447 (500)	1·4
14 d/23	7	55·8 (8092)	4·5	1·7	7·4	55·8 (8092)	4·5	1·7	7·4	3654 (530)	5·1
28 d/23	8	62·2 (9017)	4·4	2·1	6·3	62·2 (9017)	4·4	2·1	6·3	3447 (500)	3·3
56 d/23	7	64·0 (9286)	5·1	2·0	5·7	64·0 (9286)	5·1	2·0	5·7	3516 (510)	1·6
24 h/50	7	73·4 (10640)	1·8	3·7	0·8	66·6 (9663)	5·5	4·6	10·4	3240 (470)	3·1
16 h/60	6	72·9 (10570)	0·7	3·6	2·2	65·3 (9466)	3·9	4·9	7·2	3240 (470)	1·5
8 h/70	7	72·7 (10550)	0·9	3·6	5·5	67·7 (9822)	1·4	4·7	8·2	3310 (480)	10·0
Constant cure cycle, 7 h at 60°C plus 1 h at 100°C Resin-to-hardener ratio											
100:36	7	72·8 (10560)	0·8	3·6	2·5	67·0 (9713)	3·2	4·6	6·3	3240 (470)	1·3
100:39	7	69·8 (10120)	0·8	3·8	3·3	58·7 (8520)	5·9	5·9	17·7	3103 (450)	3·0
100:42	7	66·5 (9646)	0·8	4·1	1·7	54·8 (7950)	4·4	7·7	11·8	2896 (420)	1·6

[a] % coefficient of variation.

DANIEL A. SCOLA

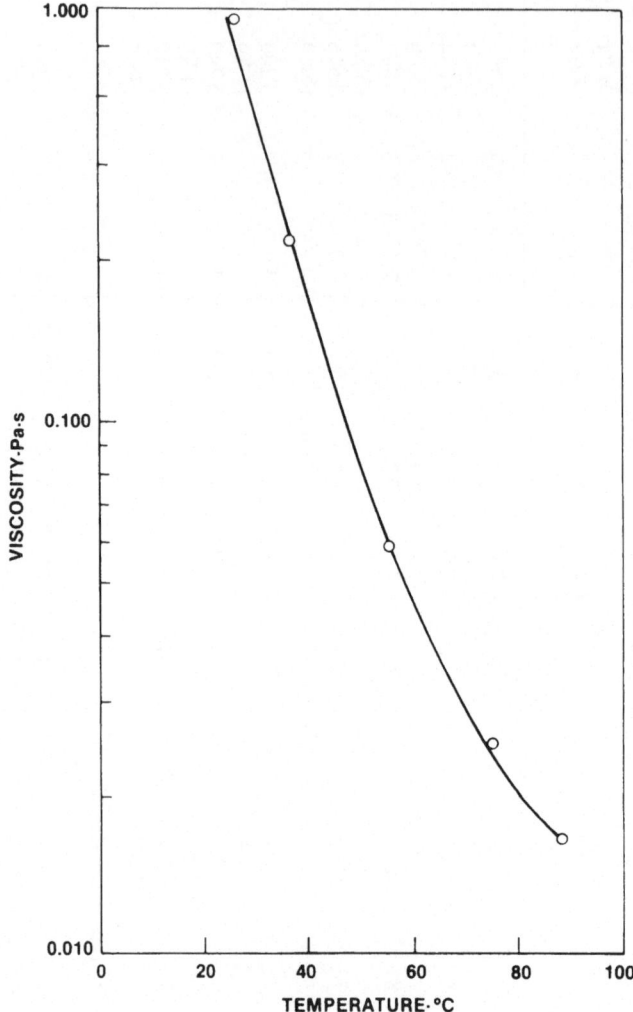

FIG. 5.4. Initial viscosity as a function of temperature for the DGEBF/T-403 polyethertriamine system.[22]

temperature and tensile properties that the best overall properties are obtained with resins containing the stoichiometric, or 5% less than the stoichiometric, quantity of triamine as curing agent. Tensile properties of the DGEBF/polyethertriamine (100/49) system as a function of cure cycle are listed in Table 5.13.

Utilisation of aliphatic amines as curing agents in filament-winding

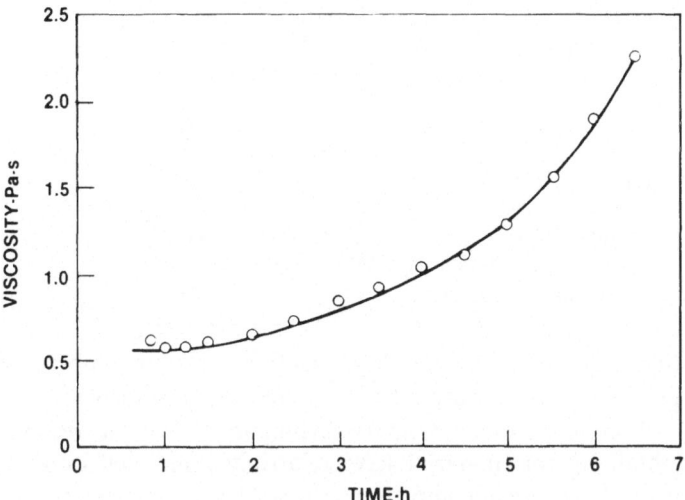

FIG. 5.5. Viscosity versus time at 20 °C (300 g sample) for the DGEBF/T-403 polyethertriamine system.[22]

TABLE 5.9

COMPRESSIVE PROPERTIES OF DGEBA (EEW 175)/POLYETHERTRIAMINE T-403 (100:36) CURED AT VARIOUS CONDITIONS[21]

Cure cycle (time/°C)	No. of specimens	At maximum				Secant modulus to 0·01 strain	
		Strength		Elongation			
		Mpa (psi)	CV %[a]	%	CV %[a]	MPa (ksi)	CV %[a]
8 d/23	4	87·6 (12 710)	2·4	2·0	3·1	3 374 (489·3)	1·7
14 d/23	4	93·9 (13 620)	0·5	2·0	2·0	3 377 (489·8)	3·5
28 d/23	4	97·7 (14 170)	1·1	1·9	0·8	3 455 (501·1)	4·0
56 d/23	4	103·1 (14 960)	0·9	1·9	2·6	3 582 (519·5)	5·1
140 d/23	3	96·5 (14 000)	1·7	2·8	1·4	4 002 (530·5)	2·0
16 h/60	9	87·7 (12 720)	0·7	3·7	1·4	3 485 (505·4)	2·2

[a] % coefficient of variation.

TABLE 5.10
PROCESSING FACTORS FOR DGEBF (EEW ~ 159)/POLY-
ETHERTRIAMINE T-403 (100/49)[22]

Factor	Quantity
Viscosity, 25 °C	0·53 Pa
Density, 25 °C	1·116 g/cm^3
Pot-life, 500 g specimens at 23 °C	14 h
Time to reach 2·0 Pa at 25 °C	6 h
Exotherm of 500 g from 25 °C	5 °C

applications is restricted by their high reactivity at room temperature. The
amine group in polyoxypropylenetriamine 16 is somewhat hindered,
because it is attached to a secondary carbon atom. For this reason epoxy
resins containing this material have relatively long shelf-lives at room
temperature. Even greater improvements in the room-temperature shelf-
life of primary amines was demonstrated by Rinde and co-workers.[23] 2,5-
Dimethyl-2,5-hexane diamine (DMHDA) (17), a sterically hindered
primary aliphatic amine, was investigated as a curing agent with DGEBA
epoxy resins. The viscosity–time curves at 25, 40 and 60 °C for the
DGEBA/DMHDA system are shown in Fig. 5.6. The shelf-life at 25 °C is
approximately 7 h and at 40 °C, approximately 4 h. A summary of
properties of the cured DGEBA/DMHDA resin is listed in Table 5.14.

Novel piperidine derivatives (18) have been introduced as curing agents

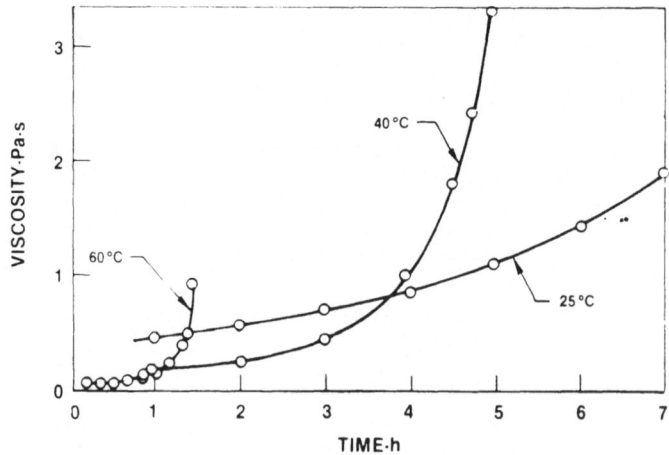

FIG. 5.6.   Viscosity as a function of time for 300 g samples of pure
DGEBA/DMHDA at 25, 40, and 60 °C.[23]

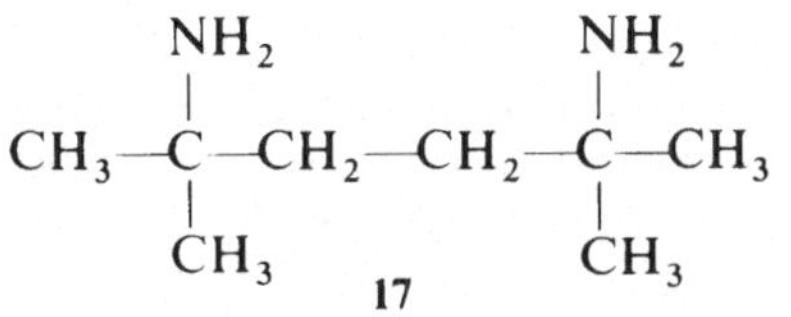

$$CH_3-\underset{\underset{CH_3}{|}}{\overset{\overset{NH_2}{|}}{C}}-CH_2-CH_2-\underset{\underset{CH_3}{|}}{\overset{\overset{NH_2}{|}}{C}}-CH_3$$

17

2,5-Dimethyl-2,5-hexanediamine (ref. 23)

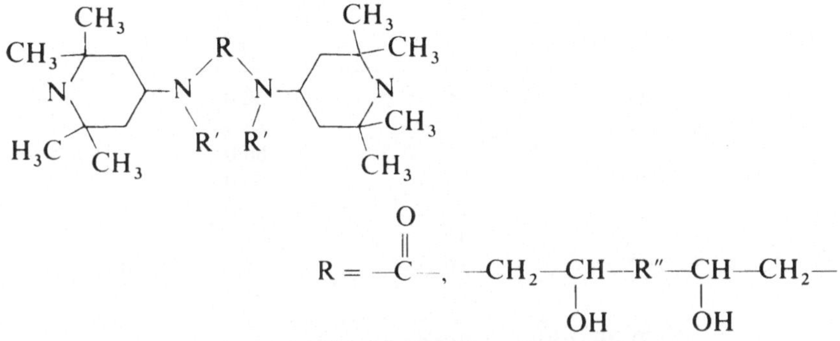

$$R = -\overset{\overset{O}{\|}}{C}-, \quad -CH_2-\underset{\underset{OH}{|}}{CH}-R''-\underset{\underset{OH}{|}}{CH}-CH_2-$$

R' = N, alkyl

R" = bivalent, aliphatic, aromatic or heterocyclic radical

18

bis(N,N'-4-Aminopiperidine derivatives) (ref. 24)

TABLE 5.11

TENSILE PROPERTIES OF VARIOUS AMOUNTS OF CURING AGENT[a]
(Cured 8 h at 80 °C)[22]

Excess amine	No. of specimens	At maximum				Modulus to 0·5% strain	
		Stress		Elongation			
		Mpa (ksi)	CV %[b]	%	CV %[b]	GPa (ksi)	CV %[b]
−23	6	75·8 (11·0)	0·3	3·5	3·1	3·39 (492)	1·9
−18	6	75·4 (10·9)	0·4	3·5	2·4	3·43 (498)	1·3
−14	9	73·8 (10·7)	0·5	3·6	2·1	3·34 (485)	2·8
−10	7	72·2 (10·5)	0·5	3·9	2·1	2·94 (427)	3·8
−5	6	74·8 (10·9)	0·6	4·1	0·7	3·08 (447)	2·6
0	9	75·8 (11·0)	0·4	4·3	2·0	2·99 (434)	2·5
+5	8	72·8 (10·6)	0·5	4·3	1·5	2·79 (404)	2·7
+10	—	71·4 (10·4)	0·3	4·1	3·1	2·78 (403)	6·3

[a] DGEBF (EEW ~ 159)/polyethertriamine T-403.
[b] % coefficient of variation.

TABLE 5.12

HEAT DISTORTION TEMPERATURE AND WATER ABSORP-
TION AT VARIOUS AMOUNTS OF CURING AGENT[a]
(Cured 80 °C/8 h)[22]

Excess amine	Water absorption[b] (%)	Heat distortion Temperature[c] (°C)
−23	1·46	30·0
−18	1·39	29·4
−14	—	60·0
−10	1·27	60·0
−5	1·20	67·0
0	1·15	67·0
+5	1·16	71·7
+10	1·26	64·4

[a] DGEBF (EEW ~ 159)/polyethertriamine (T-403).
[b] After 6 h in boiling water.
[c] Measured at 455 kPA (66 psi).

for epoxy resins intended for coating applications.[24] The curing agents were prepared by reaction of 2,2,6,6-tetramethyl-4-aminopiperidine (TAD) (12 equivalents) with DGEBA (EEW ~ 190) (4 equivalents) at 60 °C to 110 °C to produce an amine/epoxy resin adduct of unknown equivalent weight, or by reaction of TAD (6 equivalents) with urea (1 mole) to yield an amide adduct with an H-active equivalent weight of 167.

TABLE 5.13

TENSILE PROPERTIES AS A FUNCTION OF CURE CYCLE FOR DGEBF (EEW ~ 159)/POLY-
ETHERTRIAMINE (100:40)[22]

Cure Temperature (°C)	Time (h)	No. of specimens	At maximum Stress MPa (ksi)	CV %[a]	Elongation %	CV %[a]	Modulus to 0·5% strain MPa (ksi)	CV %[a]
50	14	8	71·3 (10·3)	0·5	3·6	1·5	3·18 (461)	1·8
60	11	9	73·6 (10·7)	0·5	4·2	1·2	2·85 (414)	2·0
70	9	9	74·3 (10·8)	0·3	4·1	1·0	2·91 (422)	1·5
80	8	7	74·6 (10·8)	0·3	4·3	1·6	2·87 (416)	1·5
90	6	8	75·1 (10·9)	0·4	4·4	1·9	2·81 (407)	3·4
100	3	9	75·5 (11·0)	1·0	4·3	1·4	2·87 (416)	1·3

[a] % coefficient of variation.

TABLE 5.14
PROPERTIES OF DGEBA/DMHDA EPOXY RESIN SUMMARISED[23]

Resin	DGEBA (EEW ~ 175)
Curing agent (DMHDA)	2,5-dimethyl-2,5-hexane diamine
Mix ratio	100/21·8
Cure schedule	room temperature/16 h + 60 °C/1 h
	+ 130 °C/3 h
Tensile properties	
Failure stress, MPa	84·8
Failure strain, %	8·6
Modulus, GPa	2·7
Compressive strength, MPa	105·3
Compressive strain, %	10·4
Shear strength, MPa	31·8
Shear modulus, GPa	0·7
Flex strength, at 5% strain, MPa	98·0
Tangent modulus, GPa	2·7
Water absorption, 100 °C/24 h	
(ASTM-D-570-62), % gain	1·32
T_g, °C	142
Coefficient of thermal expansion	$6·0 \times 10^{-5}$
(− 60 to 135 °C) cm/cm/ °C	
Izod impact ASTM D256,	
Method A, J/m	36·5

Coating compositions of each curing agent were prepared by reaction with a solid epoxy having an equivalent weight of 850–890 g and with filler and flow agents. Properties of the epoxy resin powder lacquers and electrostatically applied films cured at 180 °C for 30 min, are listed in Tables 5.15 and 5.16 respectively.

In a review by Kaman, several aliphatic, cycloaliphatic and aliphatic–heterocyclic amines are described (19–31).

Compounds 19, 20, 21a, 21b and 22 are commercially available. Compounds 23, 24, 25, 26 and 27 are the subject of several patents.[25–29] The polymeric diamine 28,[30–32] the cyclohexene diamines 29,[33] the cyclohexane diamines 30[34] and the polysiloxane diamines 31[35] are also the subject of patents.[30–35] The diamines 23–29 and 31 may be available commercially as curing agents, but may be sold under a trade name or designation, thereby eluding recognition. The cyclohexene diamines will effect cure of a DGEBA (EEW ~ 190) epoxy at room temperature in two days. Complete cure can be attained at 100 °C/1h. The cyclic polyamines are slower curing at room temperature than the straight-chain polyamines.

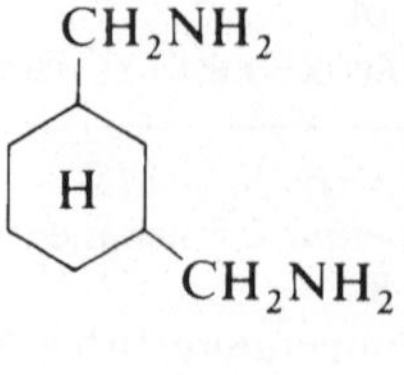

$$CH_2NH_2$$

19 (ref. 3) 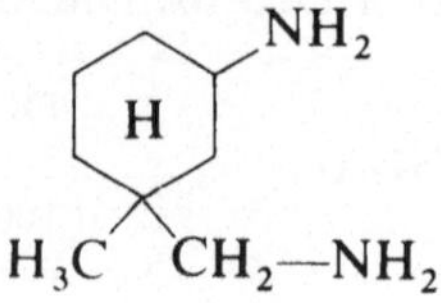   **20** (ref. 3)

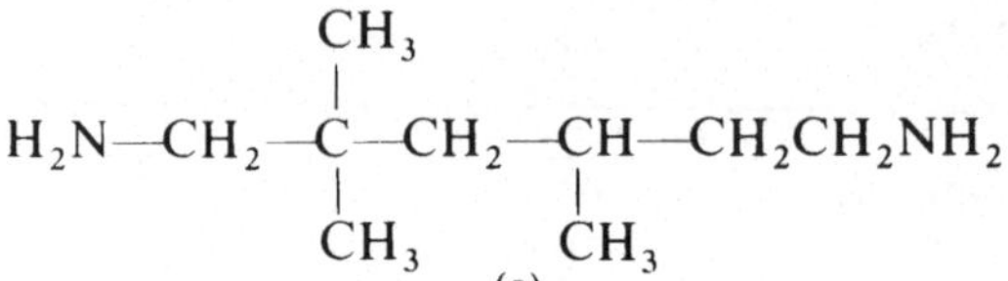

$$H_2N-CH_2-\underset{\underset{CH_3}{|}}{\overset{\overset{CH_3}{|}}{C}}-CH_2-\underset{\underset{CH_3}{|}}{CH}-CH_2CH_2NH_2$$

(a)

$$H_2N-CH_2-\underset{\underset{CH_3}{|}}{CH}-CH_2-\underset{\underset{CH_3}{|}}{\overset{\overset{CH_3}{|}}{C}}-CH_2CH_2NH_2$$

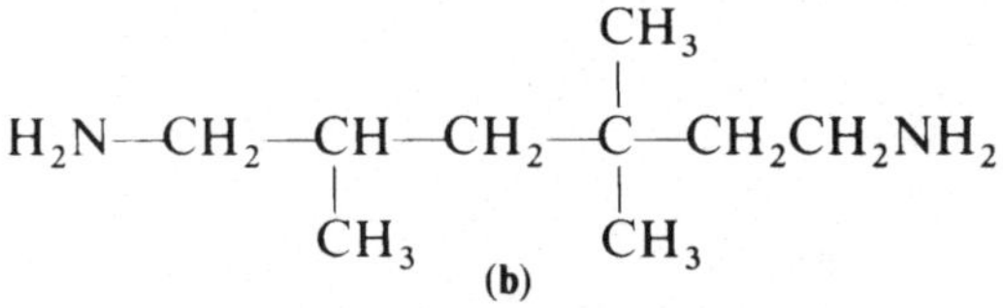

(b)

21 (ref. 3)

$$H_2N(CH_2)NH(CH_2)_2NH_2$$

22 (ref. 3)

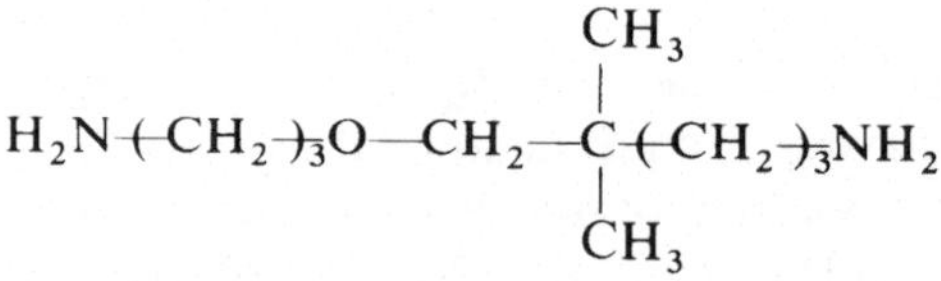

$$H_2N(CH_2)_3O-CH_2-\underset{\underset{CH_3}{|}}{\overset{\overset{CH_3}{|}}{C}}(CH_2)_3NH_2$$

23 (ref. 25)

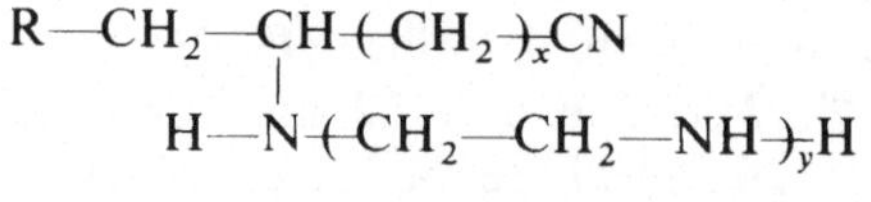

$$R-CH_2-\underset{|}{CH}(CH_2)_xCN$$
$$H-N(CH_2-CH_2-NH)_yH$$

$(x = 2-10, \ y = 1-5)$

24 (ref. 26)

$$H_2N(CH_2)_5C$$

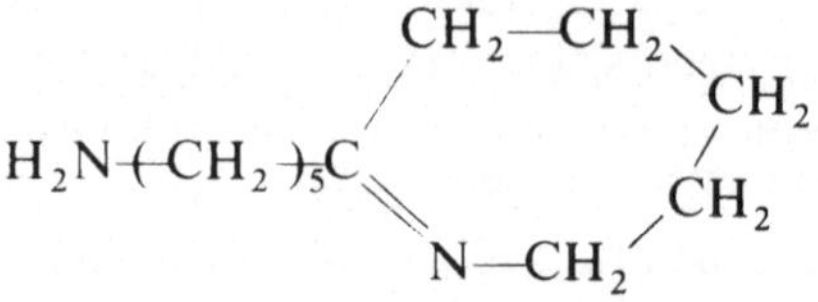

25 (ref. 27)

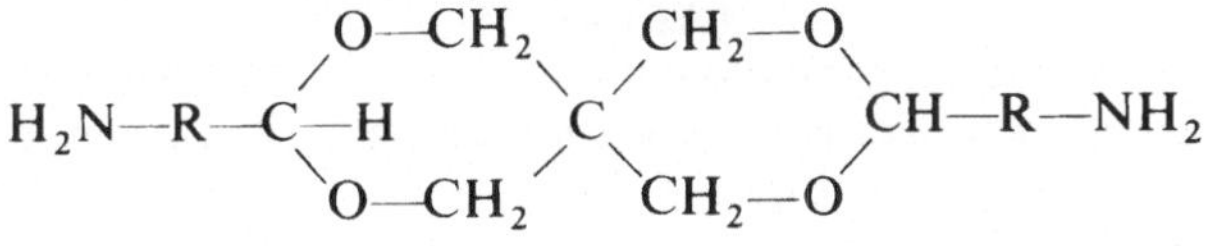

$$H_2N-R-\underset{\underset{O-CH_2}{|}}{\overset{\overset{O-CH_2}{|}}{C}}-H \quad \underset{\underset{CH_2-O}{}}{\overset{\overset{CH_2-O}{}}{C}} \quad CH-R-NH_2$$

$$R = C_5-C_7$$

26 (ref. 28)

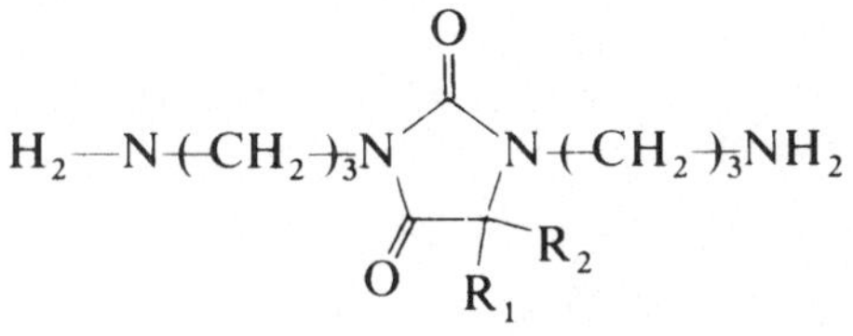

$$H_2-N+CH_2)_3N \underset{\underset{O}{\overset{\overset{O}{\parallel}}{}}}{\overset{}{}} N+CH_2)_3NH_2 \qquad H_2N-(polymer)-NH_2$$

$$\underset{R_1}{\overset{R_2}{}}$$

28 (ref. 30, 31, 32)

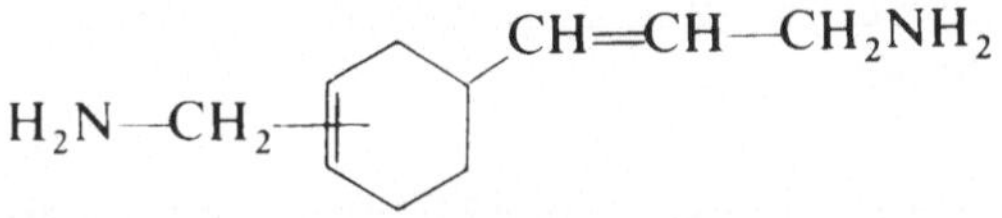

27 (ref. 29)

$$H_2N-CH_2- \quad CH=CH-CH_2NH_2$$

29 (ref. 33)

$$\underset{\overset{|}{R}}{\overset{\overset{NH_2}{}}{}} \quad CH_2NH_2$$

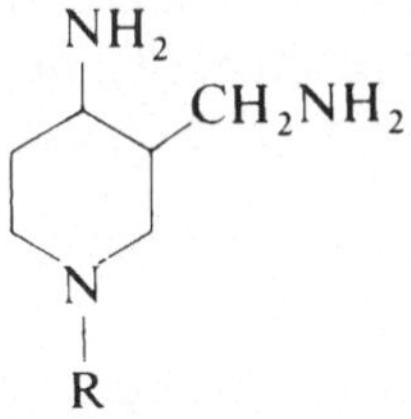

30 (ref. 34)

$$R = H, \ CH_3-, \ CH_3(CH_2)_3-,$$
$$\qquad a \qquad b \qquad\qquad c$$

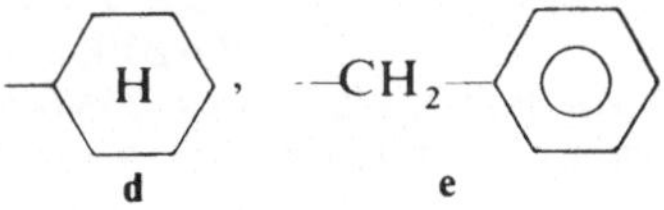

$$-\!\!\!\left\langle \ H \right\rangle, \quad -CH_2-\!\!\!\left\langle \bigcirc \right\rangle$$
$$\qquad\quad d \qquad\qquad\qquad e$$

$$H_2N-R-O-\underset{\underset{CH_3}{|}}{\overset{\overset{C_6H_5}{|}}{Si}}-O-\underset{\underset{CH_3}{|}}{\overset{\overset{C_6H_5}{|}}{Si}}-O-\underset{\underset{CH_3}{|}}{\overset{\overset{C_6H_5}{|}}{Si}}-O-R-NH_2$$

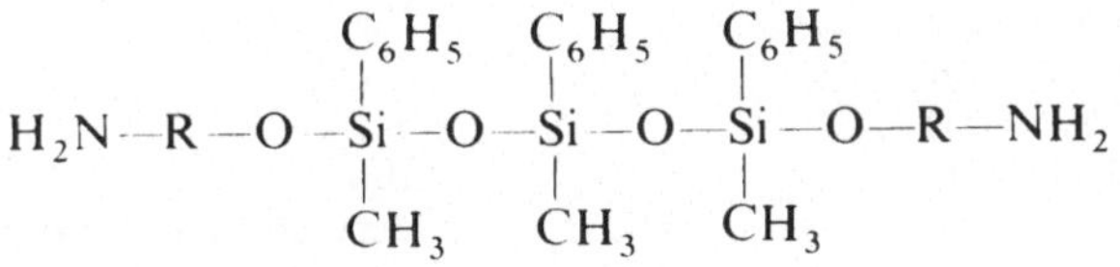

31 (ref. 35)

However, the cyclic diamines generate epoxy systems with greater thermal stability than do the straight-chain aliphatic polyamines.

Some properties of four DGEBA/diamine epoxy resin systems, using diamines **20**, **21a/21b** (50/50 mixture), **30** (R = H) and **30** (R = butyl) are listed in Table 5.17. These resin systems are characteristic of cyclic diamines and hindered aliphatic diamines.

TABLE 5.15

PROPERTIES OF POWDER LACQUER COATING[a] EPOXY (DGEBA EEW 850–940) CURED
BY 2,2,6,6-TETRAMETHYL-4-AMINOPIPERIDINE–DGEBA ADDUCT[24]

Coating thickness	60–70μm
Hardness (DIN 53157)	180–190 sec
Elasticity (DIN 53156)	8–10 mm
Adhesion (DIN 53151)	0
Gloss (ASTM-D-523)	95
Flow storage for 30 days at 40°C	Very good, no orange peel effect
Accelerated weathering in a 450 LF Xenotest apparatus (DIN 53231)	
Loss of gloss after 500 h	12%
Loss of gloss after 1 000 h	23%
Loss of gloss after 1500 h	31%
Loss of gloss after 2 000 h	35%

[a] Cured 180°C (30 min).

The low viscosity at the gel time (2 h) at room temperature for curing
agents **20** and **21a/21b** (50/50) makes them suitable for filament-winding
applications. Moreover, the epoxy resin system containing curing agent **20**
exhibits a relatively high glass transition temperature (T_g) of 130°C, and
excellent mechanical properties. This system is suitable for use at 100°C in
structural applications.

TABLE 5.16

PROPERTIES OF THE POWDER LACQUER COATING[a] EPOXY (DGEBA EEW 850–940)
CURED BY 2,2,6,6-TETRAMETHYL-4-AMINOPIPERIDINE–UREA ADDUCT[24]

Coating thickness	50–60 μm
Hardness (DIN 53157)	190–200 sec
Elasticity (DIN 53156)	9 mm
Adhesion (DIN 53151)	0
Gloss (ASTM-D-523)	100
Flow	Very good, no orange peel effect
Flow after storage for 30 days at 40°C	Unchanged, very good
Flowability after storage for 30 days at 40°C	Very good, free flowing
Accelerated weathering in the 450 LD Xenotest apparatus (DIN 53231)	
Loss of gloss after 500 h	13%
Loss of gloss after 100 h	24%
Loss of gloss after 1 500 h	34%
Loss of gloss after 2 000 h	37%

[a] Cured 220°C/15 min.

TABLE 5.17

SOME PROPERTIES OF DGEBA/DIAMINE SYSTEMS[3]

	Diamine 20	Diamine 21 (50/50)	Diamine 30 (R=H)	Diamine 30 (R=Butyl)
Composition				
DGEBA, g	100	100	100	100
Diamine, g	24	21	13·5	25
Viscosity, cP 25 °C	2 000	750–850	9 000–13 000	—
Use time, min	70	50	—	—
Gel time, min	160	60	—	—
Cure cycle	80 °C/1 h + 150 °C/4 h	80 °C/1 h + 150 °C/2 h	100 °C/24 h	100 °C/24 h
Cured properties				
HDT, °C	149	105	—	—
T_g, °C	135	92	—	123
Flexural strength, MPa	122·5	112·7	103·8	127·4
Flexural strain, mm	10·4	1·5	5·6	14·1
Flexural modulus, GPa	4·38	3·72	—	—
Izod impact, kg-cm/cm²	17	20	7·7	39·0
Tensile strength, MPa	71·5	63·7	—	73·5
Strain-to-failure	3·6	3·6	—	9·0

The polysiloxane diamines, 31, used as curing agents with DGEBA, generate flexible, moisture-resistant, weather-resistant coatings and films.[35] Mixtures of diamines 20 and 21/21b (50/50) were evaluated as diamine–DGEBA adduct curing agents in coating applications using the epoxy resin DGEBA (EEW ~ 180).[3] The diamine–epoxy adduct composition is shown in Table 5.18. The diamines 20 and 21 are dissolved in the solution of cellusolve acetate in benzyl alcohol, and then the epoxy resin is added. The solution is stirred for 2 h, giving an amine–epoxy adduct

TABLE 5.18

COMPOSITION OF DIAMINE/DGEBA ADDUCT[3]

Components	Weight (g)
Diamine 20	50
Diamine 21 (50:50 a & b)	45
Benzyl alcohol	88
Cellosolve acetate	12
DGEBA (EEW ~ 180)	40

TABLE 5.19

PROPERTIES OF DGEBA EPOXY COATINGS CURED WITH DIAMINE/DGEBA ADDUCT[3]

Exposure	Koenig degree of hardness (sec)			Buckholz harness	Impact strength $((kg\text{-}cm)/cm^2)$
	1 day	3 days	7 days		
50% relative humidity, room temperature	150	187	185	100	16·1
100% relative humidity, room temperature	76	161	167	71	14·0
5°C	30	129	149	83	12·2
120°C/30 min	170	—	—	111	14·3

solution with a viscosity of 615 cP at 25 °C. This material was evaluated as a curing agent with DGEBA (EEW ~ 180) to give the properties indicated in Table 5.19.

Bicyclic amidines (32) have been claimed as unusual curing agents for epoxy resins, but no properties of the uncured or cured resin systems were given in the patent.[36] Specific examples of two amidine compounds are

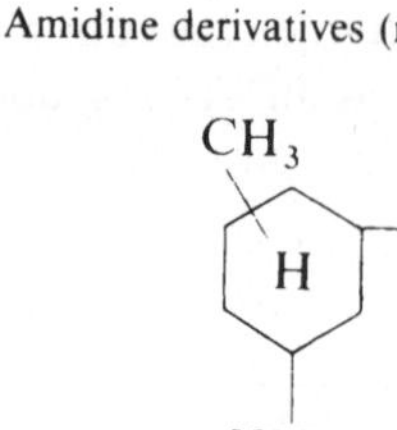

R = H or alkyl C_1–C_4 group

R' = H or alkyl C_1–C_{18}, cycloalkyl C_1–C_{18}, arylalkyl C_7–C_{10}, aryl groups C_6–C_{20}

32
Amidine derivatives (ref. 36)

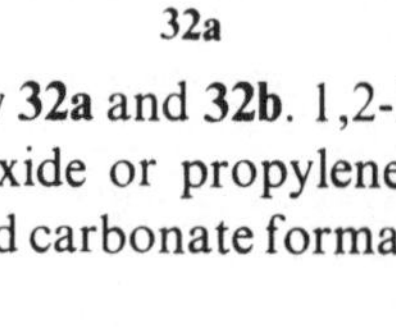

32a 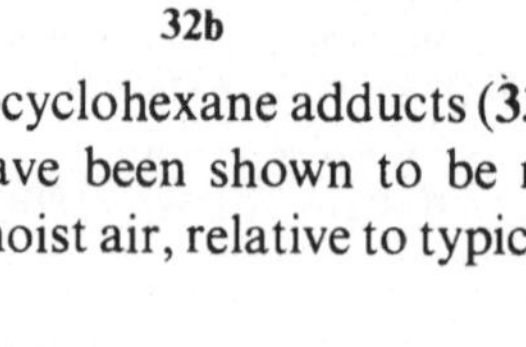   32b

illustrated by 32a and 32b. 1,2-Diaminocyclohexane adducts (33a and 33b) of ethyleneoxide or propyleneoxide have been shown to be resistant to yellowing and carbonate formation in moist air, relative to typical aliphatic

TABLE 5.20

EFFECT OF COMPOSITION ON GLASS TRANSITION TEMPERATURE (T_g) OF N-(2-HYDROXYETHYL)-1,2-DIAMINOCYCLOHEXANE CURED DGEBA (EEW ~ 190)[a][37]

Sample number	Diamine used (phr)	Equivalents diamine used	Equivalent ratio DGEBA/amine	T_g (°C)
1	27·5	0·521	1·010	128·0
2	28·0	0·531	0·990	130·0
3	28·5	0·540	0·974	132·0
4	29·0	0·55	0·956	130·5
5	29·5	0·56	0·939	130·0

[a] Cured cycle 80°C/2 h + 150°C/2 h.

or alicyclic amines such as 1,2-diaminocyclohexane.[37] The adduct curing agent 33 (MW 158·25, equivalent weight 52·75) when used in more or less stoichiometric ratios with DGEBA (EEW ~ 190) and cured at 80°C/2 h yielded resins with glass transition temperatures of 130°C (Table 5.20). This is essentially equivalent to the T_g of the resin generated from the diamine 20/DGEBA adduct and DGEBA (Table 5.17).

Amine-terminated butadiene–acrylonitrile copolymers (ATBN) (36) of varying equivalent weights have been synthesised from N-aminoethyl-piperizine (AEPZ) (35) and carboxyl-terminated butadiene copolymers (CBTN) (34).[38]

The polyamide–amine-terminated elastomers can react with epoxy resins to produce either rubber-like materials or more rigid elastomer toughened epoxy resins.

ATBNs of varying equivalent weights were prepared by altering the AEPZ/CTBN molar ratio. As the AEPZ/CTBN molar ratio decreased, the viscosity of the ATBN increased and the amine equivalent weight increased, as is shown in Table 5.21.

A summary of the ATBNs made from AEPZ and CTBN, with varying

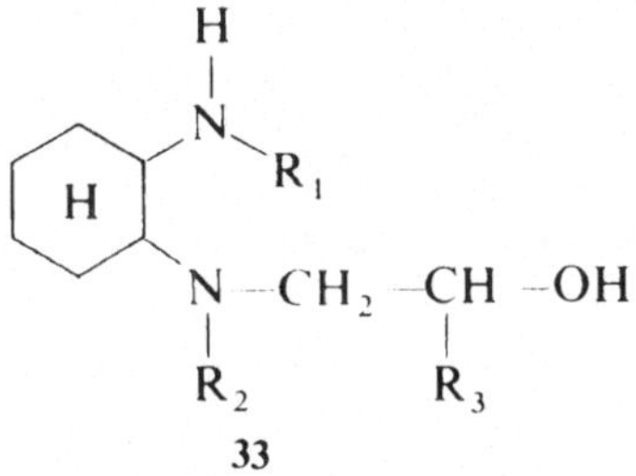

a $R_1, R_2, R_3 = H$
b $R_1, R_2 = H$
    $R_3 = CH_3—$

33

1,2-Diaminocyclohexane-epoxide adducts (ref. 37)

$$\displaystyle{+\!CH_2\!-\!CH\!=\!CH\!-\!CH_2\!)_a\!(CH_2\!-\!\underset{\underset{\displaystyle CN}{|}}{CH}\!)_b CH_2 CH_2\!-\!\overset{\displaystyle \overset{O}{\|}}{C}\!-\!OH}$$

$$\mathbf{34}$$

$$+\ H_2 NCH_2 CH_2\!-\!N\overset{\displaystyle \frown}{\underset{\displaystyle \smile}{\ }}N\!-\!H\ \longrightarrow$$

$$\mathbf{35}$$

$$\displaystyle{+\!CH_2 OH\!=\!CH_2\!-\!CH_2\!)_a\!(CH_2\!-\!\underset{\underset{\displaystyle CN}{|}}{CH}\!)_b CH_2 CH_2\overset{\displaystyle \overset{O}{\|}}{C}}$$

$$NH\!-\!CH_2 CH_2\!-\!N\overset{\displaystyle \frown}{\underset{\displaystyle \smile}{\ }}N\!-\!H$$

$$\mathbf{36}$$

Amine-terminated butadiene–acroylonitrile copolymer (ATBN) (ref. 38)

acrylonitrile content, is listed in Table 5.22. Some properties of the initial CTBNs as well as T_g values and viscosities of the ATBN products are also listed. The effects of ATBN level on the tensile properties of an epoxy resin are shown in Table 5.23. A cured composition containing from 1 to 50 parts by weight ATBN per 100 parts of DGEBA (EEW ~ 180) consists of a continuous hard epoxy phase and a particulate rubbery phase. This two-phase system has been shown to impart improved toughness of the

TABLE 5.21

ATBN MADE FROM CTBN[a] AND N-(2-AMINOETHYL) PIPERAZINE[b] AT DIFFERENT STOICHIOMETRIES[c] [38]

	A	B	C	D	E
AEPZ/CTBN molar ratio	3·0	2·0	1·75	1·5	1·25
ATBN properties					
Viscosity at 27°C, Pa	190	246	350	360	450
AEW[d]	690	858	967	1 203	1 477
AEPZ content %					
Actual[e]		5·6	4·2	2·7	2·0	1·2
Calculated[f]		10·9	6·25	4·75	3·20	1·65

[a] CTBN: Ephr COOH 0·55; viscosity at 27°C, 127 Pa.
[b] AEPZ, N-(2-aminoethyl)piperazine.
[c] Reaction temperature, 130°C.
[d] AEW, amine equivalent weight.
[e] After vacuum (266 Pa, at 100°C for 70 h).
[f] Before vacuum.

TABLE 5.22

ATBN MADE FROM VARIOUS CBTNs AND N-(2-AMINOETHYL) PIPERAZINE[a,b 38]

Acrylonitrile (wt %)	CTBN			ATBN	
	Ephr COOH	T_g (°C)	Viscosity (Pa)	T_g (°C)	Viscosity (Pa)
0	0·044	−72	44 at 27 °C	−72	76 at 27 °C
9·9	0·054	−59	75 at 23 °C	−62	89 at 23 °C
18·2	0·052	−47	177 at 23 °C	−46	187 at 23 °C
26·5	0·056	−32	1 320 at 22 °C	−33	1 408 at 22 °C

[a] AEPZ, N-(2-aminoethyl)piperazine.
[b] Reaction temperature 130 °C, AEPZ/CTBN, molar ratio = 3.

cured resin without substantial deterioration of thermal or mechanical properties. However, when the quantity of ATBN per 100 parts of DGEBA is increased from 50 to 500 parts, the composition consists of a continuous rubbery phase and a particulate rubbery phase.

It has been demonstrated that a small quantity of rubber in the form of discrete particles in a glassy thermoset, such as an aromatic-amine-cured epoxy resin, can greatly improve crack resistance and impact strength.[39] A series of ATBN-toughened epoxy resins is listed in Table 5.24. It can be seen that 2·5 to 10 phr ATBN causes a substantial increase in the toughness (fracture energy), and impact strength relative to the rubber-free epoxy.

TABLE 5.23

EFFECTS OF ATBN LEVEL ON COMPOSITIONAL PROPERTIES[38]

Parts by weight of ATBN per 100 parts by weight epoxy resin[a,b]	Compositional properties		
	Description	Tensile strength (MPa)	Approximate ultimate elongation (%)
About 1–20	Toughened plastic	40–70	1–15
About 20–100	Flexibilised plastic	20–50	10–50
About 100–250	Rigid elastomer	10–20	40–150
About 250–500	Elastomer	5–10	100–300
About 400–1 000	Soft elastomer	1–7	300–1 000

[a] DGEBA (EEW 175–200).
[b] Composition contained 24 parts of Bisphenol A chain extender per 100 parts by weight of epoxy resin.

TABLE 5.24
ATBN TOUGHENED EPOXY PLASTICS SYSTEMS[a] [38]

	A	B	C	D	E	F
Recipe by weight						
DGEBA (EEW ~ 190)	100	100	100	100	100	100
ATBN	—	—	2·5	5	7·5	10
CTBN	—	5	—	—	—	—
Bisphenol A	24	24	24	24	24	24
Piperidine	5	5	5	5	5	5
Test data						
Elastic modulus, GPa	2·7	2·6	2·7	2·6	2·7	2·6
Tensile strength, MPa	61	66	64	60	64	55
Ultimate elongation, %	4·0	8·8	4·8	5·4	4·5	4·9
Fracture energy, kJ/m^2	3·5	85·8	21·0	61·3	11·9	91·0
Gardner impact[b] J	5·7	>36	>36	>36	>36	>36
Heat distortion temperature, °C	82	82	82	82	82	82

[a] Cured 16 h at 120°C.
[b] Specimen thickness is nominal, 6·35 mm. The Gardner Impact Tester has a limit of 36 J. The >36 indicates that the sample did not break at 36 J.

ATBN copolymers have been used in the synthesis of castable elastomers by changing the ratio of epoxy to ATBN. Table 5.25 shows the mechanical properties of ATBN/epoxy castable elastomers cured at 120°C for 15 h. The data demonstrate that varying ratios of ATBN to epoxy resin generate elastomeric properties from rigid to soft.

Minato and coworkers[40] investigated the functional amines 1,3,5-tris(aminomethyl)benzene (37) and 1,3,5-tris(aminomethyl)cyclohexane (38) as curing agents for DGEBA (EEW ~ 190) epoxy resins. These two triamines yielded epoxy castings having HDTs of 147 and 167°C respectively, while the two diamines p-xylylenediamine and 1,4-bis(aminomethyl)cyclohexane yielded epoxy castings with HDTs of 112°C and 132°C respectively. The cure cycle in all cases was room temperature/60 h + 120°C/3 h + 200°C/3 h.

Trifunctional cyclic amines (ref. 40)

TABLE 5.25
PHYSICAL PROPERTIES OF ATBN/EPOXY CASTABLE ELASTOMERS[a] [38]

	A	B	C	D	E	F	G	H	I	J	K	L
Recipe by weight												
DGEBA (EEW ~ 190)	100	100	100	100	100	100	100	100	100	100	100	100
ATBN	100	150	200	250	275	300	300	325	350	400	450	500
Bisphenol A	24	24	24	24	24	24	—	24	24	24	24	24
Test data												
100% modulus, MPa	18·3	—	—	5·8	5·8	4·9	—	4·0	3·9	2·8	1·5	1·3
Tensile strength, MPa	—	14·0	11·2	8·9	7·8	12·5	4·8	6·6	6·5	5·5	3·4	2·1
Ultimate elongation, %	43	72	98	146	154	195	94	210	204	263	260	280
Gehman freezing point, °C	-72	-71	-63	-54	-55	-48	-49·5	-52	-57	-54·5	-50	-51
Compression set, %	67·7	54·1	70·9	68·0	62·3	61·5	80·2	69·0	68·6	46·7	68·5	—
Tear resistance, kN/m	37·8	39·2	37·8	31·9	28·7	27·1	13·9	27·6	27·7	23·7	22·5	10·7
Pico abrasion index	31	34	40	28	27	26	—	—	19	—	23	—
Durometer hardness, Type A	98	97	97	91	90	87	73	83	83	81	72	66

[a] Cured 16 h at 120°C.

$$R + NH-(CH_2)_{\overline{x}})_{\overline{y}} NHR'$$

39

N-alkylpolyamines (ref. 41)

where R = aliphatic hydrocarbon radical C_5–C_8, R' = H, R or R'', $x = 2$ or 3, $y = 2$ to 4.

Another category of amine curing agent is the *N*-alkylpolyamines (**39**). Recent examples of long chain aliphatic polyamine curing agents investigated as curing agents for epoxy resins[41] have R' = equal to a substituted aliphatic radical with C_2–C_{14} carbon atoms, which contains an oxygen atom or a hydroxy group or an aromatic group, such as a phenyl group. Examples of polyamines synthesised and evaluated are:

$$RNHCH_2CH_2NHCH_2CH_2NH_2$$

$$RNHCH_2CH_2NHCH_2CH_2NHR$$

$$RNHCH_2CH_2NHCH_2CH_2NHCH_2CH_2NH_2$$

$$RNHCH_2CH_2NHCH_2CH_2NHCH_2CH_2NHR$$

where R = 4-methyl-2-hexyl, 4-methyl-2-pentyl, 2-hexyl.

Several of the *N*-alkylpolyamines were evaluated with DGEBA (EEW ∼ 185) epoxy resin in adhesive and coating applications. Adhesion of the coatings to steel was found to be excellent.

5.4. AMINE–EPOXY AND THIOALCOHOL–EPOXY ADDUCTS AS CURING AGENTS

The high reactivity and volatility of amines can be reduced by reaction of excess amine with an epoxy resin. In this method, the epoxy **40** reacts completely, but the product **41** contains residual amino groups which can further react with additional epoxy resin to form a rigid or flexible thermosetting polymer.

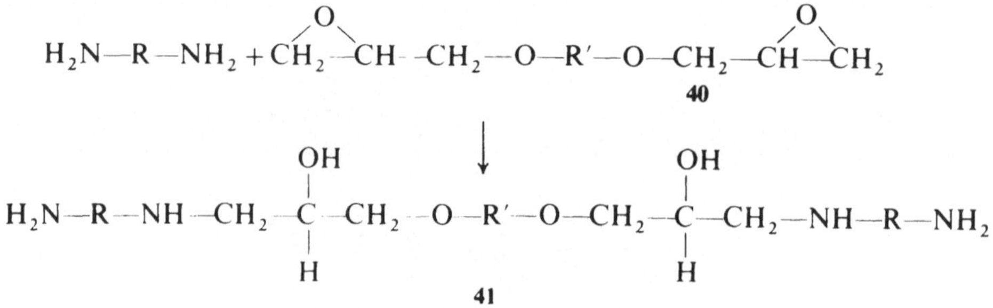

$$H_2N-R-NH_2 + \overset{O}{\overset{\diagup\diagdown}{CH_2-CH}}-CH_2-O-R'-O-CH_2-\overset{O}{\overset{\diagup\diagdown}{CH-CH_2}}$$

40

$$\downarrow$$

$$H_2N-R-NH-CH_2-\underset{H}{\overset{OH}{\underset{|}{\overset{|}{C}}}}-CH_2-O-R'-O-CH_2-\underset{H}{\overset{OH}{\underset{|}{\overset{|}{C}}}}-CH_2-NH-R-NH_2$$

41

Amine–epoxy adducts

Aliphatic, cycloaliphatic, aromatic and alkylaromatic diamines are the types of amines that have been used in preparation of adducts. Epoxies undergoing this type of adduct reaction include the standard DGEBA, epoxy phenol novalacs, epoxy cresol novalacs, hydrogenated DGEBA, and combinations of these epoxies. The synthesis of amine and polyamine adduct products and their utilisation as curing agents are described in several patents.

DiBenedetto and Gannon[42,43] describe the preparation of amine adducts **42** from several aliphatic amines, such as diethylenetriamine (DETA), 1,2-diaminocyclohexane, bis(p-aminocyclohexyl) methane, m-

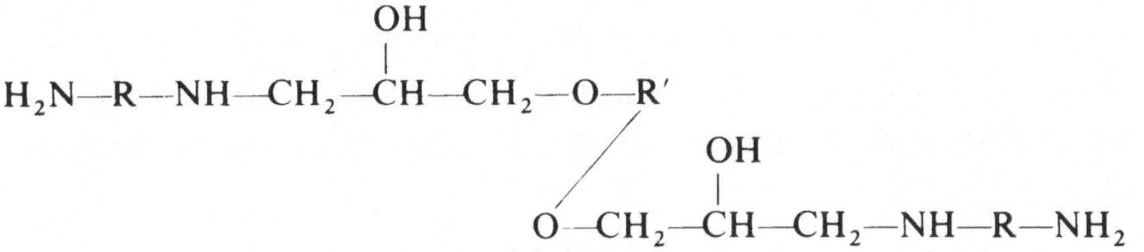

$$H_2N-R-NH-CH_2-\overset{\displaystyle OH}{\overset{\displaystyle |}{CH}}-CH_2-O-R'$$

$$O-CH_2-\overset{\displaystyle OH}{\overset{\displaystyle |}{CH}}-CH_2-NH-R-NH_2$$

42

Amine–epoxy adducts (refs. 25, 26)

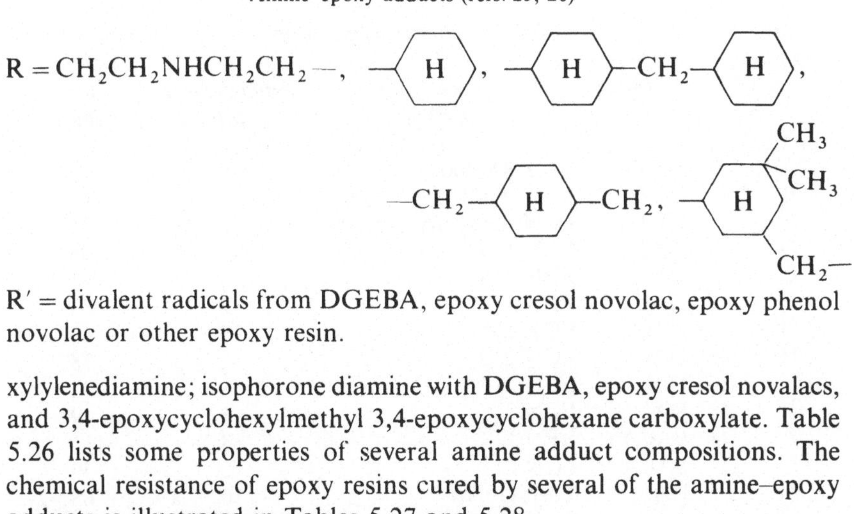

$R = CH_2CH_2NHCH_2CH_2$ —, —⟨H⟩, —⟨H⟩—CH$_2$—⟨H⟩,

—CH$_2$—⟨H⟩—CH$_2$, —⟨H⟩$\overset{\displaystyle CH_3}{\underset{\displaystyle CH_2-}{\overset{\displaystyle CH_3}{\big|}}}$

R′ = divalent radicals from DGEBA, epoxy cresol novolac, epoxy phenol novolac or other epoxy resin.

xylylenediamine; isophorone diamine with DGEBA, epoxy cresol novolacs, and 3,4-epoxycyclohexylmethyl 3,4-epoxycyclohexane carboxylate. Table 5.26 lists some properties of several amine adduct compositions. The chemical resistance of epoxy resins cured by several of the amine–epoxy adducts is illustrated in Tables 5.27 and 5.28.

Amine epoxy adducts (**43**) have been synthesised from 4,4′-diamino-diphenylsulphone and DGEBA containing Bisphenol A.[44] The compositions and hardeners prepared are illustrated in Tables 5.29 and 5.30. The four amine–epoxy adducts were evaluated as curing agents with the epoxy 4,4′-(N,N′-tetraglycidyl) methylenedianiline. The compositions and cure cycles are noted in Table 5.31. Some physical and mechanical properties are listed in Table 5.32. Impact and tensile elongation of the pure

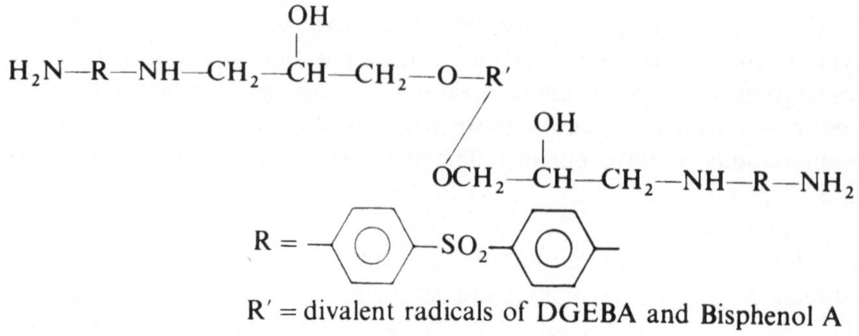

$$
\text{H}_2\text{N}-\text{R}-\text{NH}-\text{CH}_2-\overset{\overset{\displaystyle \text{OH}}{|}}{\text{CH}}-\text{CH}_2-\text{O}-\text{R}'
$$

$$
\text{OCH}_2-\overset{\overset{\displaystyle \text{OH}}{|}}{\text{CH}}-\text{CH}_2-\text{NH}-\text{R}-\text{NH}_2
$$

R = ⟨◯⟩—SO₂—⟨◯⟩—

R′ = divalent radicals of DGEBA and Bisphenol A

43
4,4′-Diaminodiphenylsulphone–epoxy adducts (ref. 44)

resins are listed in Table 5.33. Moisture absorption by the four resins is reported in Table 5.34, and the effects of this moisture on the resin mechanical properties are listed in Table 5.35.

TABLE 5.26
COMPOSITION AND PROPERTIES OF AMINE–EPOXY ADDUCTS[42,43]

Adduct designation	Adduct reactant composition			Equivalent weight of adduct
	Amine	Epoxy types/ EEWs	Amine/epoxy ratio	
A	DETA[a]	DGEBA/200	20	38
B	DETA	EPN, DGEBA/178,200	8	97
C	DETA	ECN, DGEBA/215,200	8	105
D	DAC[b]	EPN/178	15	55
E	DAC	EPN/215	10	71
F	DAC	EPN/178	8	83
G	DAC	ECN/215	15	60
H	DAC	ECN/215	10	78
I	DAC	ECN/215	8	93
J	DAC	ECN/215	6	119
K	BACA[c]	EPN/178	8	111
L	BACA	ECN/215	8	149
M	DAC	TGPAP[d]/109	10	65
N	DAC	TGPAP/109	10	46
O	DAC	TGPAP/109	7	56
P	DAC	TGPAP/109	6	61

[a] Diethylenetriamine.
[b] 1,2-diaminocyclohexane.
[c] bis(p-Amino-cyclohexyl)methane.
[d] Triglycidyl p-aminophenol.

TABLE 5.27
CHEMICAL RESISTANCE OF AMINE–EPOXY-ADDUCT-CURED DGEBA
EPOXY (ROOM TEMPERATURE CURE)[42,43]

	DETA control	Hardener adduct[c]		
		A	C	M
Composition				
Parts per hundred parts of epoxy[b]	11	20	56	35
Days to failure				
HCl (36%)	89	>14	>14	21
Acetic acid (10%)	63	>14	>14	6
NH$_4$OH (30%)	138	>14	>14	>180
Cellosolve	<3	<3	>14	>180
Acetone	<3	<3	<3	>180
MEK	<3	<3	>14	>180
Butyl acetate	>180	5	>14	>180
Methylene chloride	<3	<3	<3	<3
Trichloroethylene	>180	<3	>14	87
Ethanol (95%)	<3	<3	>14	>180
Methanol	<3	<3	10	>180

[a] Cure conditions 23 °C/10 days before immersion.
[b] DGEBA (EEW ~ 190) was used throughout as the resin component.
[c] See Table 5.26 for composition.

Dante[45] showed that amine–epoxy adducts (44) can be stabilised by post-treatment with sufficient sodium or potassium hydroxide to remove chlorine impurities. The chlorine impurities are present initially in the epoxy resin used to synthesise the adduct; they lead to an increase in the adduct viscosity on storage at room temperature.

Dante[46] further showed that amine–epoxy adducts could be prepared

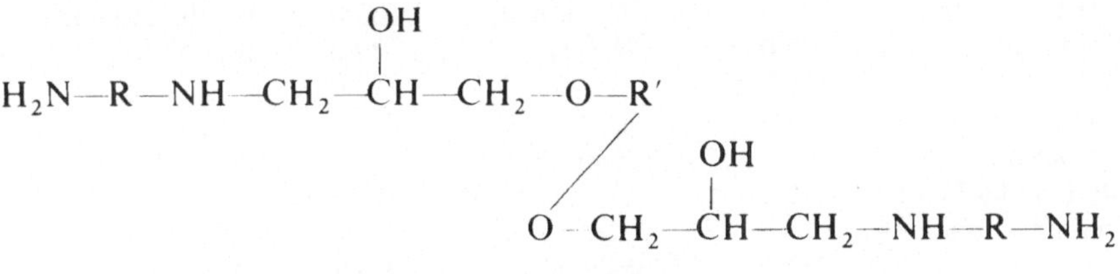

$$H_2N-R-NH-CH_2-\overset{\overset{\displaystyle OH}{|}}{CH}-CH_2-O-R'$$

$$O-CH_2-\overset{\overset{\displaystyle OH}{|}}{CH}-CH_2-NH-R-NH_2$$

R = –CH$_2$–CH$_2$NH–CH$_2$CH$_2$–
R' = divalent radical of hydrogenated DGEBA

44
DETA-hydrogenated DGEBA adducts (ref. 45)

TABLE 5.28

CHEMICAL RESISTANCE OF AMINE–EPOXY-ADDUCT-
CURED DGEBA[a] EPOXY (65 °C/10 h cure)[b 42,43]

	DETA control	M^c
Hardener adduct, phr[d]	11	35
Days to failure		
HCl (36 %)	90	28
Acetic acid (10 %)	150	53
NH_4OH (30 %)	>180	>180
Cellosolve	60	>180
Acetone	7	>180
MEK	<3	65
Butylacetate	>180	>180
Methylene chloride	<3	<3
Trichloroethylene	>180	163
Ethanol (95 %)	>180	>180
Methanol	90	>180

[a] DGEBA epoxy (EEW 190).
[b] Cure conditions 65 °C/10 h before immersion.
[c] See Table 5.26 for composition.
[d] Parts per hundred of DGEBA.

TABLE 5.29

COMPOSITION OF AROMATIC AMINE–EPOXY ADDUCTS[44]

	Hardener No. 1 (parts)		Hardener No. 2 (parts)	
DGEBA (EEW ~ 190)	100·0	0·530 equivalent	100·0	0·530 equivalent
Bisphenol A	29·0	0·254 OH	29·0	0·254 OH
50 % NaOH	0·008		0·008	
After complete reaction		0·276 epoxy		0·276 epoxy
Methylethylketone	86·0		86·0	
Diaminodiphenyl-sulphone	160	2·581 NH	129·0	2·081 NH
Theoretical yield	289	2·305 NH	258·0	1·805 NH
Calculated equivalent weight	125		143	

TABLE 5.30

COMPOSITION OF AROMATIC AMINE–EPOXY ADDUCTS[44]

	Hardener No. 3 (parts)		Hardener No. 4 (parts)	
DGEBA				
(EEW ~ 190)	100·0	0·530 epoxy	100·0	0·530 epoxy
Bisphenol A	51·8	0·454 OH	51·8	0·454 OH
50 % NaOH	0·008		0·008	
After complete reaction		0·076		0·076
Methylisobutyl-ketone	50		50	
Methylethylketone	51·2		51·2	
Diaminodiphenyl-sulfone	159·8	2·577 NH	126·5	2·041 NH
Theoretical yield	311·6	2·501 NH	278·3	1·964 NH
Calculated equivalent weight	125		142	

by reacting a DGEBA (EEW ~ 180) or a hydrogenated DGEBA (EEW ~ 243) with a primary monoamine such as ethanolamine, n-butylamine, or aniline and then reacting the resulting condensate with a poly functional amine such as polypropyleneoxide diamine, diethylene triamine. The amine–epoxy adducts were evaluated as curing agents in a water-borne coating composition containing hydrogenated DGEBA, 2-ethoxyethanol, acetic acid and water. The coatings were cured at room temperature for 7 days, and then evaluated with regard to impact,

TABLE 5.31

COMPOSITION AND CURE CYCLE[a] OF AMINE–EPOXY-ADDUCT-CURED EPOXY (PURE RESIN SYSTEMS)[44]

Component (pbw)	System				
	Control	1	2	3	4
MY-270	100	100	100	100	100
Control (4,4'-DDs)	44				
Hardener No. 1		85			
2			100		
3				85	
4					100

[a] Resin cure: 80 °C/2 h + 100 °C/1 h + 150 °C/4 h + 200 °C/7 h.

TABLE 5.32

PHYSICAL, THERMAL AND MECHANICAL PROPERTIES OF AMINE–EPOXY-ADDUCT-CURED EPOXY[44]

Property	System				
	Control 4,4'-DDS	1	2	3	4
Heat deflection temperature, $°C^a$	238	201	204	200	199
Glass transition temperature, $°C^b$	177	192	180	174	175
Room temperature					
Tensile strength, MPa	58·9	63·3	48·0	83·8	79·3
Tensile modulus, GPa	3·7	3·5	3·4	3·7	3·5
Tensile elongation, %	1·8	2·0	1·5	2·7	2·6
Flexural strength, MPa	91·7	131	116	139	135
Flexural modulus, GPA	3·4	3·4	3·2	3·7	3·5
Compressive strength, MPa	270	202	210	230	248
Compressive yield, MPa	201	182	176	167	163
Compressive modulus, GPa	1·95	1·89	1·98	2·17	2·16
150°C					
Tensile strength, MPa	44·5	55	61	42	42
Tensile modulus, GPa	2·6	2·4	2·5	1·3	1·2
Tensile elongation, %	1·9	3·9	3·6	5·9	6·0
Flexural strength, MPa	85	92	92	62	62
Flexural modulus, GPa	2·7	2·4	2·6	1·3	1·4

^{a} Determined by ASTM D-648.
^{b} Determined by the thermomechanical analysis (TMA) method in the penetration mode at a heat-uptake of 10°C/min and a 5-g loading.

TABLE 5.33

IMPACT AND TENSILE ELONGATION OF AMINE–EPOXY-ADDUCT-CURED EPOXY RESIN[44]

Property	System				
	Control 4,4'-DDS	1	2	3	4
Room temperature					
Modified Gardner, kg-cm	8·0	8·0	6·9	13·8	9·2
Unnotched Charpy, kg-cm/cm²	11·9	17·2	19·7	17·0	39·9
Tensile elongation, %	1·8	2·0	1·5	2·7	2·6
150°C					
Tensile elongation, %	1·9	3·9	3·6	5·9	6·0

TABLE 5.34

HUMIDITY AGEING OF RESIN AGED UNTIL EQUILIBRIUM
(25 DAYS) AT 80 °C AND 100 % RELATIVE HUMIDITY[44]

System	% Weight gain
Control, 4,4'-DDS	4·29
1	3·28
2	3·17
3	3·00
4	2·80

TABLE 5.35

EFFECTS OF HUMIDITY EXPOSURE AT ROOM TEMPERATURE ON AMINE–EPOXY-CURED
EPOXY RESIN MECHANICAL PROPERTIES[44]

	Control 4,4'-DDS	1	2	3	4
% Weight gain	4·29	3·28	3·17	3·00	2·80
HDT, °C	141	148	133	117	110
Flexural strength, MPa	94·3	95·8	114	107	133
Flexural modulus, GPa	3·4	3·3	3·3	3·4	3·4
Tensile strength, MPa	26·7	37·2	31·7	44·8	60
Tensile modulus, GPa	3·0	3·1	3·0	3·3	3·2
Elongation, %	0·9	1·2	1·1	1·4	2·1

flexibility, pencil hardness, and resistance to methylisobutylketone and to
water. Several formulations exhibited excellent properties. The
amine–epoxy adduct of a polyamide (amine value 370–400) and a
hydrogenated DGEBA (EEW ~ 243) was evaluated[47] as a room-
temperature-curable epoxy coating material in a composition consisting of
the adducts, titanium dioxide, cellulose acetate butyrate, and an organic
solvent (glycol ether, alcohol, ketone, or toluene).

Amine adducts **45** and **46** of *o*-hydroxy benzoic acid or dihydroxy

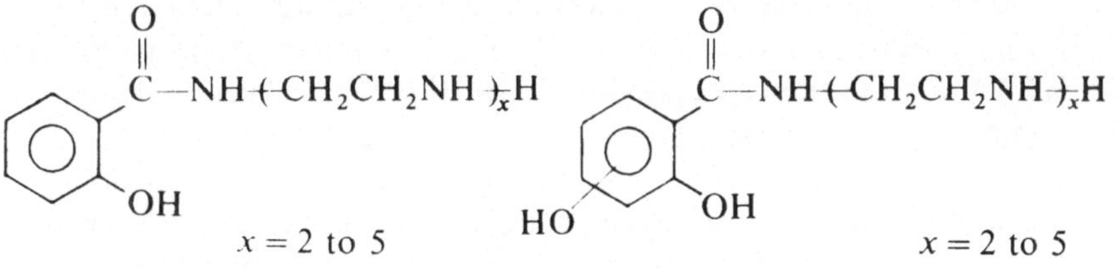

$x = 2$ to 5

45

Phenolic amidoamines

46

$x = 2$ to 5

TABLE 5.36
EPOXY RESINS–MERCAPTAN ADDUCTS[49]

	Adduct				
	A	B	C	D	E
Dimercaptan					
2,2-dimercapto diethylether	48·42	53·10	—	—	—
1,3-dimercaptopropanol-2	—	—	56·28	56·9	—
bis(2-mercaptoethyl sulphide)	—	—	—	—	49·87
Epoxy resin					
DGEBA (EEW ∼ 190)	40·95	44·91	24·43	24·7	42·18
p-tertiary butylphenol					
glycidylether (EEW ∼ 225)	—	—	13·16	13·3	—
Catalyst–accelerator					
Dabco	0·5	—	1·08	—	0·45
DMP-30	10·13	1·99	5·05	5·1	7·50
Physical properties					
Viscosity, cP, 21 °C	2 300	2 400	3 200	2 500	60 000
Density, g/cm³, 25 °C	1·21	1·20	1·20	1·22	1·23
Equivalent weight	163	153	97	109	—

benzoic acid with diethylenetriamine, triethylenetetramine, tetraethylene-pentamine and pentaethylenehexamine have been shown to be effective room temperature curing agents.[48]

For example, a film prepared from DGEBA (EEW ∼ 185) and stoichiometric quantities of the salicyclic acid/diethylenetriamine (DETA) or triethylenetetramine (TETA) adducts exhibited a film hardness B after 24 h, and film hardness 2H after 72 h cure at room temperature. Without the amidoamine adduct, the pencil hardness was approximately 5B (soft).

Mercaptan-epoxy adducts (**48**), prepared by reaction of a dimer captan, such as 2,2'-dimercaptodiethyl ether (**47**), with a diepoxide (**40**), have been shown to cure epoxies rapidly at temperatures down to −20 °C.[49]

The polymercaptan–epoxy adducts of 2,2'-dimercaptodiethyl ether (DMDE), 1,2-dimercaptopropane (DMCP), 1,3-dimercaptopropanol-2 (DMHP), and bis-(2-mercaptoethylsulphide) BMES have been evaluated. The compositions of resins evaluated with the polymercaptan–epoxy adducts are listed in Table 5.36. Table 5.37 is a summary of some physical, mechanical and chemical resistance properties of the mercaptan–epoxy-adduct-cured epoxy resins, compared with those of two typical amine-cured epoxies. The resin systems were cured at 25 °C for 1 week. The amine-

TABLE 5.37

PHYSICAL STRENGTH AND CHEMICAL RESISTANCE PROPERTIES OF MERCAPTAN-ADDUCT-CURED EPOXY RESIN SYSTEMS[a][49]

	Composition (parts by weight)				
	1	2	3	4	5
Resin portion					
Epoxy novalac resin (EEW 192)	60	—	60	—	—
DGEBA (EEW 192)	—	60	—	60	100
PGETA (EEW 146)	40	40	40	40	—
Converter portion					
Mercaptan Adduct A[b]	92	91	—	—	—
Mercaptan Adduct B[b]	—	—	87	—	—
Modified Aliphatic amine[c]	—	—	—	19	—
Modified Aromatic Amine[d]	—	—	—	—	70
Properties of cured resin[e]					
Tensile strength, MPa	59·0	41·3	4·6	72·2	59·8
Tensile elongation, %	3·3	27	82	3·9	8·5
Flexural strength, MPa	84·7	59·3	1·37	126	104
Flexural modulus, GPa	2·89	1·44	—	4·0	2·9
Hardness (Shore D)	81	77	50	85	83
Izod (kg-cm/cm notch)	8·0	13·4	42·5	4·09	2·56
Water absorption, 24 h, %	0·69	1·20	0·44	0·21	0·12
5% Acetic acid, absorption, 24 h, %	5	2·63	0·57	1·90	0·14
Xylene (isopropanol 50/50), absorption, 24 h, %	0·23	0·60	0·89	0·13	0·54

[a] Castings were not de-aired.
[b] See Table 5.36 for adduct A & B composition.
[c] Epi-cure 874.
[d] Epi-cure 8491.
[e] 1/8 in (3·2 mm) thick castings cured 25 °C/1 week.

cured epoxies exhibited higher ultimate strength, but were not as tough as the mercaptan–epoxy-adduct-cured epoxies, as indicated by the Izod impact values.

Another type of room temperature curing agent developed for films and coatings is the poly(thioxyalkanoic acids).[50,51] These are adducts of mercaptoacetic acid or 3-mercaptopropionic acid and a butadiene–styrene liquid polymer. Table 5.38 lists a comparison of the chemical resistance of films prepared from a poly(thioxyalkanoic acid) adduct, BMAA36,

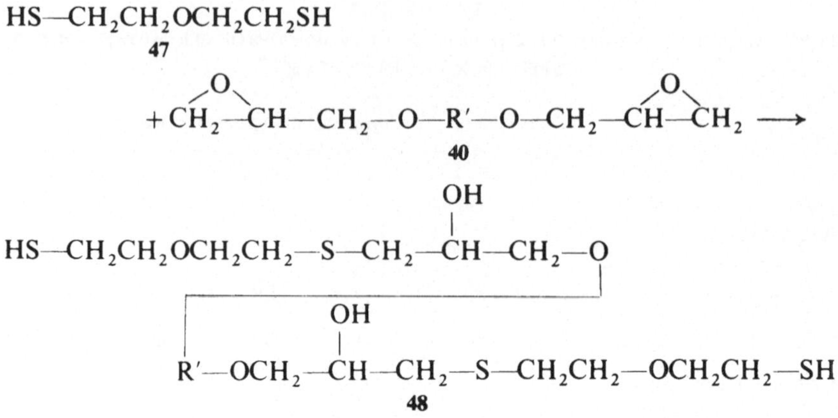

$$HS-CH_2CH_2OCH_2CH_2SH$$
47

$$+ CH_2\overset{O}{\overset{\diagup\diagdown}{-}}CH-CH_2-O-R'-O-CH_2-CH\overset{O}{\overset{\diagup\diagdown}{-}}CH_2 \longrightarrow$$
40

$$\overset{OH}{\overset{|}{HS-CH_2CH_2OCH_2CH_2-S-CH_2-CH-CH_2-O}}$$

$$\overset{OH}{\overset{|}{R'-OCH_2-CH-CH_2-S-CH_2CH_2-OCH_2CH_2-SH}}$$
48

Mercaptan–epoxy adduct

R′ = divalent radical of DGEBA or other eposy resins

TABLE 5.38

CHEMICAL RESISTANCE OF EPOXY RESIN CURED BY MERCAPTAN–BUTADIENE–STYRENE ADDUCT[50,51]

	Temperature (°C)	Hours for breakdown[a]	
		DGEBA (EEW 450–550) Versamide 115	DGEBA (EEW 190) Mercaptan-adduct[b]
H_2O	25	72 +	72 +
10% NaH	25	72 +	6·25
50% H_2SO_4	25	72 + but whitened immediately	72 +
50% H_2SO_4	100	0·5	24 +
Glacial acetic acid	25	0·0	72 +
Dimethylformamide	25	0·0	0·0
Acetone	25	24	72 +
MEK	25	2·0	72 +
MIBK	25	72 +	72 +
Toluene	25	24	72 +
Mineral spirits	25	72 +	72 +
Concentrated NH₄OH	25	72 +	3·5

[a] + = no breakdown in 72 h.
[b] Mercaptoacetic and butadiene–styrene adduct, BMMA36, molecular weight ~ 14 000.

(molecular weight ~ 14 600, 56 carboxyl groups per molecule 36 wt %
mercaptoacetic acid), and DGEBA and with that of a polyamide/epoxy
(65/35) mixture. The superior chemical resistance of the poly(thioxyalkanoic)-cured epoxy film over the polyamide-cured epoxy film is noted.
The disadvantage of the sulphur containing film is the bad odour.

5.5. AROMATIC AMINES AND MODIFICATIONS

Aromatic amines are less reactive than the aliphatic amines and for this
reason are used as latent curing agents with epoxy resins. 4,4'-
Diaminodiphenylmethane, 4,4'-diaminodiphenylsulphone, m- and p-
phenylenediamine, 4,4'-oxydianiline and eutectic mixtures of aromatic
amines are typical of commercially available aromatic amines used as
curing agents for epoxy resins. The lower reactivity of the aromatic amines,
relative to aliphatic amines requires elevated temperatures for complete
cure. The more complete cure and the rigid structure of the aromatic rings
of the aromatic amines leads to cured epoxy resins with much higher glass
transition temperatures (T_g values) than can be obtained with aliphatic
amines. Therefore, aromatic amines are the choice of curing agents for
epoxy resins, used as the matrices for advanced composites requiring
elevated temperature capabilities.

The polyamines[52] derived from condensation of N-methylaniline with
formaldehyde in molar ratios of 1·3, 1·38, 1·45 and 1·6 were investigated for

49

N-methylaniline–formaldehyde adduct (ref. 52)

steric effects in epoxy resin cure reactions (Table 5.39). Polyamines **49**
derived from the 1·3 and 1·6 molar ratios of N-methylaniline to
formaldehyde were evaluated as curing agents with the epoxy 2,2-
bis(epoxy propoxyphenyl) propane. The results are shown in Table 5.40.
The properties of resins cured with these aromatic amine epoxies do not
equal those of resins cured with 4,4'-MDA cured DGEBA. Other physical

TABLE 5.39
N-METHYLANILINE FORMALDEHYDE CONDENSATION PRODUCTS[52]

Mole ratio N-methylaniline/H_2CO	Product designation	% Yield	Melting point (°C)
1·30	H	88	53–69
1·38	G	85	40–45
1·38[a]	J	84	~40–45
1·59[a]	A	91	Amber liquid

[a] From uninhibited formaldehyde.

test data on the 1·60 mole ratio of N-methylaniline/CH_2O product included the following:

Elongation, 10%
Compressive yield strength, 38·4 MPa
Modulus of elasticity, 0·8 GPa
Thermal conductivity, $1·45 \times 10^{-3}$ J sec^{-1} m^{-2}/°C/m
$\qquad\qquad (3·46 \times 10^{-6}$ cal/sec/cm^2/°C/cm)
Coefficient of thermal expansion
$\qquad\qquad$ at 25–100°C, $0·639 \times 10^{-4}$ in/in (mm/mm) °C
$\qquad\qquad$ at −75–25°C, $0·502 \times 10^{-4}$ in/in (mm/mm) °C
Weight loss at 2×10^{-6} torr (>178 h total), 73 h at 150°C, 0·11%

Investigations of aromatic aminoesters **50**, **51** and **52** have revealed that these materials undergo partial reaction with epoxies at 80°C to a 'B' staged intermediate.[53] These 'B' staged resins were shown to be stable for several months at room temperature, and to be soluble in esters and ketones after

TABLE 5.40
SOME MECHANICAL PROPERTIES OF POLYAMINE CURED EPOXY[52]

Amine/epoxy equivalent[a] ratio	Curing agent derived from methylaniline formaldehyde reaction ratio of	Compression strength (MPa)	Tensile strength (MPa)	Elastic modulus (GPa)
1:1	1·30	106·2	97·2	0·76
1:1	1·6	100	84·1	2·55

[a] DGEBA (EEW ~ 180).

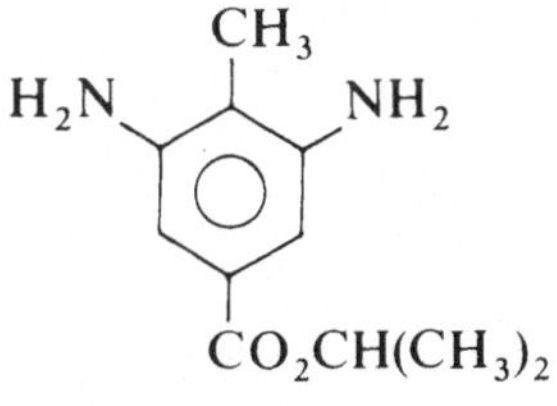

50

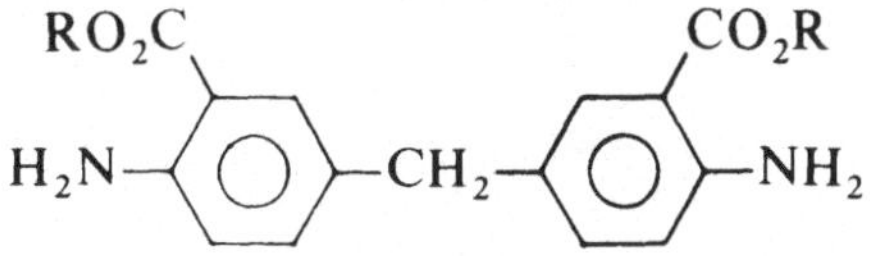

51

a R = CH$_3$—
b R = CH$_3$CH$_2$—

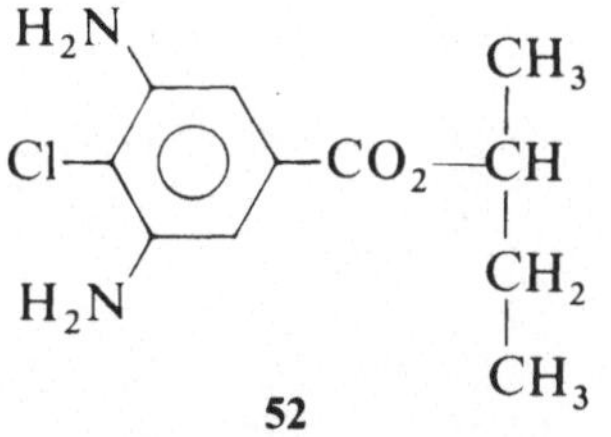

52

Aromatic aminoesters (ref. 53)

this time period.[53] The amino esters have been tested as curing agents with epoxies of several equivalent weights. Conditions of cure and the properties of aromatic-amine-ester-cured epoxies are listed in Table 5.41. The excellent properties are comparable with those of aromatic-amine-cured epoxies, such as 4,4'-diaminodiphenylsulphone and 4,4'-diaminodiphenylmethane.

Recent attempts to increase the toughness of epoxy resins have been made by increasing the distance between tl amine end groups, and by incorporating a hexafluoroisopropylidene t veen aromatic rings of the diamine. A series of bisimide amines (BIA 53 having these structural

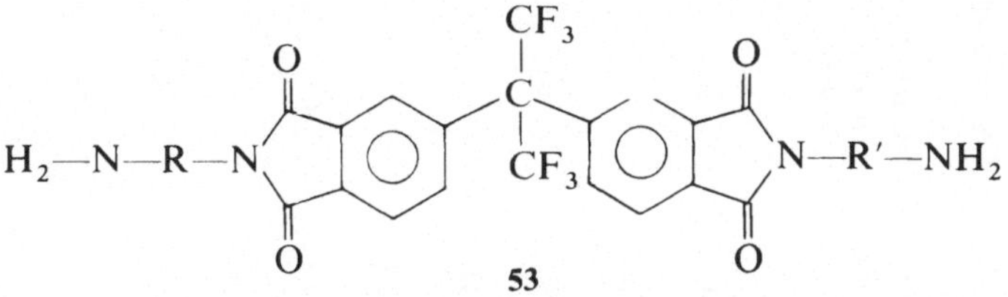

53

Bisimide amines (BIAs) (refs. 54–58)

features were synthesised by Serafini, Delvigs and Vanucci,[54,55] Scola and Pater[56,57] and Scola.[58] Other objectives of this research effort were to develop moisture resistant, high char yield epoxy resins. The general structure of the bisimide amines synthesised and evaluated is shown in **53**.

TABLE 5.41

CURE CONDITIONS AND PROPERTIES OF AROMATIC AMINE ESTERS AS CURING AGENTS FOR EPOXY RESINS[53]

Example	Diamine curing agent[a]	Curing agent (phr)	Epoxy[a] (EEW)	Cure cycle	Flexural strength (MPa)	Properties modulus (MPa)	Tensile strength (MPa)	Ball pressure hardness (MPa)
1	50	28	200	120°C/15h +160°C/5h	122	0·26	—	172
3	50	45	125	100°C/15h +160°C/6h	—	—	78·9	—
4	50	46	125	120°C/15h +170°C/30 min	63·7	3·9	—	—
6	51b	41	200	120°C/15h +160°C/5h	145	3·1	—	—
8	51a	39	200	120°C/15h +160°C/5h	89·7	5·0	49·2	277
9	51a	65	125	140°C/8h +170°C/7h	159	3·3	94·7	180
10	51a	62	125	120°C/15h +160°C/5h	178	3·7	96·8	209
13	52	30	200	120°C/15h +160°C/5h	138	3·3	86·2	182
14	52	49	125	100°C/15h +160°C/6h	132	3·1	84·8	173

[a] Epoxy resins with epoxy equivalent weights 200 and 125 were used in this study.

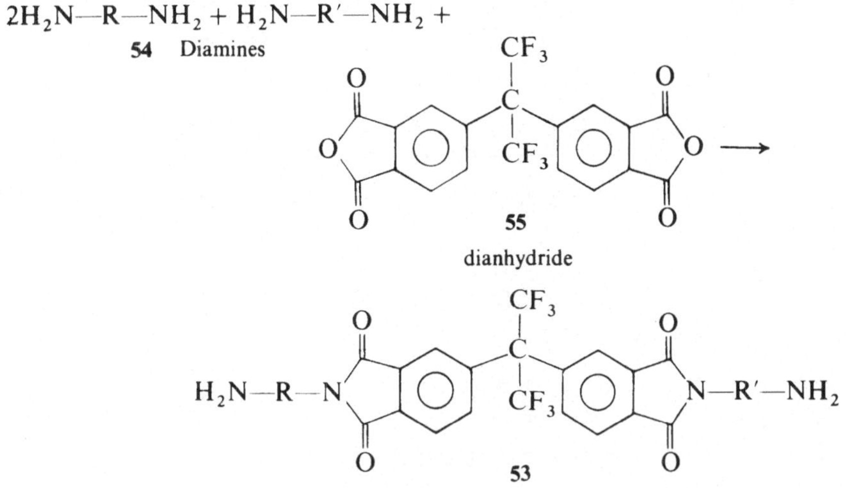

$$2H_2N\text{---}R\text{---}NH_2 + H_2N\text{---}R'\text{---}NH_2 +$$

54   Diamines

55

dianhydride

$\longrightarrow$

$H_2N\text{---}R\text{---}N \qquad\qquad N\text{---}R'\text{---}NH_2$

53

R	R'	BIA
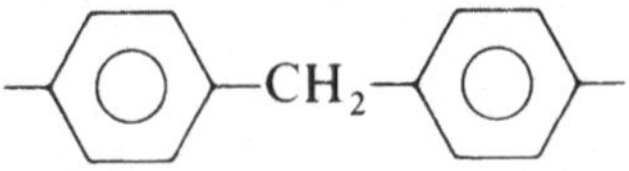	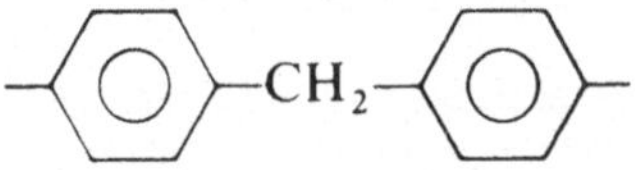	6F—4,4'—MDA
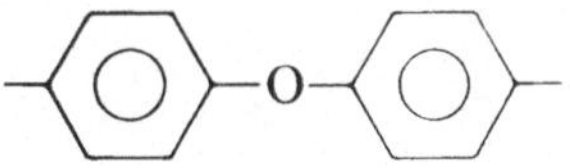	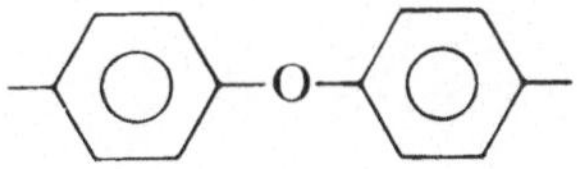	6F—4,4'—ODA
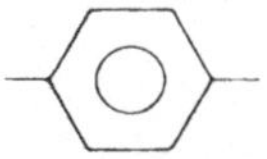	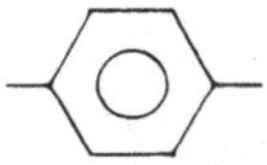	6F—PDA
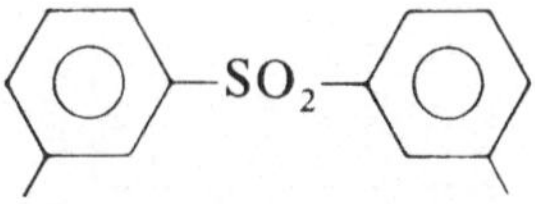	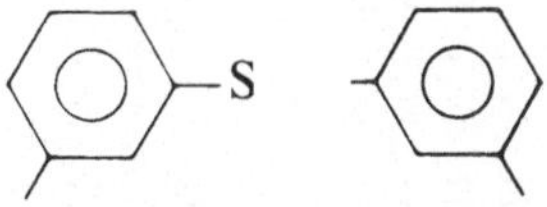	6F—3,3'—DDS
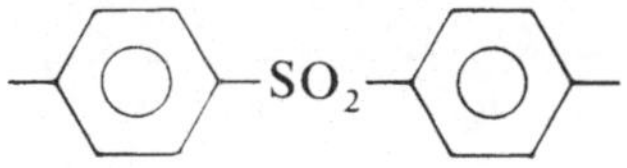	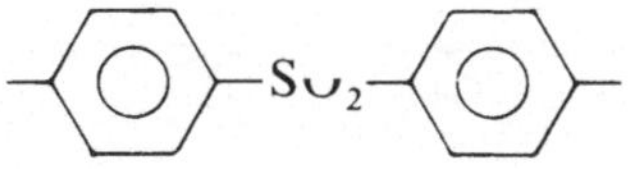	6F—4,4'—DDS
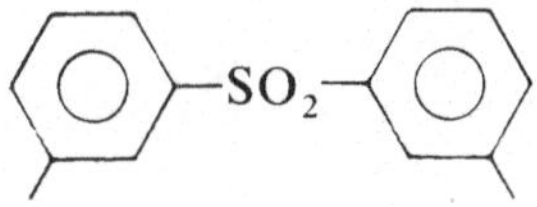	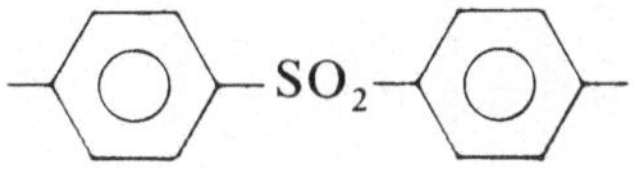	6F—3,3'—DDS—4,4'—DDS
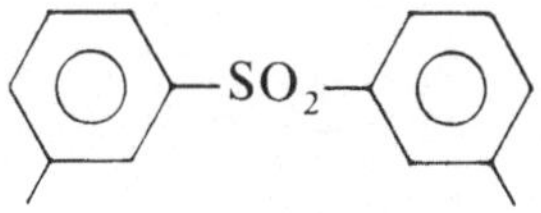	$-(CH_2)_{12}-$	6F—3,3'—DDS—1,12—DDA

Bisimide amines (BIAs) (ref. 57, 58)

TABLE 5.42

YIELDS AND ELEMENTAL ANALYSIS OF BISIMIDE AMINES[57,58]

Bisimide amine (BIA)	Yield (%)	Empirical formula	Found					Calculated				
			C	H	N	F	S	C	H	N	F	S
6F-4,4'-MDA (oligomeric)	80	$C_{45}H_{30}N_4F_6O_4$	64·50	4·00	7·09	14·51	—	67·16	3·73	6·97	14·18	—
6F-4,4'-ODA (oligomeric)	90	$C_{43}H_{26}N_4F_6O_6$	64·27	3·86	6·69	13·80	—	63·86	3·22	6·93	14·11	—
6F-PDA (oligomeric)	97	$C_{31}H_{18}N_4F_6O_4$	59·79	3·26	8·57	17·99	—	59·62	2·88	8·97	18·27	—
6F-3,3'-DDS (oligomeric)	93	$C_{43}H_{26}N_4F_6S_2O_8$	56·95	2·96	6·15	12·54	6·89	57·08	2·88	6·20	12·61	7·08
6F-4,4'-DDS (oligomeric)	99	$C_{43}H_{26}N_4F_6S_2O_8$	55·87	3·42	6·19	12·46	5·39	57·08	2·88	6·20	12·61	7·08
6F-3,3'-DDS-4,4'-DDS (oligomeric)	93	$C_{43}H_{26}N_4F_6S_2O_8$	56·87	3·02	6·11	12·48	6·98	57·08	2·88	6·20	12·61	7·08
6F-3,3'-DDS-1,12-DDA NaOH treated (polymeric) (Method A)	24	$C_{43}H_{42}N_4F_6SO_6$ as monomer	58·33	3·62	4·45	17·65	2·49	60·28	4·94	6·54	13·30	3·37
		$C_{62}H_{49}N_4F_{12}SO_{16}$ as polymer						58·56	3·48	4·94	18·03	2·53
6F-3,3'-DDS-1,12-DDA NaOH treated (polymeric) (Method B)	29	$C_{43}H_{42}N_4F_6SO_6$ as monomer	57·52	3·73	4·33	17·29	2·48	60·28	4·44	6·54	13·30	3·37
		$C_{62}H_{49}N_4F_{12}SO_{16}$ as polymer						58·56	3·48	4·94	18·03	2·53
6F-3,3'-DDS-1,12-DDA monomer & polymer mixture (Method D)	95	$C_{43}H_{42}N_4F_6SO_6$ as monomer	55·60	3·57	3·73	18·19	1·89	60·28	4·94	6·54	13·30	3·37
		$C_{62}H_{49}N_4F_{12}SO_{16}$ as polymer						58·56	3·48	4·44	18·03	2·53
6F-3,3'-DDS-1,12-DDA monomeric (Method E)	84	$C_{43}H_{42}N_4F_6SO_6$ as monomer	60·34	4·90	6·55	13·60	3·01	60·28	4·94	6·54	13·30	3·37
		$C_{62}H_{49}N_4F_{12}SO_{16}$ as polymer						58·56	3·48	4·44	18·03	2·53

The introduction of the hexafluoroisopropylidene group between the aromatic rings has led to the development of thermally stable, tough polyimides,[59,60] and in increased moisture resistance and strain-to-failure epoxies[56,57,58] without sacrifice in the use-temperature of the epoxy.

Several novel bisimide amines (BIAs) were synthesised, characterised and then evaluated as curing agents with a standard state-of-the-art epoxy, 4,4'-(N,N'tetraglycidyl)-methylenedianiline. The BIAs were synthesised by reaction of 4,4'-hexafluoroisopropylidene (biphthalic acid anhydride) (6F-anhydride) with aromatic or aliphatic diamines in dimethylformamide at reflux temperature in yields ranging from 24–99 %. The synthetic route used to prepare the bisimide amines (BIAs) is represented by the general equation shown below. The BIA synthesised, and the designation for each BIA, are also indicated. The yields and elemental analysis of the BIAs synthesised are listed in Table 5.42. The monomeric, oligomeric and polymeric nature of the BIAs was determined by elemental analysis and by high pressure liquid chromatography (HPLC).

The general composition of each resin system evaluated, and the bisimides epoxy resin (IME) designation, are listed in Table 5.43. 'Stoichiometric' quantities of epoxy and bisimide amine components were used. The equivalent weights were calculated assuming that all bisimide amines are monomeric. Homogeneous powders of the IME resins were prepared by dissolving the resin components in tetrahydrofuran (100 ml), followed by solvent removal to give a solventless homogeneous one-part powder mixture. Vacuum heat treatment at 60–80 °C for 8 h was required to get complete removal of the solvent.

For the resin systems under study, only the control resin was capable of being fabricated into neat resin specimens by pouring the liquid resin mix at 150 °C into preheated (150 °C) moulds. All the other resin systems are solid, stable 'one-part' mixtures at room temperature, and were compression moulded at 150 °C/1 h + 177 °C/2 h + 204 °C/24 h (6·9 MPa).

The density, the coefficient of thermal expansion, and the shrinkage due to cure of each IME resin were determined. The control resin C-2 has a density of 1·24 g/cm^3; the IME densities are higher because of the presence of the hexafluoroisopropylidene group, and ranged from 1·29 to 1·35 g/cm^3. The coefficient of thermal expansion of the control resin C-2 was 8·6 × 10^{-5} cm/cm/°C, while for the IME resins it ranged from 4·5 to 5·3 × 10^{-5} cm/cm/°C. The shrinkage of the control resin was 0·80 percent, while for the IME resins, it ranged from 0·44 to 0·80 percent, depending on the system. Moisture absorption properties are listed in Table 5.44. A comparison of the moisture absorbed by the IME resins for the 95 percent

TABLE 5.43
COMPOSITION OF MY720/IME EPOXY RESINS[57,58]

Resin No.	Epoxy/hardener(s)	Weight (g)
Control C-1	MY720/4,4'-DDS	
Control C-2	MY720/3,3'-DDS	120/60
IME-1	MY720/6F-4,4'-DDS	3·20/5·79
IME-2	MY720/6F-4,4'-MDA	3·45/5·52
IME-3	MY720/6F-4,4'-ODA	3·45/5·58
IME-4	MY720/6F-PDA	4·00/4·99
IME-5	MY720/6F-4,4'-DDS/4,4'-DDS	4·00/3·80/1·04
IME-6	MY720/6F-4,4'-MDA/4,4'-MDA	4·50/3·62/0·89
IME-7	MY720/6F-4,4'-ODA·4,4'-ODA	4·40/3·56/0·88
IME-8	MY720/6F-PDA/PDA	5·20/3·25/0·52
IME-9	MY720/6F-3,3'-DDS	32·0/57·9
IME-10	MY720/6F-3,3'-DDS/3,3'-DDS	42·5/34·3/10·4
IME-11	MY720/6F-3,3'-DDS-1,12-DDA (Method D)	65·0/111·4
IME-11-3	MY720/6F-3,3'-DDS-1,12-DDA (Method E)	32·5/41·8
IME-11-D	MY720/6F-3,3'-DDS-1,12-DDA (Method D)	32·5/144·6
IME-12	MY720/6F-3,3'-DDS-1,12-DDA/3,3'-DDS (Method D)	85·0/72·8/21·1
IME-12-3	MY720/6F-3,3'-DDS-1,12-DDA/3,3'-DDS (Method E)	42·5/27·3/11·9
IME-12-D	MY720/6F-3,3'-DDS-1,12-DDA/3,3'-DDS	42·5/29·1/11·9
IME-13	MY720/glyamine 200/6F-3,3'-DDS-4,4'-DDS	46·8/11·9/135·7
IME-14	MY720/glyamine 100/6F-3,3'-DDS-4,4'-DDS	46·8/5·6/135·7
IME-15	MY720/glyamine 200/6F-4,4'-DDS	46·8/11·8/135·7
IME-16	MY720/glyamine 100/6F-4,4'-DDS	46·8/5·6/135·7
IME-18-1	MY720/6F-3,3'-DDS-1,120DDA (Method B)	12·0/22·2

relative humidity and boiling water conditions with the moisture absorbed by the control resins, C-1 and C-2 and typical epoxy resins under the same conditions shows that several IME resins absorb considerably less moisture than these systems.

The glass transition temperatures derived from thermal mechanical analysis (TMA) thermograms are listed in Table 5.45 along with the tensile and compression properties. Most IME resins exhibited one transition in the TMA curve; this was assigned as the glass transition temperature (T_g). Resins IME-11, IME-12-3, and IME-12D exhibited two transitions (Table 5.45) suggesting that these systems may be heterogeneous, thereby behaving as two-phase systems.

Because compression moulding methods were used in fabricating resin

TABLE 5.44
MOISTURE ABSORPTION OF IME EPOXY RESINS[57,58]

Resin system	After 24 h at room temperature in water (%)	Saturation at 82°C, 95% relative humidity (wt%)	After 72h water boil (wt%)
C-1	—	3·4	—
C-2	0·20	0·82	3·20
IME-1	—	2·4	—
IME-2	—	1·6	—
IME-3	—	1·4	—
IME-4	—	1·8	—
IME-5	—	1·8	—
IME-6	—	1·8	—
IME-7	—	2·0	—
IME-8	—	2·5	—
IME-9-1	0·22	0·75	2·22
IME-10	0·24	0·48	2·61
IME-11-3	0·17	0·00	2·81
IME-11-D	0·49	0·50	—
IME-12-3	0·22	0·30	—
IME-12-D	0·61	0·95	2·52
IME-13	0·35	0·61	2·39
IME-14	0·39	0·65	3·17
IME-15	0·39	0·80	4·43
IME-16	0·54	0·90	—

specimens, miniature tensile specimens 5·08 cm long by 0·635 cm wide by ~1·0 cm thick, having a 0·32 cm reduced section in the 0·32 cm gauge length were fabricated from compression moulded rectangular specimens. A limited number of specimens were fabricated and this is reflected both in the lack of data for several IME systems, and in the low tensile strengths and strain-to-failure values of the control resin and several of the IME resins. The relatively high tensile strengths (31·7–54·2 MPa) (4600–7870 psi), Table 5·45, and strain-to-failure values (1·20–2·20%) of several IME resins (IME-1, IME-10, IME-11-3, IME-12 and IME-12-3) compared with the control resin suggest that these resin systems are tougher than the control system and state-of-the-art epoxy resins.

The results of the compression properties are listed in Table 5.45. The compression strengths of the IME resins compare favorably with those of a typical epoxy and of the control resin. The high compressive strength 232 MPA (33·6 psi) for the control resin C-2, and the low tensile strength,

TABLE 5.45
GLASS TRANSITION TEMPERATURES AND MECHANICAL PROPERTIES OF IME RESINS[57,58]

IME resin system	T_g (°C)	Tensile properties			Compression properties	
		Strength MPa (psi)	Modulus GPa (10^6 psi)	Strain-to-failure (%)	Strength MPa (ksi)	Modulus GPa (10^6 psi)
C-1	227	41·1 (5966)	4·13 (0·60)	0·99	232 (33·6)	4·41 (0·64)
C-2	222	18·7 (2710)	4·07 (0·59)	0·54		
IME-1	215	48·7 (7870)	3·76 (0·55)	1·25		
IME-2	215	—	—	—		
IME-3	220	—	—	—		
IME-4	—					
IME-5	220	44·8 (6500)	4·82 (0·65)	0·99		
IME-6	217	—	—	—		
IME-7	212	—	—	—		
IME-8	—					
IME-9	197	—	—	—	131 (19·0)	3·38 (0·49)
IME-10	200	36·0 (5230)	3·72 (0·54)	0·83	270 (39·2)	4·76 (0·69)
IME-11	97 219					
IME-11-3	82	31·7 (4600)	4·24 (0·61)	1·20	143 (20·7)	—
IME-11-D	149				82·8 (12·0)	5·45 (0·79)
IME-12	210	37·9 (5480)	1·93 (0·28)	2·20	204 (29·5)	3·38 (0·49)
IME-12-3	85 120	38·2 (5540)	2·55 (0·37)	1·22	262 (38·0)	4·55 (0·66)
IME-12-D	106 160			—	124 (17·9)	3·93 (0·57)
IME-13	217	20·0 (2900)	4·14 (0·60)	0·50	206 (29·8)	4·89 (0·71)
IME-14	212	17·2 (2500)	3·93 (0·57)	0·45	214 (31·0)	4·76 (0·69)
IME-15	180			—	206 (29·8)	4·76 (0·69)
IME-16	216	16·7 (2420)	3·72 (0·54)	0·49	210 (30·5)	4·60 (0·67)

TABLE 5.46

SHORT BEAM SHEAR STRENGTHS[a] OF CELION 6000/EPOXY RESIN COMPOSITES[b,c] IN DRY AND HUMIDITY EXPOSED[d] CONDITIONS[57]

Composite no.	Short beam shear strength, MPa (ksi)					
	Dry condition			Humidity exposed		
	RT	150°C	177°C	RT	150°C[e]	177°C[e]
Control	94·5 (13·70)	68·0 (9·86)	58·2 (8·44)	67·1 (9·73)	40·8 (5·92)	28·2 (4·09)
20-IME-1	102 (14·80)	73·1 (10·60)	64·6 (9·37)	72·8 (10·55)	40·8 (5·91)	34·5 (5·00)
32-IME-2	84·8 (12·30)	39·3 (5·71)	31·1 (4·52)	66·4 (9·63)	29·0 (4·21)	17·3 (2·51)
28-IME-3	84·1 (12·20)	46·9 (6·80)	39·9 (5·79)	70·0 (10·15)	28·4 (4·12)	22·8 (3·31)
29-IME-4	72·4 (10·50)	57·4 (8·32)	53·7 (7·79)	65·6 (9·51)	32·8 (4·76)	26·9 (3·91)
24-IME-5	111 (16·15)	72·0 (10·45)	52·6 (7·63)	93·8 (13·60)	50·1 (7·27)	30·9 (4·49)
31-IME-6	84·8 (12·20)	46·9 (6·80)	36·3 (5·26)	65·9 (9·56)	30·3 (4·39)	21·9 (3·18)
26-IME-7	77·2 (11·20)	46·9 (6·80)	41·1 (5·95)	66·3 (9·61)	33·8 (4·90)	26·2 (3·80)
27-IME-8	89·3 (12·95)	43·4 (6·30)	36·0 (5·23)	78·6 (11·40)	35·5 (5·15)	21·6 (3·13)

[a] Span-to-depth ratio, 4/1.
[b] Composites were compression moulded, for cure cycle see ref. 57.
[c] Composites were post-cured at 477 K (240 °C) for 24 h.
[d] 87% relative humidity at 355 K (82 °C) to saturation.
[e] After 20 min soak at temperature.

TABLE 5.47

TENSILE PROPERTIES OF $10°$ OFF-AXIS CELION 6000/IME COMPOSITES[58]

Composite no.	Resin formulation	Tensile properties		Intralaminar shear properties		Composite intralaminar shear strain-to-failure (%)	Shear strain-to-failure in resin (%)
		Strength MPa (ksi)	modulus GPa (10^6 psi)	Strength MPa (ksi)	Modulus GPa (10^6 psi)		
C-2-2B (control)	MY720/3,3'-DDS	316 (45·8)	84·8 (12·3)	54·0 (7·84)	7·79 (1·13)	1·08	4·65
IME-9-2B	MY720/6F-3,3'-DDS	433 (62·8)	80·0 (11·6)	73·8 (10·7)	7·52 (1·09)	1·93	10·6
IME-10-2B	MY720/6F-3,3'-DDS	453 (65·8)	88·3 (12·3)	77·2 (11·2)	7·24 (1·05)	1·79	9·07
IME-18-1	MY720/3,3'-DDS-1,12-DDA (Method B)	388 (56·2)	72·4 (10·5)	66·3 (9·61)	6·82 (0·99)	1·71	9·56
IME-11-3A	MY720/6F-3,3'-DDS-1,12-DDA (Method D)	275 (39·8)	73·8 (10·7)	46·8 (6·80)	6·20 (0·90)	1·51	6·80
IME-12-3A	MY720/6F-3,3'-DDS-1,12-DDA (Method B)	166 (24·1)	90·3 (13·1)	28·4 (4·12)	6·14 (0·89)	0·90	3·73
IME-13-2	MY720/Gly200/6F-3,3'-DDS-4,4'-DDS	391 (56·7)	95·8 (13·9)	66·8 (9·69)	8·14 (1·18)	1·35	6·15
IME-14-2	MY720/Gly100/6F-3,3'-DDS-4,4'-DDS	416 (60·3)	82·8 (12·0)	71·0 (10·3)	8·14 (1·18)	1·36	6·53
IME-15-2	MY720/Gly200/6F-4,4'-DDS	263 (38·1)	93·8 (13·6)	44·9 (6·52)	7·59 (1·10)	0·75	3·60
IME-16-2	MY720/Gly100/6F-4,4'-DDS	338 (49·6)	82·8 (12·0)	58·5 (8·48)	7·59 (1·10)	1·20	6·08

18·7 MPa (2710 psi) suggest that the latter values are not representative of the system. This points out the difficulties in using miniature tensile specimens in measuring tensile properties. The high compressive strength values for IME-10, IME-12 and IME-12-3 were also reflected in relatively high tensile strengths. This suggests that these properties more nearly reflect the true properties of these resins than do the C-2 properties.

The performance of the IME resins in a unidirectional Celion 6000 graphite/IME epoxy resin composite is illustrated in Table 5.46. The short beam shear strengths of two of the Celion 6000/IME epoxy resin composites, namely the IME-2 and IME-5 systems are clearly superior to those of the composite containing the control system.

Data for the 10° off-axis tests[61] for Celion 6000 graphite/IME resin composite samples tested in the 'as fabricated' (dry) condition are listed in Table 5.47. Four 10° off-axis properties, the uniaxial tensile strength, the intralaminar shear strength, composite shear strain-to-failure, and the calculated shear-strain-to-failure in the resin of each composite system were compared with the same properties of the control composite C-2. For the control composite, the tensile strength, intralaminar shear strength, shear strain-to-failure in the composite and shear strain-to-failure in the resin were as follows: $45·8 \times 10^6$ psi (316 MPa), 7·84 ksi (54·0 MPa), 1·08 percent and 4·65 percent. Comparison of these properties with those exhibited by the IME composites shows that five systems exhibited better properties in all categories than the control resin. These are listed as follows, in order of decreasing values for each property: IME-9-2B > IME-18-1 > IME-10-2B > IME-14-2 > IME-13-2. Two systems, IME-11-3-A and IME-16-2, exhibited similar tensile strengths and intralaminar shear strengths to those of the control composite, but much greater composite shear strain-to-failure and resin shear strain-to-failure than the control composite C-2.

TABLE 5.48
PROPERTIES OF ED EPOXY RESINS[62]

Epoxy	Molecular weight	Epoxy content	Softening point (°C)
ED-5	400	22–23	—
ED-12	500	14–15	20–25
ED-2000	1 500	10–11	73–82
ED-15	2 000	4–5	48–100

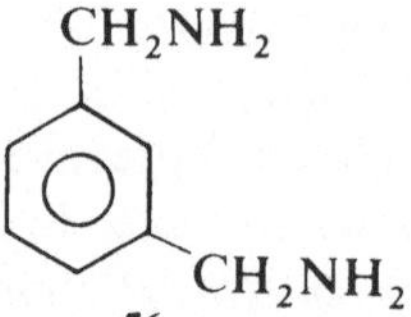

CH$_2$NH$_2$

56

m-Xylylenediamine (MXDA)

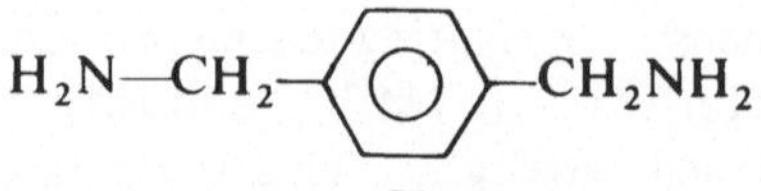

H$_2$N—CH$_2$—⟨O⟩—CH$_2$NH$_2$

57

p-Xylylenediamine (PXDA)

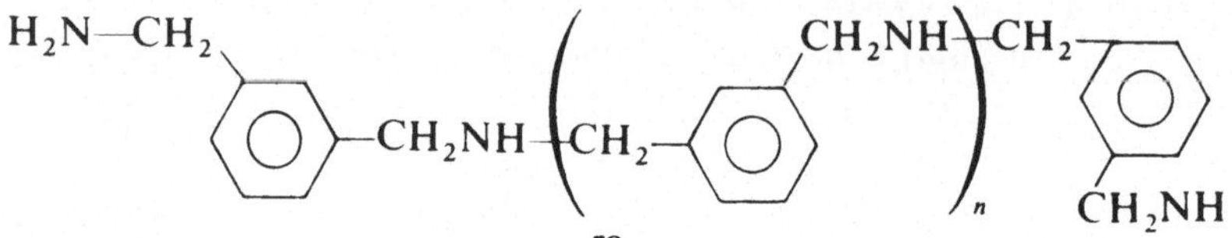

58

m-Polyxylylenepolyamine (MPA)

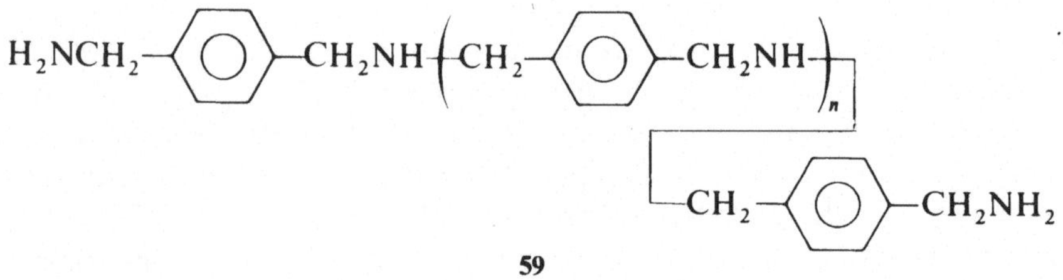

59

p-Polyxylylenepolyamine (PPA)

m-Xylylenediamine and *p*-xylylenediamine (**56** and **57**) and *m*-polyxylylenepolyamine (**58**) and *p*-polyxylylenepolyamine (**59**) have been investigated[62] as curing agents for epoxy resins having molecular weights varying from 400 to 2000 (Table 5.48). The thermomechanical properties (% strain) of the resins were studied as a function of temperature (20–180 °C). The results are shown in Figs 5.7 and 5.8. The deformation resulting from uniaxial compression was investigated at constant load (49 N, 5 kg, 10 sec) on specimens (10·5 mm diameter × 7 mm thick) prepared by casting unfilled epoxy compound in a mould. The specimens were cured first at room temperature and then 100 °C/3 h. The lowest strain was exhibited by the epoxy resin with lowest EDS epoxy/polyxylylene diamine ratio (0·91) (Fig. 5.7). With the ED-5 epoxy the *p*-polyxylylenepoly-amine (PPA) exhibited a lower strain as a function of temperature than the *m*-polyxylylenepolyamine- (MPA) cured epoxy (Fig. 5.8). Also, the *p*-polyxylylenepolyamine (PPA) exhibited minimum elastic deformation at a 45–50 % amine content with the ED-5 epoxy, indicating a maximum crosslink density for this composition relative to resins with 20, 30 and

MOLES EPOXY/ACTIVE H: (1) 2.0; (2) 1.33; (3) 1.25; (4) 1.6; (5) 0.91

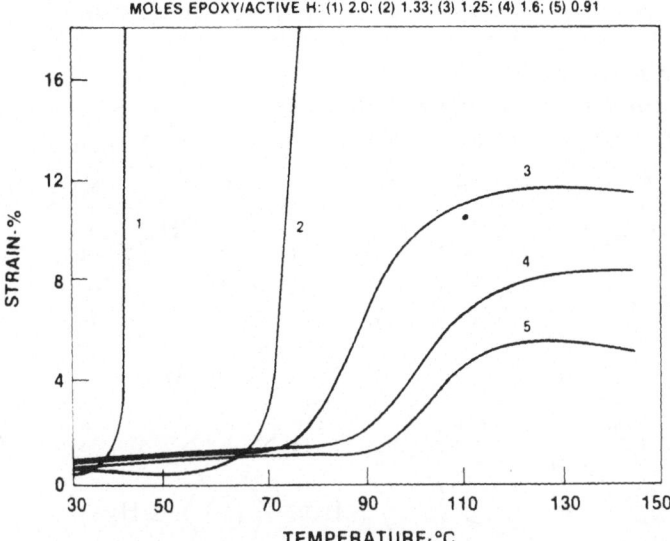

FIG. 5.7.   Strain as a function of the temperature of ED-5 epoxy resin cured with *p*-xylylenediamine (PXDA).[62]

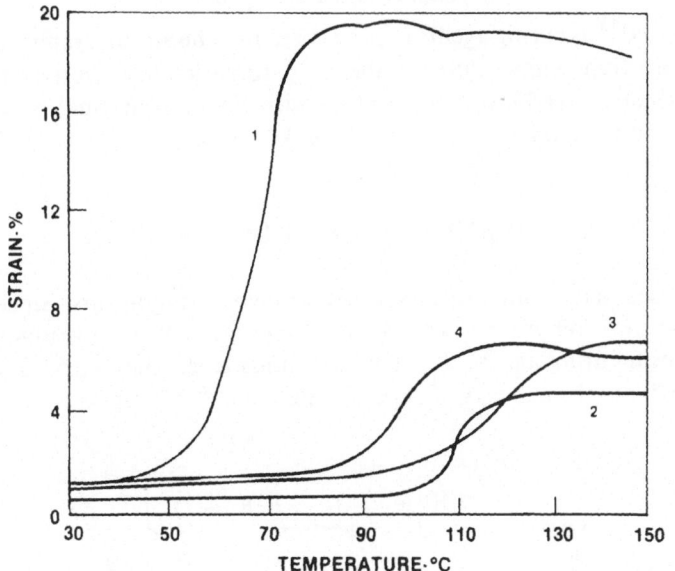

FIG. 5.8.   Strain as a function of temperature for ED-5 epoxy resin cured with (1) *p*-polyxylylenepolyamine (PPA), 20 phr; (2) PPA, 40 phr; (3) PPA, 50 phr; (4) PPA, 60 phr.[62]

60 wt % amine. The polyamine- and *p*-xylylenediamine-cured resins exhibited lower swelling in dichloroethane vapour relative to the *m*-xylylenediamine-cured resin.

Polyaromatic amines (**60–63**) prepared by reaction of a mononuclear diamine, such as metaphenylene diamine, with formaldehyde were

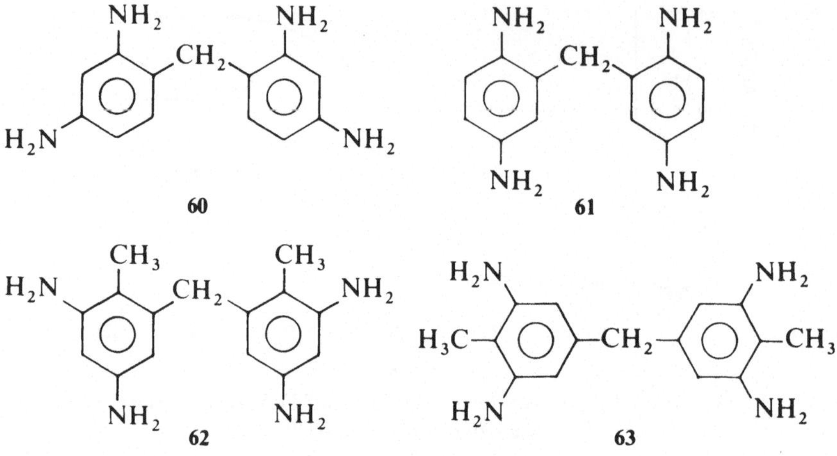

Polyaromatic amines (ref. 63)

investigated[63] as curing agents for epoxy resins. The authors report that the heat distortion temperature of the polyaromatic-amine-cured epoxy is higher than the HDT of epoxy cured with the conventional product of aniline and formaldehyde (4,4'-methylenedianiline).

5.6.   DIANHYDRIDE CURING AGENTS

Anhydrides and dianhydrides are well known curing agents. Since 1966, very few new anhydrides and dianhydrides have been investigated for epoxy resin curing agents. Kaman[3] mentions three dianhydrides (**64, 65, 66**), which are the subjects of Japanese patents,[64–66] but no properties of

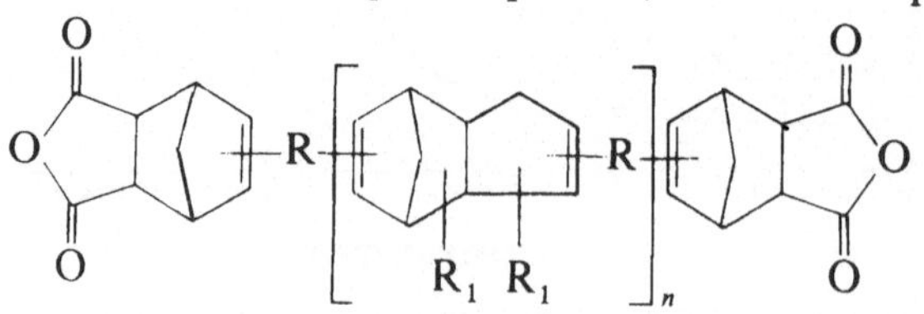

64
(ref. 64)

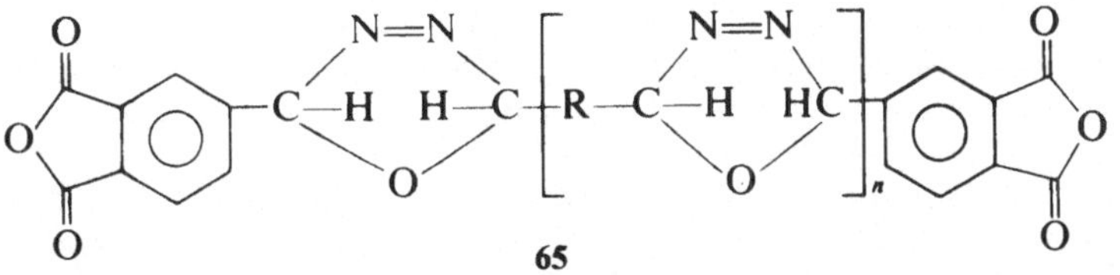

65
(ref. 65)

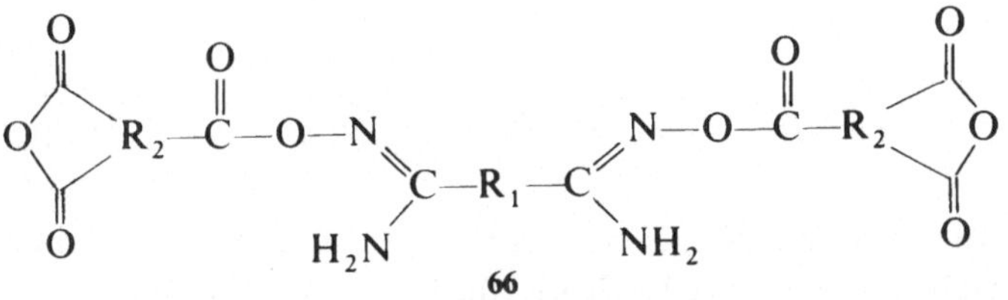

66
(ref. 66)

epoxies cured by these materials were given. The literature does not reveal their use as curing agents for epoxy resins.

The dianhydride **67** prepared by a photochemical addition of maleic anhydride to ethylbenzene was evaluated as a curing agent with DGEBA epoxy.[67] Thus DGEBA (1 equivalent) mixed with the dianhydride (0·4

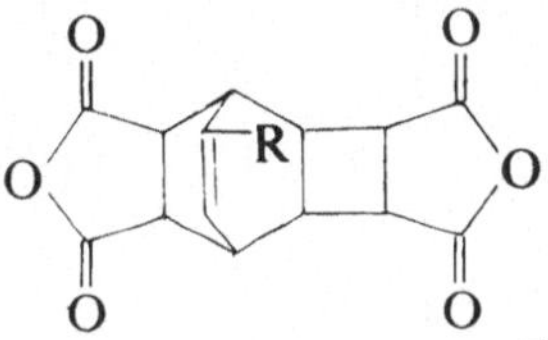

R = ethyl or isopropyl

67

Maleic anhydride–ethylbenzene adduct (ref. 67)

equivalent), and phthalic anhydride (0·4 equivalent), cured at 160°C/24 h gave a cured epoxy with flexural strength 124 MPa and HDT, 110°C. The values for nadic methylanhydride/(DGEBA) 1:1 were 110 MPa and 60°C respectively.

Polycyclic aromatic polycarboxylic anhydrides containing aromatic nitro groups have been prepared by nitric acid (70%) oxidation of coal or lignite.[68,69] The reaction conditions and neutralisation equivalents of the anhydride mixture for each run are listed in Table 5.49. The polycyclic anhydride mixture (10·5 g) from run No. 3 mixed with diglycidyl ether of Bisphenol A (25 g) having an epoxy equivalent of 900, was compression moulded at 160°C/1600 psi (11·0 MPa) for 2 h, and then post-cured at

TABLE 5.49

REACTION CONDITIONS AND PRODUCT ANALYSIS OF NITRIC ACID OXIDATION PRODUCT OF LIGNITE[68]

Run No.	Temperature (°C)	Reaction time (h)	Acetone water insoluble product (g)	Neutral equivalent of anhydride mixture
1	50	5	67·0	129
2	70	2	51·1	119
3	90	2	52·5	119, 105

175 °C for 16 h to give a Barcol hardness of 16. A resin sample prepared from polycyclic anhydride (20 g) from run No. 1 and the same resin (9 g) and 0·9 g tin octoate as accelerator and cured as above gave a product with a Barcol hardness of 25 to 28.

A polysulphide tetracarboxylic acid dianhydride (68) has been evaluated with an epoxy resin.[70] The complex dianhydride mixed with a low molecular weight epoxy resin, and cured at 70 °C/2 h + 130 °C/20 h

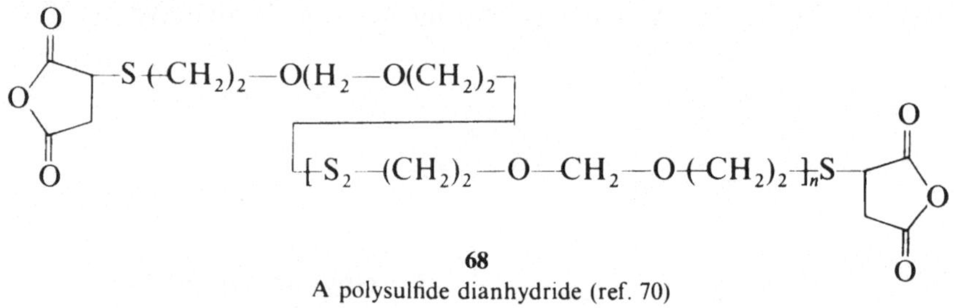

68
A polysulfide dianhydride (ref. 70)

gave a flexible product having a $T_g \approx -5\,°C$. Diels–Alder adducts of maleic anhydride and high molecular weight dienes from the residue of butadiene synthesis as curing agents for epoxies are reported,[71] but no reference to curing action with epoxies is made.

Isoimidylphthalic anhydrides (69 and 70) have been patented[72,73] for use

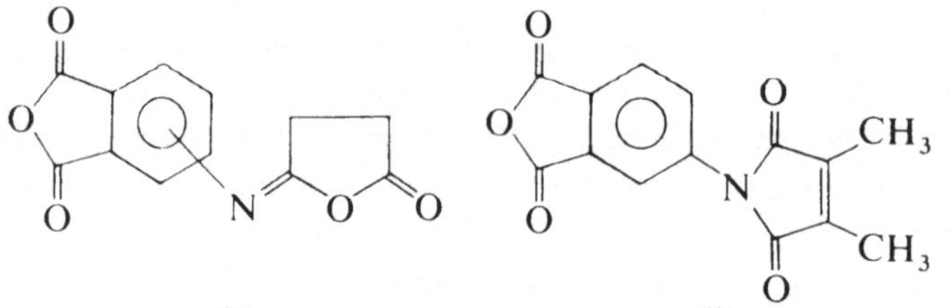

69a   **69b**

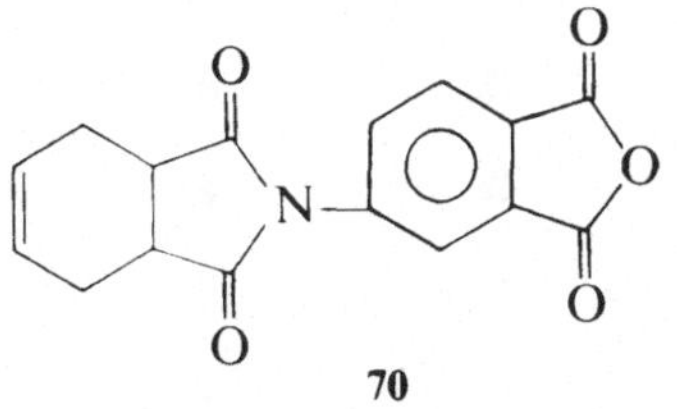

70

Isoimidylphthalic anhydrides (refs. 72, 73)

as epoxy curing agents, but no properties of the cured epoxies were mentioned.

5.7.   LATENT CURING AGENTS

Efforts to improve latency of one-part amine/epoxy resin systems have been accomplished by two methods: (1) development of aromatic poly(diamines) such as polymethylenedianilines, poly(m- and p-xylylenedi-amines) and aromatic amine–epoxy adducts; and (2) development of steric hindered amines such as 2,2'-diethyl-4,4'-methylenedianiline (2,2'-Et-4,4'-MDA). Many of these materials are commercially available products (for example, 2,2'-Et-4,4'-MDA) and are sold under tradenames and other designations.

One-part epoxy formulations which exhibit long-time latent characteristics are desirable from a cost and a handling viewpoint. The latent curing agent used in adhesives, coating, moulding resin, and prepreg formulations must also provide cured epoxy resins with the specific physical, thermal and mechanical properties required for the application. Considerable research over the years has been directed towards developing latent curing agents which lead to cured epoxies with the required characteristics.

An unusual approach towards development of latency in a curing agent is illustrated by the aminimide compounds **71**,[74] Table 5.50. The aminimides **71** contain the ylide structure

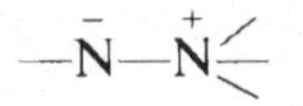

$$-\overset{-}{N}-\overset{+}{N}\overset{\diagup}{\diagdown}$$

which is responsible for the latent properties of these compounds. The aminimides are prepared by a trimolecular condensation of an ester, a disubstituted hydrazine and an epoxide.[75,76]

The latency of these compounds is derived from the room temperature

TABLE 5.50
PROPERTIES OF AMINIMIDE COMPOUNDS[74]

Name	Structure	Appearance	Solubility[a]							Toxicity LD_{50} (mouse)
			1	2	3	4	5	6	7	
YPH-103	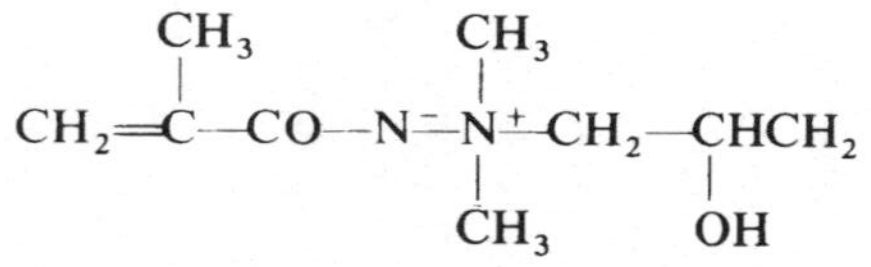 $CH_2=\overset{\underset{\displaystyle CH_3}{\displaystyle \vert}}{C}-CO-N^--N^+\!\!\!\underset{\displaystyle CH_3}{\overset{\displaystyle CH_3}{\vert\vert}}\!\!\!CH_2-\underset{\displaystyle OH}{\overset{\displaystyle \vert}{C}}HCH_2$	White crystal mp 144–146°C	●	○	×	×	●	●	●	>6·0
YPH-201	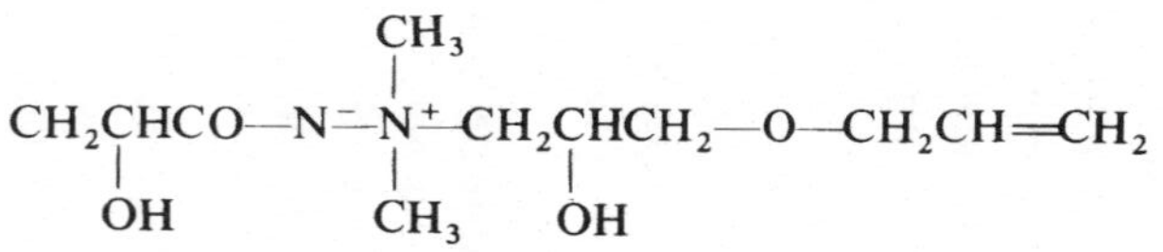 $CH_2\underset{\displaystyle OH}{\overset{\displaystyle \vert}{C}}HCO-N^--N^+\!\!\!\underset{\displaystyle CH_3}{\overset{\displaystyle CH_3}{\vert\vert}}\!\!\!CH_2\underset{\displaystyle OH}{\overset{\displaystyle \vert}{C}}HCH_2-O-CH_2CH=CH_2$	Viscous liquid n_D^{20} 1·4828	●	●	●	×	●	●	●	>13·3
YPH-208	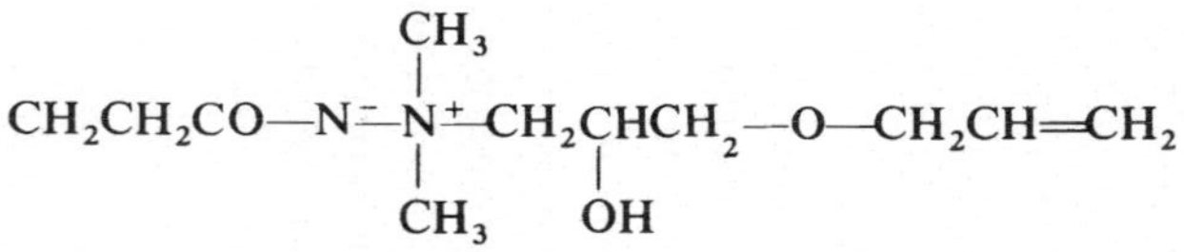 $CH_2CH_2CO-N^--N^+\!\!\!\underset{\displaystyle CH_3}{\overset{\displaystyle CH_3}{\vert\vert}}\!\!\!CH_2\underset{\displaystyle OH}{\overset{\displaystyle \vert}{C}}HCH_2-O-CH_2CH=CH_2$	Viscous liquid n_D^{20} 1·4871	●	●	●	●	●	●	●	>12·7

[a] (1) H_2O; (2) n-BuOH; (3) MEK; (4) xylene; (5) cellosolve; (6) DMF; (7) DMSO; (●) freely soluble; (○) soluble; (×) insoluble.

$$a \quad R = CH_2\!=\!\overset{\overset{\displaystyle CH_3}{|}}{C}\!-\!$$

$$R' = CH_3\!-\!\overset{\overset{\displaystyle }{|}}{CH}\!-\!CH_2\!-\!$$
$$OH$$

$$\overset{\overset{\displaystyle O}{\|}}{R}\!-\!C\!-\!N\!\overset{\text{-}}{=}\!\overset{\overset{\displaystyle CH_3}{|}}{N}{}^+\!-\!R'$$
$$\overset{|}{CH_3}$$

$$b \quad R = CH_3\!-\!\overset{\overset{\displaystyle OH}{|}}{CH}\!-\!$$

$$R' = CH_2\!=\!CH\!-\!CH_2\!-\!O\!-\!CH_2\!-\!\overset{\overset{\displaystyle OH}{|}}{CH}\!-\!CH_2\!-\!$$

$$c \quad R = CH_3\!-\!CH_2\!-\!$$

71
Aminimide (ref. 74)

$$R' = CH_2\!=\!CH\!-\!CH_2\!-\!O\!-\!CH_2\!-\!\overset{\overset{\displaystyle OH}{|}}{CH}\!-\!CH_2\!-\!$$

stability of the ylide structure. These compounds exhibit accelerator or curing properties with an epoxide by the tertiary amine (**72**), which is generated by thermal decomposition of the aminimide, for example

$$CH_2\!=\!\overset{\overset{\displaystyle CH_3}{|}}{C}\!-\!\!-\!\overset{\overset{\displaystyle O}{\|}}{C}\!-\!N\!\overset{\text{-}}{=}\!\overset{\overset{\displaystyle CH_3}{|}}{N}{}^+\!-\!CH_2\!-\!\overset{\overset{\displaystyle OH}{|}}{CH}\!-\!CH_3 \overset{heat}{\longrightarrow}$$
$$\overset{|}{CH_3}$$

71a

$$CH_3\!-\!\overset{\overset{\displaystyle OH}{|}}{CH}CH_2\!-\!\overset{\overset{\displaystyle CH_3}{|}}{N} + CH_2\!=\!\overset{\overset{\displaystyle CH_3}{|}}{C}\!-\!N\!=\!C\!=\!O$$
$$\overset{|}{CH_3}$$

72   **73**

The tertiary amine promotes ring opening of the epoxide ring. The isocyanate **73** generated in the presence of epoxides and anhydride also appears to promote the epoxide anhydride cure reaction, although this has not been clearly established.

In the presence of amine curing agents, the isocyanate generated from the aminimide would react to form ureas, thereby preventing the normal amine–epoxide reaction. Therefore these materials have not been investigated with amines in amine/epoxy systems.

The toxicological and some physical properties of three aminimides are listed in Table 5.50. The toxicity of these compounds is very low, as is indicated by the low LD_{50} values on mouse; human skin is not irritated, as evidenced by skin patch tests. The pot-life of these compounds with a DGEBA (EEW ~ 180) epoxy is illustrated in Table 5.51 which shows a

TABLE 5.51

POT-LIFE AND GELATION TIME OF DGEBA (EEW 190) CONTAIN-
ING AMINIMIDE (10 phr)[74]

YPH	Pot-life (days)	Gelation time (min)		
		185°C	155°C	125°C
103	30	25	60	180
201	50	15	20	30
208	30	25	60	120

room temperature 30–50 day pot-life; while at 125 °C, a 2–3 h gelation time
for compounds **71a** and **71c** is noted.

Some thermal, mechanical, and electrical properties of two one-part
epoxy formulations are compared with three latent catalysts, namely
tris(dimethylaminomethyl)phenyl 2-ethylhexanoic salt, 2-ethyl-4-imid-
azole and BF_3-monoethylamine (Table 5.52). The aminimide-cured

TABLE 5.52

MECHANICAL AND ELECTRICAL PROPERTIES OF ONE-PART EPOXY RESIN COMPOSITIONS[74]

Item	Comp. 1[a]	Comp. 2[b]	Comp. 3[c]	Comp. 4[d]	Comp. 5[e]
Heat deformation temperature (°C)	84	92	80	141	170
Mechanical properties					
Flexural strength, MPa	138	121	99·1	107	114
Izod impact strength, kg-cm/cm[f]	6·0	8·9	1·9	—	3·5
Electrical properties					
Volume resistivity, Ω-cm $\times 10^{-15}$	2·0	—	1·0	10	10
Dielectric constant (1 kHz)	4·5	—	3·5	3·7	3·7
tan δ (1 kHz)	0·0377	—	0·003	0·004	0·008

[a] DGEBA (EEW 190) (100 parts) and YPH-208 (8·0 parts) are cured at 120 °C for 5 h, and
then at 180 °C for 8 h.
[b] DGEBA (EEW 190) (100 parts) and YPH-201 (6·0 parts) are cured at 110 °C for 3 h, and
then at 150 °C for 8 h.
[c] DGEBA (EEW 190) (100 parts) and trisdimethylaminomethylphenyl-2-ethylhexoic acid
salt (10 parts) are cured at 80 °C for 2 h. The pot-life of this resin was 7 h and 23 °C.
[d] DGEBA (EEW 190) (100 parts) and 2-ethyl-4-methylimidazole (4 parts) are cured at 60 °C
for 2 h. The pot-life of this resin was 20 h at 30 °C.
[e] DGEBA (EEW 190) (100 parts) and BF-monoethanolamine (3 parts) are cured at 120 °C for
4 h, and then at 200 °C for 4 h. The pot-life of this resin was 6 months at 23 °C.
[f] Izod: 1 kg-cm/cm of notch = 9·81 J/m.

TABLE 5.53
ADHESIVITY OF VARIOUS EPOXY COMPOSITIONS[74]

Hardener or catalyst (phr)		Curing conditions		Adhesivity (MPa)	
		Temp (°C)	Time (h)	Al–Al[a]	Fe–Fe[b]
YPH-103	(6)	155	5	11·5	19·3
	(8)	185	1·5	23·5	
YPH-208	(8)	155	5·0	23·1	
Epicure Z	(20)	110	1·5	9·5	
	(20)	170	1·5	10·9	
BF₃-MEA[c]	(5)	130 + 150	2 + 2	4·9	No adhesion

[a] Aluminium test pieces (thickness 2·0 mm, width 10 mm, adhesion area 1 cm²) were immersed in phosphoric acid/n-BuOH/iso-PrOH/water mixture (1:4:3:1) for 10 min, washed with aqueous acetone, and dried in air.
[b] Iron test pieces (thickness 1 mm, width 10 mm, adhesion area 1 cm²) were burnished with a sandpaper, washed with acetone, and dried in air.
[c] MEA, mono(ethanolamine).

epoxy systems exhibit lower T_g values and higher flexural and impact strengths than the 2-ethyl-4-methylimidazole and BF₃-monoethylamine cured systems. The uses of the aminimides as hardeners in adhesive applications are compared in Table 5.53 with Epicure Z and BF₃-MEA. The strong adhesive strengths of these adhesive systems relative to Epicure Z and BF₃-MEA was attributed to the toughness of the aminimide cured resin. The mechanical properties of anhydride-cured epoxies using aminimide as accelerator are noted in Table 5.54. The three aminimide-accelerated epoxy systems are compared to benzyldimethylamine as accelerator.

Ketimines (**76**), the reaction products of ketones and primary amines (**74**) or ammonia (**75**), have been shown to exhibit latent properties in epoxy

$$\underset{74}{\underset{R_2}{\overset{R_1}{\diagdown}}} C{=}O + \underset{75}{H_2NR_3} \rightleftharpoons \underset{76}{\underset{R_2}{\overset{R_1}{\diagdown}}} C{=}N{-}R_3 + H_2O$$

Ketimines (ref. 77)

resins.[77] The reaction is reversible, and the ability of a ketimine to function as a curing agent depends on this reversibility; that is, the presence of water. Table 5.55 lists ketimines prepared and their boiling points. Compared with free amines, the pot-life of the ketimine systems is two to three times longer.[77] A series of clear films cured at 23 °C and 57 % relative humidity

TABLE 5.54

MECHANICAL   AND   ELECTRICAL   PROPERTIES   OF   EPOXY–ACID   ANHYDRIDE
COMPOSITIONS[74]

Item	Compositions[a]			Reference[b] BDMA
	A	B	C	
Heat deformation temperature, °C	98	142	127	128
Mechanical properties				
Bending strength, MPa	84·9	88·9	124·6	124·8
Modulus of elasticity, GPa	104	104	35	27
Tensile strength, MPa	5·18	6·3	—	—
Izod impact strength,				
kg-cm^2	1·2	1·1	2·0	2·1
Electrical properties				
Volume resistivity,				
(Ω-cm) $\times 10^{-15}$	1·7	3·9	170	20
Dielectric constant, 1 MHz	4·0	4·1	2·3	3·2
tan δ, 1 MHz	0·013	—	0·012	—
Arc resistance time, s	187	187	125	134
Water absorption				
Room temperature, %	0·05	0·06	0·11	—
Boiling water absorption, %	0·59	0·15	—	—

[a] Formulation and curing condition:
  DGEBA (EEW 450–550)   100 parts ⎫ Composition A: 150 °C × 2 min
  pyromellitic anhydride   17 parts ⎬ Composition B: 150 °C × 2 min
  YPH-208   2·0 parts⎭   + 180 °C × 1 h
  DGEBA (EEW 190)   100 parts ⎫
  hexahydrophthalic anhydride   80 parts ⎬ Composition C: 90 °C × 2 h
  YPH-208   1·0 parts⎭   + 150 °C × 8 h
[b] Formulation and curing condition:
  DGEBA (EEW 190)   100 parts ⎫
  hexahydrophthalic anhydride   80 parts ⎬ Reference   100 °C × 2 h
  BDMA (benzyldimethylamine)   1·0 parts⎭   + 150 °C × 4 h

after 7 days were prepared from ketimine-cured DGEBA epoxy, which was catalysed with 5 phr of phenol as an accelerator. The properties of this 0·005 in thick film are listed in Table 5.56. The ketimine derived from meta-xylylenediamine (MXDA) and methyl isobutylketone (MIBK) exhibited the best overall film properties of those tested. The extended room-temperature latency and low viscosity of the uncured resin systems and colourless films produced from ketimine-cured epoxy films are some of the advantages of these systems. The disadvantages are: that these curing

TABLE 5.55
KETIMINES[77]

Ketone[a]	Amine[b]	Boiling point of ketimine ($°C$)	(mm Hg)
DMK	EDA	44–46	1·0
DMK	DTA	75	1·0
MEK	PDA	91–95	1·0
MEK	DTA	100–104	0·5
MIBK	EDA	91–92	1·0
MIBK	HMDA	140	0·4
MIBK	MXDA	180	1·0
MIBK	DTA	138	1·0
MIBK	TEPA	215–220	1·0
DIBK	MXDA	200	0·8
MPK	EDA	114–116	Melting point

[a] DMK = acetone; MEK = methyl ethyl ketone, MIBK = methyl isobutyl ketone, DIBK = diisobutyl ketone, MPK = acetophenone.
[b] EDA = ethylene diamine, DTA = diethylenetriamine, PDA = propylene diamine, HMDA = hexamethylene diamine, MXDA = methaxylylene diamine, TEPA = tetraethylene pentamine.

systems are limited to films and surface coatings; and that the cures are slower than with the corresponding amines.

Metal acetylacetonates have been evaluated as latent accelerators for epoxy–anhydride solventless resin systems.[78] The types of metal acetylacetonates evaluated are listed in Table 5.57. These were evaluated with a

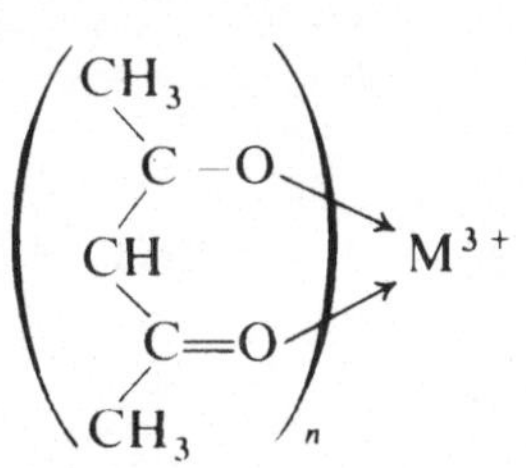

$$\left(\begin{array}{c} CH_3 \\ | \\ C-O \\ | \\ CH \\ | \\ C=O \\ | \\ CH_3 \end{array} \right)_n M^{3+}$$

77
Metal acetylacetonates (ref. 78)

where M = Ti(IV), Al(III), Mn(II), Fe(III), Mg^{+2}, Mn(III), Co(II), Cu(II), Cr(III), Ni(II), Zn(II), Co(III), Va(III), UO_2(IV), Zr(IV), Th(II), Sr(II), K^+, Pb(II) or Be(II).

TABLE 5.56
PROPERTIES OF COATING PREPARED FROM KETIMINES[77]

Ketimine[a]	phr (25% excess)	Cure rate		Pencil hardness after 7 days	Film appearance	Solvent resistance after 7 days cure[b]					
		Set to touch (h)	Hard through (h)			Toluene 15 min	MEK 15 min	MIBK 15 min	Boiling H$_2$O 15 min	Cold H$_2$O 24 h
DMK–EDA	25	6	<22	H	Trace of blush	NA	Soft	NA	NA	NA
DMK–DTA	24	2	4	2H	Trace of blush	NA	Soft	NA	NA	NA
MEK–PDA	30	5	<22	HB	Excellent	Dissolved	Peeled	Peeled	Whitened	Whitened
MEK–DTA	27	2	4	H	Blushed	Dissolved	Peeled	Peeled	Whitened	Whitened
MIBK–EDA	36	3	6	B	Trace of blush	Peeled	Peeled	Peeled	Whitened	NA
MIBK–HMDA	45	–	<22	B	Trace of blush	Peeled	Peeled	Softened	NA	Whitened
MIBK–MXDA	49	4	6	H	Excellent	NA	Peeled	NA	NA	NA
MIBK–DTA	35	2	4	F	Trace of blush	NA	Soft	NA	NA	Whitened
MIBK–TEPA	33	3	7	H	Blushed	NA	NA	NA	NA	NA
DIBK–MXDA	63	>6	77	B	Excellent	Peeled	Peeled	Peeled	Whitened	NA
MPK–EDA	43	>6	<22	H	Excellent	NA	NA	Peeled	NA	NA

[a] Same code as Table 5.55.
[b] NA = not affected.

TABLE 5.57

VARIOUS METAL ACETYLACETONATES EVALUATED AS LATENT ACCELERATORS FOR EPOXY–ANHYDRIDE RESINS[78]

Metal acetylacetonate	Formula
Titanium oxyacetylacetonate	$TiO(C_5H_7O_2)_2$
Aluminium acetylacetonate	$Al(C_5H_7O_2)_3$
Cerous acetylacetonate	$Ce(C_5H_7O_2)_3$
Manganese(II) acetylacetonate	$Mn(C_5H_7O_2)_2$
Iron(III) acetylacetonate	$Fe(C_5H_7O_2)_3$
Magnesium acetylacetonate	$Mg(C_5H_7O_2)_2$
Manganese (III) acetylacetonate	$Mn(C_5H_7O_2)_3$
Cobalt(II) acetylacetonate	$Co(C_5H_7O_2)_2$
Copper(II) acetylacetonate	$Cu(C_5H_7O_2)_2$
Chromium(III) acetylacetonate	$Cr(C_5H_7O_2)_3$
Nickel acetylacetonate	$Ni(C_5H_7O_2)_2$
Zinc acetylacetonate	$Zn(C_5H_7O_2)_2$
Cobalt(III) acetylacetonate	$Co(C_5H_7O_2)_3$
Vanadium(III) acetylacetonate	$V(C_5H_7O_2)_3$
Uranyl acetylacetonate	$UO_2(C_5H_7O_2)_2$
Zirconium acetylacetonate	$Zr(C_5H_7O_2)_4$
Thorium acetylacetonate	$Th(C_5H_7O_2)_4$
Strontium acetylacetonate	$Sr(C_5H_7O_2)_2$
Sodium acetylacetonate	$C_5H_7O_2Na$
Potassium acetylacetonate	$C_5H_7O_2K$
Lead acetylacetonate	$Pb(C_5H_7O_2)_2$
Beryllium acetylacetonate	$Be(C_5H_7O_2)_2$

1:1 stoichiimetric DGEBA/1-methyltetrahydrophthalic anhydride formulation at a level of 0·05 to 0·10 weight percent. The gel time and storage data are listed in Table 5.58 and the electrical properties are listed in Table 5.59. The metal acetylacetonates which gave the best combination of latency, fast gel time and low power factor values are those of titanium(IV), chromium(III), cobalt(III), zirconium(IV), aluminium(III), manganese(III), and cobalt(II). These are all transition metals, and this fact may be related to their effectiveness. The solubility of the acetylacetonate in the resin solution may also be a major factor in the fast gel times. The most effective latent accelerators undergo thermal decomposition in the temperature range 90–170 °C. It is therefore possible that the decomposition products may, in fact, be the active species responsible for the polymerisation of epoxy–anhydride resin mixtures.

Stark[79] described a latent one-component epoxy resin consisting of, for example, DGEBA (EEW ~ 185) tetraisopropyltitanate initiator, and a

TABLE 5.58

GEL TIME AND STORAGE DATA FOR EPOXY–ANHYDRIDE RESIN WITH DIFFERENT METAL
ACETYLACETONATE ACCELERATORS[78]

Sample No.	Metal acetylacetonate used (0·10% on impregnant)	Gel time data (min)[a] 150°C	Gel time data (min)[a] 175°C	Catalysed storage lifetime at 25°C (days)[b]
1	Titanium oxyacetylacetonate	35–35	30–35	110
2	Aluminium acetylacetonate	35–40	30–35	95
3	Cerous acetylacetonate[c]	—	50–55	—
4	Manganese(II) acetylacetonate[c]	—	55–65	—
5	Iron(III) acetylacetonate	—	<15	<10
6	Magnesium acetylacetonate[c]	—	50–55	—
7	Manganese(III) acetylacetonate	80–90	40–45	160
8	Cobalt(II) acetylacetonate	50–55	35–40	130
9	Copper(II) acetylacetonate[c]	—	90–100	—
10	Chromium(III) acetylacetonate	40–50	30–40	>200
11	Nickel acetylacetonate	—	45–50	>90
12	Zinc acetylacetonate[c]	—	20–25	50
13	Cobalt(III) acetylacetonate	80–90	25–35	>200
14	Vanadium(III) acetylacetonate	70–80	40–45	>90
15	Uranyl acetylacetonate	<10	<10	<4
16	Zirconium acetylacetonate	50–55	30–35	>90
17	Thorium acetylacetonate[c]	60–65	50–55	—
18	Strontium acetylacetonate[c]	100–110	60–65	—
19	Sodium acetylacetonate	35–40	20–25	>90
20	Potassium acetylacetonate	25–30	15–20	>90
21	Lead acetylacetonate[c]	100–110	70–80	—
22	Beryllium acetylacetonate	100–110	60–65	>90

[a] On 10-g impregnant samples in 2-in diameter aluminium dish.
[b] Time for viscosity to reach 1 000cPs at 25°C (some values obtained by extrapolation of data).
[c] Poor or partial (<50%) solubility in resin.

silicone accelerator **78**. Examples of the silicone accelerators described are given in Scheme **78**. Several titanate or zirconate initiators were also described as initiators in this patent. For example, a mixture of 100 parts of DGEBA (EEW ~ 185), 1·5 parts of initiator and 4·89 parts of accelerator, where R = CH$_3$ and X = H, gelled in 1·3 min at 150 °C, while replacement of the accelerator with 5·2 parts of catechol yielded a resin which gelled in 1·9 min at 50 °C, a much lower temperature. By variations in R and X, and

TABLE 5.59

ELECTRICAL PROPERTIES OF CURED SAMPLES OF EPOXY–ANHYDRIDE RESIN WITH DIFFERENT METAL ACETYLACETONATE ACCELERATORS[78]

Sample No.	Metal acetylacetonate used ($\sim 0.10\%$ on resin)	Electrical properties at $150°C$ and $60\,Hz^a$	
		$100 \times \tan \delta$	e'
1	Titanium oxyacetylacetonate	2·4	6·7
2	Aluminium acetylacetonate	2·5	6·9
3	Cerous acetylacetonate^b	8·0	6·9
4	Manganese(II) acetylacetonate^b	43·0	6·8
5	Iron(III) acetylacetonate	1·6	6·4
6	Magnesium acetylacetonate	1·8	7·1
7	Magnesium(III) acetylacetonate	5·3	7·4
8	Cobalt(II) acetylacetonate	2·8	6·6
9	Copper(II) acetylacetonate	310·0	19·0
10	Chromium(III) acetylacetonate	2·0	7·4
11	Nickel acetylacetonate	2·2	6·7
12	Zinc acetylacetonate^b	112·0	9·8
13	Cobalt(III) acetylacetonate	2·2	6·9
14	Vanadium(III) acetylacetonate	1·8	6·5
15	Uranyl acetylacetonate	15·0	7·9
16	Zirconium acetylacetonate	4·7	7·6
19	Sodium acetylacetonate	12·0	7·9
20	Potassium acetylacetonate	27·0	8·4

^{a} Samples cured 16 h at 150 °C (~ 0.200 in (5 mm) thick).
^{b} Metal acetylacetonate concentration in impregnants probably $<0.05\%$.

in initiators, gelation times varied from 12 to 36·2 min at 150 °C and from 64 min to 6 days at 50 °C.

where $R = CH_3$—, phenyl—

$X = -OCH_3, -CH(CH_3)_2, -C(CH_3)_3, H$

78

Dialkylsiloxane bis(substituted catechol) derivatives (ref. 79)

Utilisation of extra-coordinate siliconate salts (**79, 80** and **81**) as latent accelerators and curing agents for epoxy resin has been demonstrated by Vincent et al.[80] The benzyldimethylammonium penta-coordinate salt (**79**) (8 parts) mixed with solid DGEBA (EEW $\sim$ 925) (100

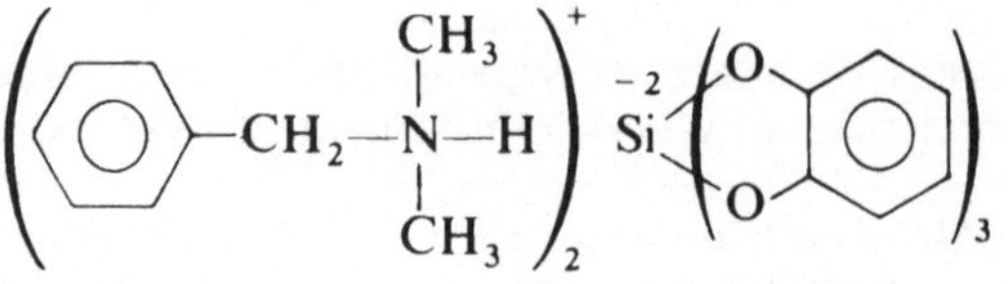

$$\left(\left(\bigcirc\!\!\!-CH_2-\underset{\underset{CH_3}{|}}{\overset{\overset{CH_3}{|}}{N}}-H\right)^{+}_{2}\overset{-2}{Si}\left(\overset{O}{\underset{O}{}}\bigcirc\right)_3\right)$$

79

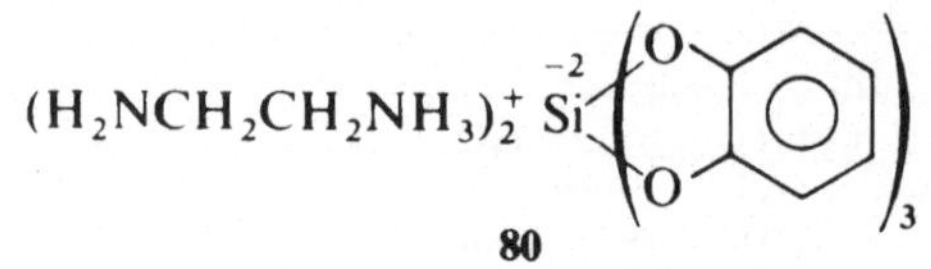

$$(H_2NCH_2CH_2NH_3)^{+}_2\;\overset{-2}{Si}\left(\overset{O}{\underset{O}{}}\bigcirc\right)_3$$

80

$$\left(\left(\bigcirc\!\!\!-CH_2-\underset{\underset{CH_3}{|}}{\overset{\overset{CH_3}{|}}{N}}-H\right)^{-}\bigcirc\!\!-\overset{-1}{Si}\left(\overset{O}{\underset{O}{}}\bigcirc\right)_2\right)$$

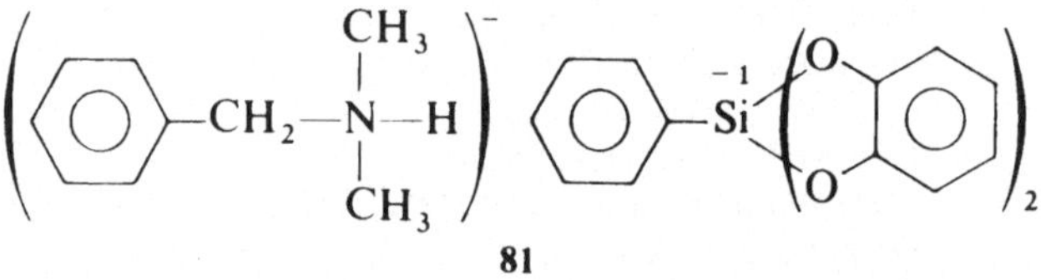

81

Extra-coordinate siliconate salts (ref. 80)

parts) and trimellitic anhydride in catalytic amounts (2·0 g phr), gelled in 8·5 sec at 200 °C and maintained a shelf-life of 5 months at room temperature. The ability of the ethylene salt to function as a curing agent in various concentrations was demonstrated by tensile lap shear strengths (Table 5.60). In this case, the salt was used with DGEBA in 35 phr and cured at 150 °C/1 h.

Piperazine salts (**82**) of polycarboxylic acids have been shown to function as curing agents for epoxy resins.[81] The composition, gel time at 120 and

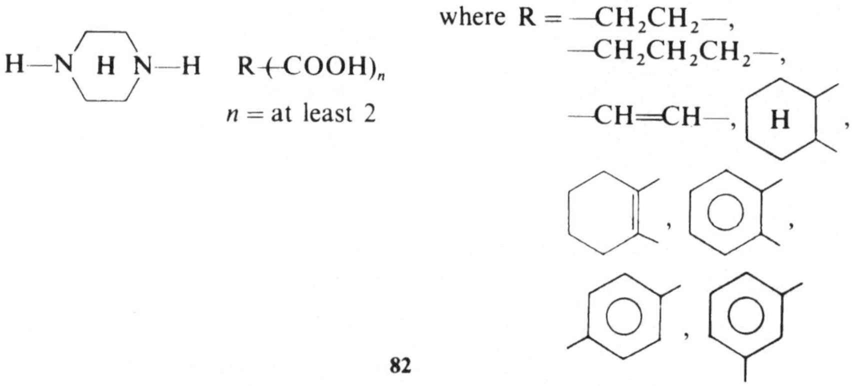

$$H-N\!\!\begin{array}{c}\diagup\!\!\diagdown\\ H\end{array}\!\!N-H \quad R\!\!-\!\!(COOH)_n$$

$$n = \text{at least } 2$$

where R = —CH$_2$CH$_2$—,
—CH$_2$CH$_2$CH$_2$—,
—CH=CH—,

82

Piperazine polycarboxylic acid salts (ref. 81)

TABLE 5.60

DEPENDENCE OF ADHESIVE STRENGTH ON CURING AGENT[a]
CONCENTRATION[80]

Curing agent concentration (phr)[b]	Adhesive lap-shear (MPa)
10	4·13–4·82
15	13·8–16·6
20	13·8–16·6
25	14·4–15·1
35	15·8–17·2

[a] $(H_2NCH_2CH_2N^+H_3)_2 Si^{-2}(C_4H_4O_2)_3$.
[b] Parts per hundred of epoxy (EEW 190).

160 °C, and room-temperature shelf-life of several piperazine salt/DGEBA compositions are listed in Table 5.61. Table 5.62 lists the composition of the piperazine salt/DGEBA resin, cure schedule and tensile lap shear strength of stainless steel/stainless steel joints ($1·27\,cm \times 0·33\,cm \times 0·16\,cm$). The cure time for dicyandiamide is shown for comparison, but the tensile shear strength value for the dicyandiamide-cured adhesive was not reported. The piperazine salts cure more rapidly than dicyandiamide.

TABLE 5.61

PIPERAZINE–ACID SALT/DGEBA GEL TIMES[81]

Curing agent		Gel time		Shelf-life (day)
Piperazine–polycarboxylic acid salt	Amount (phr)[a]	120°C (min)	160°C (sec)	
Piperazine–succinic	27	23	156	60–65
Piperazine–adipic	30	35	180	50–55
Piperazine–sebacic	38	9	100	50–55
Piperazine–maleic	27	9·5	100	40–45
Piperazine–hexahydrophthalic	34	10	170	60–70
Piperazine–tetrahydrophthalic	34	5·5	108	35–40
Piperazine–phthalic	33	10·5	120	50–55
Piperazine–isophthalic	33	25	180	60–65
Piperazine–trimellitic	31	30	480	140–150
Piperazine (comparative)	20	5	60–75	1/12

TABLE 5.62

TENSILE PROPERTIES OF PIPERAZINE SALT/DGEBA EPOXY RESINS[a][81]

Curing agent		Cure time 120°C (h)	Tensile shear strength (MPa)
Piperazine–polycarboxylic acid salt	Amount (phr)		
Piperazine–succinic	27	1	46·9
Piperazine–adipic	30	1	39·2
Piperazine–sebacic	38	1	37·2
Piperazine–maleic	27	1	39·2
Piperazine–hexahydrophthalic	34	1	22·7
Piperazine–tetrahydrophthalic	34	1	45·6
Piperazine–phthalic	33	1	39·2
Piperazine–isophthalic	33	1	46·9
Piperazine–trimellitic	31	1	37·2
Dicyandiamide (comparative)	8	3	

[a] DGEBA (EEW 180).

A polyoxypropylenediamine salt of zinc-pyrrolidone carboxylic acid (83) has been shown to function as a latent curing agent.[82] The complex is a clear yellow liquid, with an amine content of 0·825 equivalents/kg and a viscosity of 80 000 cP at 25 °C. The shelf-life at 40 °C of various ratios of DGEBA/complex A are shown in Table 5.63. The range of properties of the uncured mixtures and cured resin systems are summarised in Table

TABLE 5.63

GY250/COMPLEX A: PROCESSING AND FINAL PROPERTIES AS A FUNCTION OF THE GY250[a]/COMPLEX A RATIO[82]

GY250	100	100	100	100	100
Complex A	100	75	50	25	10
Initial viscosity at 40 °C, cP	8 850	7 800	6 370	4 240	2 640
Usable life at 40 °C to 15 000 cP	7 h 30 min	10 h	17 h 30 min	64 h	≈70 days
Gel time (hot plate)					
80 °C	100 min	136 min	173 min	—	—
100 °C	26 min 30 s	37 min 30 s	48 min 30 s	—	—
120 °C	8 min 30 s	13 min	17 min 30 s	—	—
140 °C	3 min 10 s	4 min 45 s	6 min 45 s	126 min	350 min
160 °C	—	—	—	54 min	161 min
180 °C	—	—	—	26 min 30 s	78 min

[a] DGEBA (EEW 180).

$$\left[\begin{array}{c} H_2C-CH_2 \\ O=\underset{\underset{H}{N}}{}\quad CO_2^- \end{array}\right]_2 \overset{+2}{Zn}\cdot H_2N\!\!-\!\!(CH_2\!-\!\underset{\underset{CH_3}{|}}{CH}\!-\!O)_n CH_2\!-\!\underset{\underset{CH_3}{|}}{CH}\!-\!NH_2$$

83

Zinc pyrrolidinone carboxylate (polyoxypropylene diamine salt) (Complex A) (ref. 82)

5.64. As expected, the variation of processing and end properties can be controlled by the resin/complex ratio. A resin sample DGEBA/complex A (100 g/250 g) cured at 100 °C/2 h + 140 °C/4 h yielded a tensile strength of 1·9–2·3 N/mm², and 4·7 % water absorption after 4 days in water at room temperature.

The salt of succinic acid and dodecamethylenediamine **84** (29 parts) mixed with DGEBA (10 parts) was considered to have a good shelf-life after 1 week at room temperature.[73] The shear strength of steel–steel adhesive

$$\begin{array}{l} CH_2-CO_2^- \\ | \qquad\qquad\qquad H_3\overset{+}{N}\!\!-\!\!(CH_2)_{\overline{12}}\overset{+}{N}H_3 \\ CH_2-CO_2^- \end{array}$$

84

Dodecamethylene diammonium succinate (ref. 83)

joints cured at 130 °C/2 h was 23 MPa, compared with 16·2 MPa for dodecamethylene-diamine-cured epoxy joints.[83]

Imidazole complexes (**85**) have been shown[84] to be useful in latent curable epoxy systems at 100 °C. The metal perfluoromethyl sulphonate imidazole remained latent for about 55 min. Table 5.65 lists cure times at 150 °C using the epoxy resin 3,4-epoxycyclohexylmethyl 3,4-epoxycyclohexane carboxylate. Barcol hardness is also given. The imidozole

where M = Cu, Co, Ni, Cd, Zn

$M\cdot L_nSO_3R_f$

$$L = \underset{}{N}\!\!\diagdown\!\!\underset{}{NH} \qquad \underset{}{N}\!\!\diagdown\!\!\underset{}{N}-CH_3 \qquad \underset{}{N}\!\!\diagdown\!\!\underset{\underset{CH_3}{|}}{N}-CH_3$$

$$\underset{\underset{CH_3}{|}}{N}\!\!\diagdown\!\!N\!-\!\!\bigcirc$$

$$R_f = -CF_3,\ -CF_2CF_3,\ -CF_2CF_2CF_3$$

85

Metal imidazole perfluoromethylsulphonate complexes (ref. 84)

TABLE 5.64
PROPERTY RANGE OF UNCURED AND CURED RESIN SYSTEMS[82]

Property	Numerical value range
Viscosity (40 °C)	8 850–2 640 cP
Usable life (40 °C) (time to 15 000 cP)	7·5–1 680 h
Gel time (140 °C)	3–350 min
Flexural strength	0·078–1·18 MPa
Elongation at break	35–1 %
Martens heat distortion temperature	<25–124 °C
Water absorption (23 °C)	1·7–0·2 %
Power factor tan δ, 1 % value	<23–150 °C
Weight loss 28 days at 180 °C	7·2–3·3 %

TABLE 5.65
CURE TIME AND HARDNESS OF IMIDAZOLE CURED EPOXIES[84]

Curing agent[a] (10 wt %)	Cure time at 150 °C (min)	Barcol hardness
$ZnL_4(SO_3CF_3)_2$	10	84
$CoL_6(SO_3CF_3)_2$	10	84
$NiL_6(SO_3CF_3)_2$	10	86
$CuL_4(SO_3CF_3)_2$	10	85

[a] L = imidazole.

complexes/DGEBA or 3,4-epoxy cyclohexylmethyl-3,4-epoxycyclohexane carboxylate systems cure efficiently and with low exotherm.

Sawa, Nomoto and Suzuki[85] suggest that the imidazole derivatives **86**, **87** and **88** are useful as curing agents for epoxy resins. No properties of epoxy resins cured with these materials are reported.

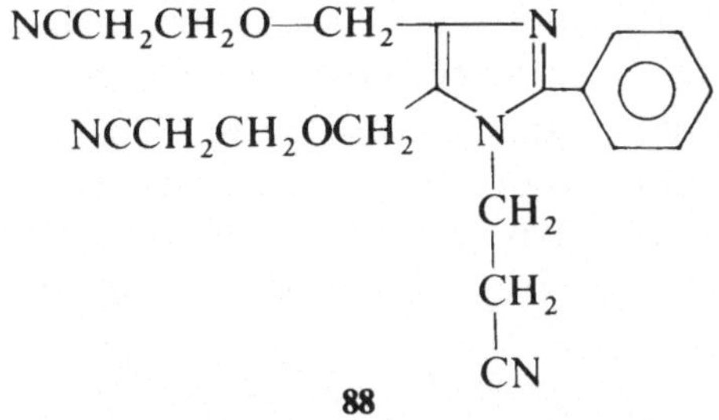

$$NCCH_2CH_2O—CH_2$$

88

Cyanoethyloxymethyl imidazole derivatives

5.8. COMPLEX AND SALT CURING AGENTS

A novel approach to the cure of an epoxy resin utilises a dialkylhydroxy-arylsulphonium salt (**89, 90, 91, 92**) with an organic oxidant, such as organic peroxides, azonitriles, or quinones. This procedure has been described by Crivello.[86-88] For example, three parts of dimethyl-4-hydroxy-3,5-dimethoxyphenylsulphonium hexafluoroarsenate **90a** and 3 parts of benzoylperoxide, mixed with 100 parts of a DGEBA resin (EEW ~ 180) gave a gelled and hardened composition in 2·5 min at 160 °F.

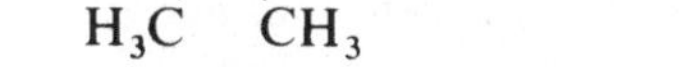

89

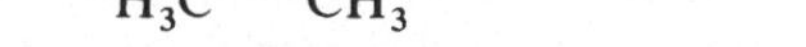

a AsF_6^-
b SbF_6^-

90

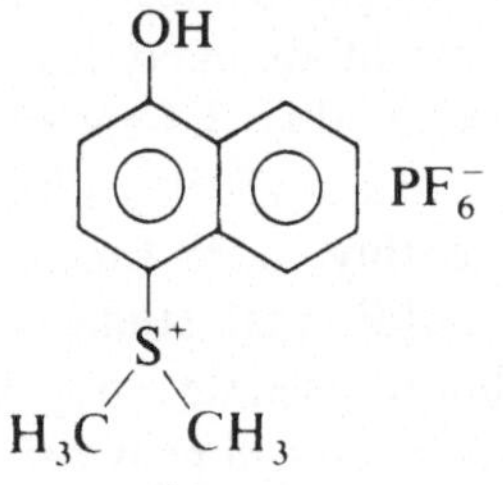

a AsF_6^-
b SbF_6^-

91 PF_6^-

92 AsF_6^-

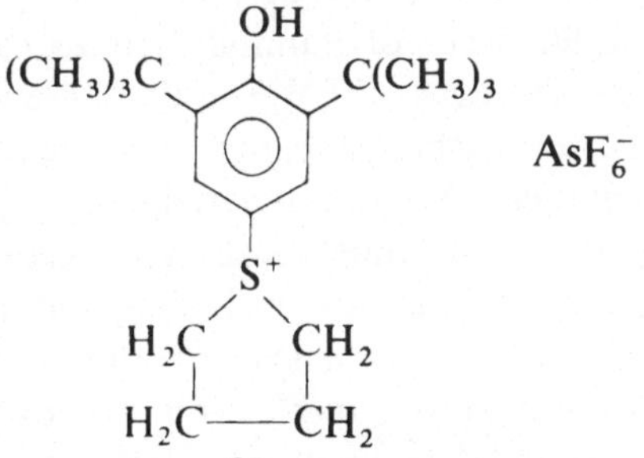

Hydroxyaryl disubstituted sulphonium salts (refs. 86–88)

The gel time with no peroxide catalyst was greater than 7 min. Other epoxy resins such as 3,4-epoxycyclohexylmethyl 3,4-epoxycyclohexane carboxylate, epoxy novalac resin, 1,4-butadiendiol diglycidylether exhibited similar reactivity with this novel curing agent. The author indicated that the curable compositions could be used for coatings, potting compounds, printing inks, sealants, adhesives, moulding compounds, laminates and varnishes.

The use of iodosobenzene diacetate as an oxidant[88] at 6 wt % in combination with the sulphonium salts **89a** or **90a** (3 wt %) with 3,4-epoxycyclohexylmethyl 3,4-epoxycyclohexane carboxylate also cuased gelation to a hard mass to occur within 3 min. Other oxidants such as peracetic acid containing cobalt naphthenate caused hardening within 5 min. Utilisation of heat sensitive onion salts (**93**, **94** and **95**) combined

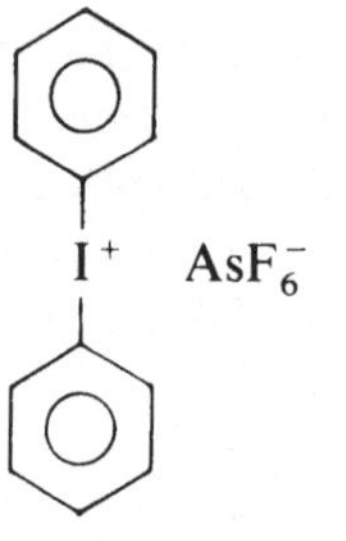

93

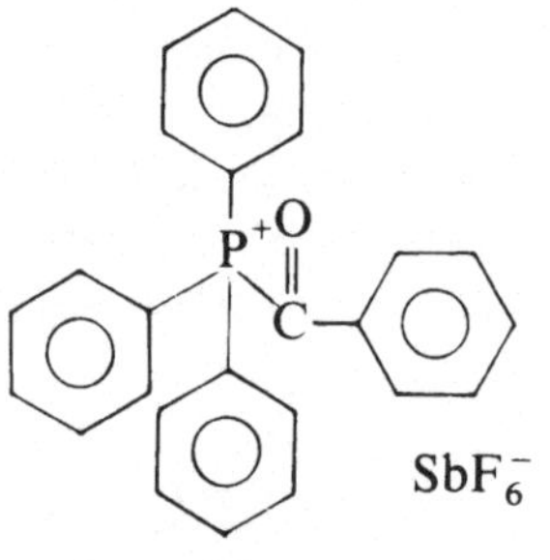

94

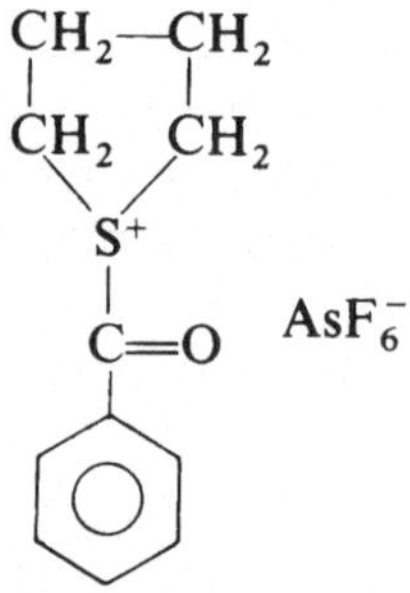

95

Onion salts (refs. 89, 90)

with certain reducing agents as curing agents for epoxy resins, and radiation sensitive aromatic onion salts for polymerisation of epoxy resin materials, have also been described by Crivello.[89] For example the onion salts shown above with certain reducing agents such as pentachlorothiophenol, ferrocene or stannous octoate can initiate cure of an epoxy resin such as DGEBA (EEW ~ 180), 3,4-epoxycyclohexylmethyl 3,4-epoxycyclohexane carboxylate, to a tack-free film between 120 and 150 °C within five minutes. Without the reducing agent, considerably more time is required or, in some cases, no reaction occurs. Additional studies[90] revealed that a mixture of diphenyliodonium hexafluoroarsenate, copper benzoate (0·1 to 10 parts) and stannous octoate or benzoin will provide a room-temperature-curable resin composition when mixed with DGEBA or 3,4-epoxycyclohexylmethyl 3,4-epoxycyclohexane carboxylate. Crivello[91] further showed that certain radiation sensitive onion salts, such as

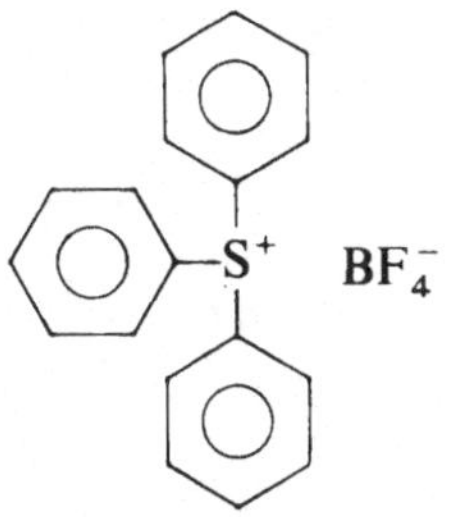

96
Triphenylsulphonium fluoroborate (ref. 91)

triphenylsulphonium fluoroborate (**96**), can effect polymerisation of epoxy resins, such as 3,4-epoxycyclohexylmethyl 3,4-epoxycyclohexane carboxylate or DGEBA (EEW ∼ 180), 4-vinylcyclohexene dioxide, and an epoxy novalac (EEW ∼ 206), in the presence of ultraviolet light radiation from 30 sec to 5 min depending on the onion salt, epoxy resin and other components present. Other anion counterions for the sulphonium atoms included PF_6^-, AsF_6^-, SbF_6^-, $FeCl_4^-$, $SnCl_4^-$, $SbCl_6^-$ and PCl_6^-.

Payne[92] revealed that certain triaryl sulphonium salts, such as triphenylsulphonium chloride, function as accelerators with a polycarboxylic acid anhydride in the cure of epoxy resins. For example DGEBA (EEW ∼ 185) can be gelled at 100 °C by a mixture of nadic methyl anhydride and triphenylsulphonium chloride (1·34 milliequivalents) within 17 min. Benzyldimethylamine or dimethylaminomethylphenol (DMP-10) require 52 and 72 min respectively to affect gelation.

TABLE 5.66
GELATION TIME FOR ALUMINIUM SALTS WITH 3,4-EPOXY CYCLOHEXYL-
METHYL 3,4-EPOXYCYCLOHEXANE CARBOXYLATE RESIN[93]
(containing an epoxy/silane ratio of 3·5/1)

Catalyst	Gel time at 160 °C
Al diisopropoxide phenyl salicylate chelate	90 sec
Al tris(phenyldimethylsilanolate)	2 min
Al benzoate	160 sec
Al tris(hexafluoroacetylacetonate)	45 sec
Al oleate	2–7·5 min
Al stearate	2–3 min
Al resinate	60 sec
Al triisopropoxide[a]	60 sec

[a] Temperature, 150 °C.

TABLE 5.67

GELATION TIME FOR SILANES WITH 3,4-EPOXYCYCLOHEXYLMETHYL 3,4-EPOXYCYCLO-
HEXANE-CARBOXYLATE RESIN (EPOXY/SILANE RATIO 3·5/1) CONTAINING 0·1 wt %
ALUMINIUM ISOPROPOXY METHOXY(ETHYLACETOACETONATE)[93]

Organosilicon compound	Gel time at 160°C
Diphenyl methylethoxy silane	40 sec
Dimethyl dimethoxy silane	3–4·5 min
Phenyl methyl dimethoxy silane	50 sec
Diphenyl dimethoxy silane	30 sec
Methyl trimethoxy silane	120 sec
Methyl triethoxy silane	110 sec
Propyl trimethoxy silane	50 sec
3,3,3-Trifluoropropyl trimethoxy silane	45 sec
Methacryloxypropyl trimethoxy silane	75 sec
Tetraethoxy silane	75 sec
$(C_6H_5Si(OCH_3)_2)_2O$	40 sec
Methoxylated copolymer of 33 mol % $C_6H_5SiO_{3/2}$ and 67 mol % $(CH_3)_2SiO$ containing 17 wt % methoxy groups	50 sec

The combined effects of aluminium alkoxide or acetylacetonates and organosilanes such as phenyltrimethoxysilane in the rapid cure of epoxy films between 160 and 200 °C is noted.[93] Variations in the type of aluminium complex (Table 5.66) and organosilane (Table 5.67) with 3,4-epoxycyclohexylmethyl 3,4-epoxycyclohexane carboxylate to form a solvent-resistant film at 160 °C within 2–7·5 min were also investigated. For example, when the epoxy resin 3,4-epoxycyclohexylmethyl 3,4-epoxycyclohexane carboxylate was mixed wtih 0·5 % by weight of phenyltrimethoxy-silane based on the weight of epoxy resin and silane and 0·1 % by weight aluminium tris(acetylacetonate), coated onto aluminium, generated a hard, solvent-resistant coating after heating at 200 °C for 2 min.

The salt **97** of salicylic acid (SA) and N,N-bis(3-aminopropyl) methylamine (also called methyliminobispropylamine, MIBPA) was found to be an effective curing agent with DGEBA (EEW ~ 185).[94] The most effective weight ratios of MIBPA to SA were found to be 60:40 and 50:50 in

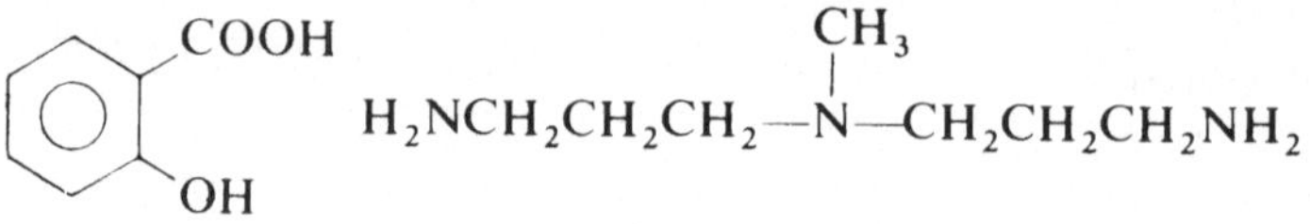

$$H_2NCH_2CH_2CH_2-\overset{\overset{\displaystyle CH_3}{\displaystyle |}}{N}-CH_2CH_2CH_2NH_2$$

97

Methyliminobisisopropyl amine salt of salicylic acid (ref. 94)

TABLE 5.68

PROPERTIES OF REACTION INJECTION MOULDING (RIM) OF DGEBA[a]/MIBPA[b]: SA[c] BLENDS[d94]

Curing agent weight ratio	Parts of salt complex per 100 g DGEBA	Viscosity at 25°C (cPs)	Gel time, 200 g mass (min.)	Peak temp. (°C)	Time to peak temp. (min.)	HDT 264 psi/ 66 psi	Tensile properties		
							Strength (MPa)	Modulus (GPa)	Elongation to break (%)
MIBPA	21	500	29·5	267·4	37·5	—	51·7	2·34	14·1
MIBPA/SA (95:5)	20	700	16·8	283·4	25·0	78/87	64·2	2·61	14·2
MIBPA/SA (80:10)	20	850	11·4	260·2	18·0	84/96	57·2	2·30	16·7
MIBPA/SA (75:25)	20	2 400	6·7	240·2	14·0	96/103	65·5	2·39	10·1
MIBPA/SA (60:40)	26	6 500	4·8	226·5	7·0	93/105	74·5	2·87	7·5
MIBPA/SA (50:50)	30	13 500	5·0	220·4	7·5	91/102	—	—	—

[a] DGEBA (EEW ~ 185).
[b] Methyliminobispropylamine.
[c] Salicylic acid.
[d] Cured 150°C/3 min.

terms of the highest glass transition temperature (T_g) and flexural and tensile properties of DGEBA/MIBDA:SA cured castings. For a neutralised acid–base salt, the theoretical weight ratio would be $1{\cdot}428/1{\cdot}000$ (60/42). However, higher concentrations of salicylic acid in the MIBPA/SA combination also decreased the time to peak temperature and gelation time at 150 °C. The most effective weight ratios of DGEBA/MIBPA:SA were 100/24 to 100/30 based on glass transition temperatures and mechanical properties. The author suggested the use of this epoxy/curing system for the production of reaction injection moulded (RIM) epoxy products. Some properties of the RIM-prepared DGEBA/MIBPA:SA mouldings are given in Table 5.68.

Electron charge transfer complexes **98** and **99**, consisting of a mono- or polyfunctional anhydride, such as methylendomethylenetetrahydro-phthalic anhydride or 1-methyl-tetrahydtophthalic anhydride and a Lewis

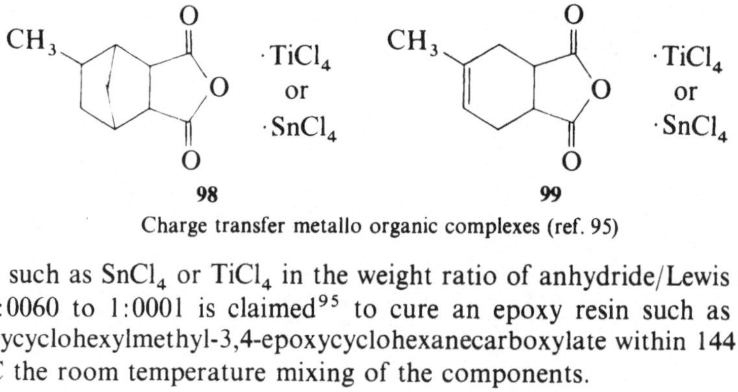

Charge transfer metallo organic complexes (ref. 95)

acid, such as $SnCl_4$ or $TiCl_4$ in the weight ratio of anhydride/Lewis acid of 1:0060 to 1:0001 is claimed[95] to cure an epoxy resin such as 3,4-epoxycyclohexylmethyl-3,4-epoxycyclohexanecarboxylate within 144 h at 25 °C the room temperature mixing of the components.

5.9. MISCELLANEOUS CURING AGENTS AND ACCELERATORS

The toxicological problems associated with handling volatile polyamine curing agents have led to the development of less volatile and toxic materials. Diacetone acrylamide (DAA) (**100**), when complexed with diethylenetri-amine (DETA) is an example of a nonvolatile room temperature curing

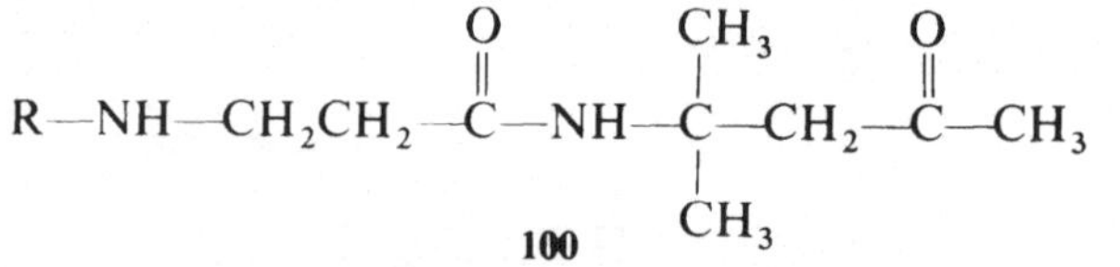

Diacetone acrylamide (ref. 96)

TABLE 5.69

PROPERTIES OF DIACETONE ACRYLAMIDE/DETA, DETA AND POLYAMIDE CURED DGEBA RESIN[a][96]

	DAA[b]/DETA added		Diethylene-triamino	Polyamide[c]
	1:2 *18 phr*	*1:1* *30 phr*		
Tensile strength, MPa	127·4	117·6	98·0	78·4
Tensile elongation, %	4	8	1	4
Izod (notched), ft-lb/in (notch) (Nm/m)	0·7 (37·4)	0·8 (42·8)		

[a] DGEBA (EEW 180–195) annealed castings, cure cycle not given.
[b] DAA is diacetone acrylamide.
[c] amine no. 345.

agent which does not exhibit skin irritating tendencies.[96] The performance of diacetone acrylamide as a curing agent was investigated as a 1:1 complex with DETA. The results listed in Table 5.69 show that the tensile properties are superior to the two controls, DETA and polyamide. Weathering tests on DAA/DETA cured DGEBA coatings indicated that these coatings exhibited improved weathering characteristics and water tolerance over the conventional DETA- and polyamine-cured coatings.[96]

Boron trifluoride–triethylphosphate (1:1) and boron trifluoride–triisopropylphosphate adducts (1:1) were investigated as epoxy curing agents. Optimum curing agent levels were established from heat capacity changes at the glass transition temperature.[97]

The rate of cure of trimethylolphosphine, methyldimethylolphosphine and benzyldimethylolphosphine with a DGEBA was investigated. The heat resistance of the epoxy resins was found to be similar to the heat resistance of polyethylenepolyamine-cured epoxy.[98] It has been demonstrated that the physicomechanical properties of 1,8-diamino-4-methyl-2-nonadiene (20 parts) or 1,8-diamino-2,3,6,7-tetramethyl-6-octadiene (20 parts) cured with DGEBA (ED 20) were similar to those properties of an epoxy (ED 20) cured with polyethylenepolyamine.[99] The adduct of neopentanediamine and polyfunctional epoxy of fatty acid epoxide has been claimed to be useful as a curing agent.[100] Phenylalkylamines are described as curing agents for epoxy resins for high solids epoxy resin coatings,[101] but the structures of the amines were not given.

The Diels–Alder adducts of maleic anhydride with 2-phellandrene or α-terpinene were esterified with glycerol and the ester acid was then evaluated as a curing agent with DGEBA. The cured epoxy resins were considered to have good thermal stability and impact strength.[102] A blend of poly(glycidyl p-hydroxybenzoate) with a diamine or polyamine, a polymer mercaptan and an epoxy resin was found to exhibit a fast cure between 100–200 °C within minutes.[103]

3-Aminophenyl 4-chloro-3-aminobenzenesulphonate (101) has been suggested as a curing agent for epoxies.[104]

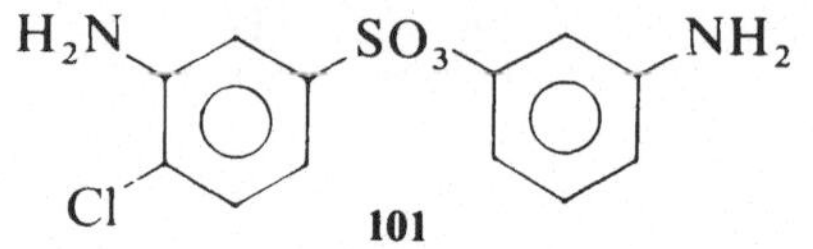

101

3-Aminophenyl 4-chloro-3-aminobenzene sulphonate (ref. 104)

O-Substituted N-(aminobenzyl) anilines (102) have been designated as curing agents for epoxies, but no properties of the cured epoxies are reported.[105]

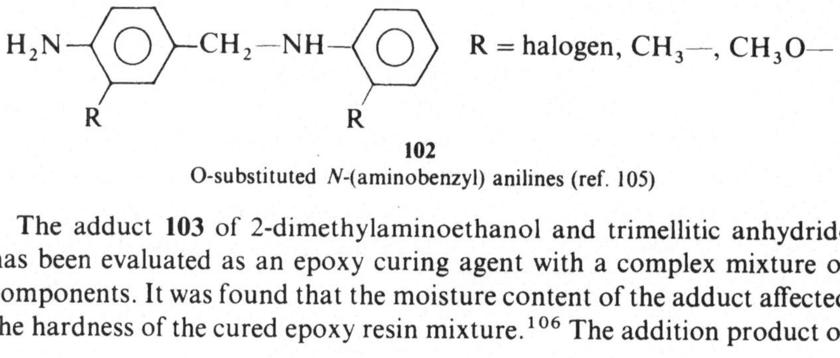

R = halogen, CH_3—, CH_3O—

102

O-substituted N-(aminobenzyl) anilines (ref. 105)

The adduct 103 of 2-dimethylaminoethanol and trimellitic anhydride has been evaluated as an epoxy curing agent with a complex mixture of components. It was found that the moisture content of the adduct affected the hardness of the cured epoxy resin mixture.[106] The addition product of

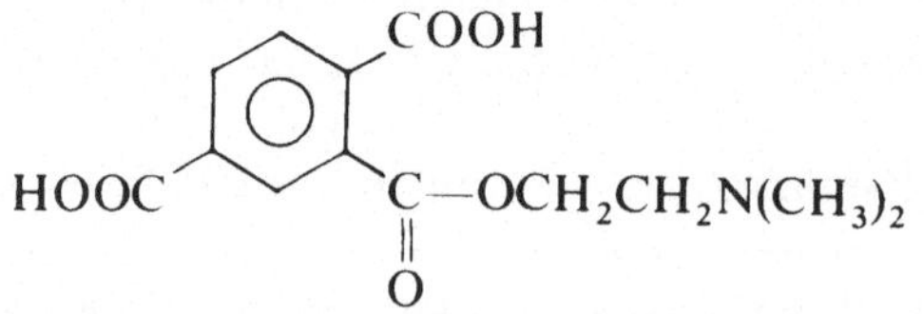

103

2-Dimethylaminoester of trimellitic anhydride (ref. 106)

1,2-diaminocyclohexane and acrylonitrile was evaluated as an epoxy curing agent with DGEBA. The coating prepared from the adduct and DGEBA hardened in seven days, and was considered to have good adhesion and appearance.[107]

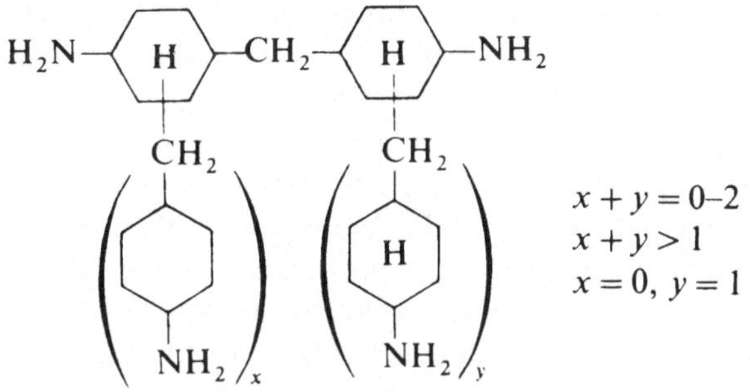

104

Polycycloaliphatic polyamines (ref. 108)

Polycycloaliphatic polyamines (**104**) (20 parts) mixed with 1,2-diaminocyclohexane (80 parts) when used as curing agents with DGEBA yielded a resin with a T_g of ~ 161 °C.[108] Amine-cured epoxy resins containing 0.001 to 3 % of $H_n ZF_m$ catalyst (where Z = Si, As, Sb, P, B, BOH, PO$_2$ or PO$_3$, $n = 1$–2 and $m = 1$–6) are considered fast-cured resins.[109] For example DGEBA (50 g) cured with trimethylhexamethylene-diamine (8 g) and HBF$_4$ (0.005 %) at 15 °C gelled in 35 min, compared with 135 min without HBF$_4$.[109] A mixture of the diamine, **105**, (125.5 g) with Epikote 154 (100 g) was cured at 170 °C/2 h + 200 °C/3 h to yield a product having a T_g of 225 °C.[110]

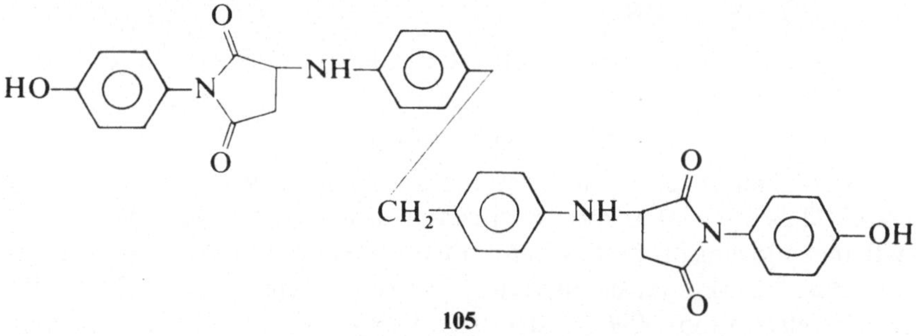

105

Phenolic-amine succinimide derivative (ref. 110)

Isopropoxytris[2-(2-aminoethyl)aminoethoxy] titanium–phenolic Novolac adduct is considered a curing agent for epoxy resins. No properties of a cured epoxy were given.[111] Salts of 2-phenyl-2-imidazoline and pyromellic acid, trimellitic acid or trimeric acid were tested as curing agents with DGEBA epoxy resins in coating applications.[112] Coatings prepared from fine salt resin particles (< 100 μm) exhibited better properties than coatings prepared from larger salt resin particles (27–304 μm).

Nylons[113] prepared from polyamines, $H_2NCH_2CH_2NH_2$, $HN(CH_2$-$CH_2NH_2)_2$, $H_2N(CH_2)_6NH_2$, piperazine, piperazine diamine, m-C_6H_4-$(NH_2)_2$, and dicarboxylic acids (malonic, succinic, adipic, sebacic, isophthalic, phthalic, and terephthalic) were tested as curing agents for epoxy resins. The nylon, poly(piperazine sebacate) was found useful as a latent curing agent, and with DGEBA gave better adhesive properties, heat resistance and impact strength than polyamine-cured DGEBA resins. This nylon had a 3 month room-temperature shelf-life, and could be cured between 50 to 100 °C. Rabkina and Potekhina[114] found that the diamine 106 (22 parts) mixed with DGEBA (EEW = 210) (100 parts) had a pot-life

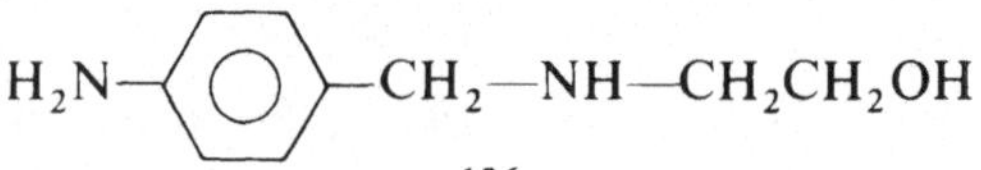

$$H_2N-\!\!\left\langle\bigcirc\right\rangle\!\!-CH_2-NH-CH_2CH_2OH$$

106

p-Hydroxyethylaminomethyl aniline (ref. 114)

of 3 days at 20 °C, and could be cured at 100 °C within 10 h. When the amine was hexamethylenediamine (50 wt %) the pot-life was decreased to 3·5–4 h and cure time to 4–5 h. The mechanical and electrical properties of the cured composition were similar to a polyethylene polyamine cured resin.

Tris-(alkylamino) silanes 107 have been tested as curing agents for epoxy

$$R'-Si\!+\!NHR)_3$$

a $R = CH_3-$, $R' = -\!\!\left\langle\bigcirc\right\rangle$

b $R = CH_3CH_2-$, $R' = -\!\!\left\langle\bigcirc\right\rangle$

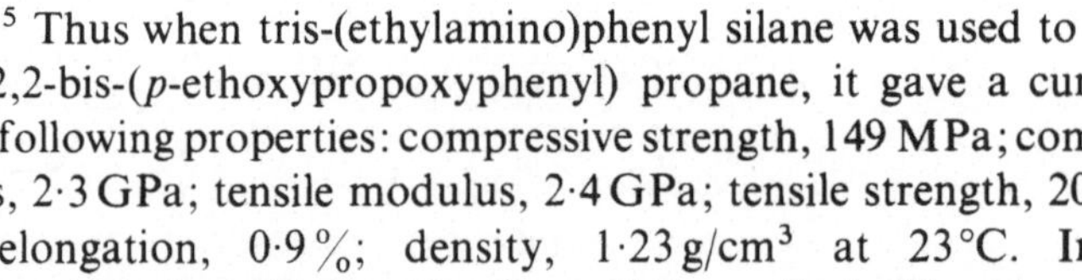

107

Tris-(alkylaminosilanes) (ref. 115)

resins.[115] Thus when tris-(ethylamino)phenyl silane was used to cure the epoxy, 2,2-bis-(p-ethoxypropoxyphenyl) propane, it gave a cured resin with the following properties: compressive strength, 149 MPa; compressive modulus, 2·3 GPa; tensile modulus, 2·4 GPa; tensile strength, 20·8 MPa; tensile elongation, 0·9%; density, 1·23 g/cm³ at 23 °C. Imidazole derivatives 108 where R, R_1, R_2, R_3 and R_4 = alkyl, alkenyl, acyl, aryl,

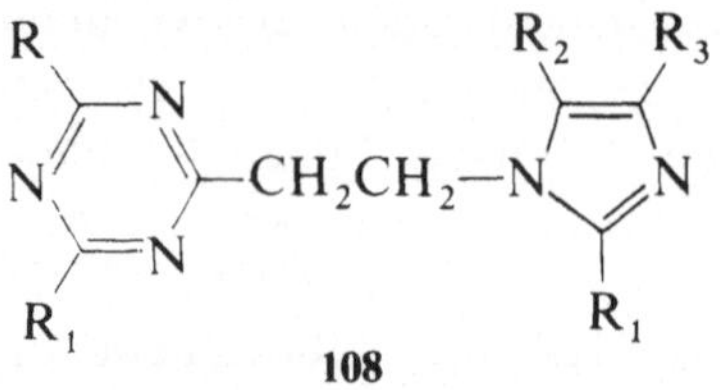

$$\text{(imidazole structure)}-CH_2CH_2-N$$

108

Imidazole derivatives (ref. 116)

cycloalkyl, cyclohexyl, amino, cyano or nitro have been evaluated with amines to determine effects on cure rate.[116] For example, the time to cure an epoxy resin at 170°C with methylenebis(o-chloroaniline) containing catalytic amounts of the imidazole derivative where R, R_1, $R_2 = H$, $R_3 = CH_3$ and $R_4 = CH_2CH_3$ and phthalic anhydride was decreased from 180 to 40 min, compared to compositions without the imidazole derivative and phthalic anhydride.

Beitchman et al.[117] describe Bisphenol A amine salt curing agents **109** comprising a dialkyl substituted polyamine and Bisphenol A. These salts

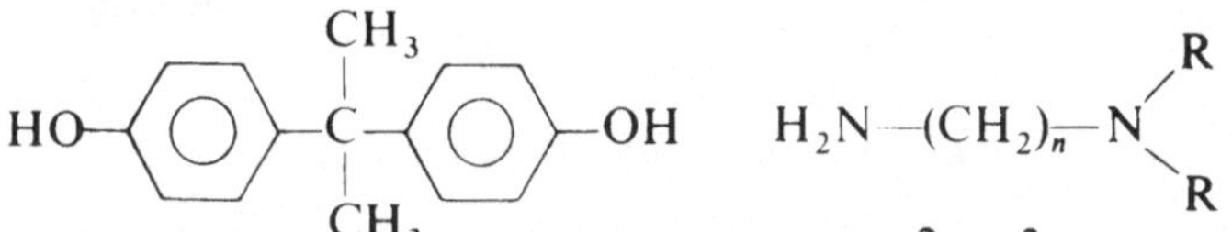

$$HO-\langle O \rangle - \underset{\underset{CH_3}{|}}{\overset{\overset{CH_3}{|}}{C}} - \langle O \rangle - OH \qquad H_2N-(CH_2)_n-N\overset{R}{\underset{R}{\big<}}$$

$$n = 2 \text{ or } 3$$
$$R = CH_3-, CH_3CH_2-$$

109

Amine salts of Bisphenol A (ref. 117)

are latent curing agents with solid state 1,2-epoxy resins having a softening point of not less than 40°C. The amine salt/epoxy powder blends can be cured in the range 132–204°C.

A phenolic curing agent **110**, consisting of a dihydric phenol, a diepoxide and dicarboxylic acid, which reacted in the molar ratio dihydric phenol:diepoxide:dicarboxylic acid of $\geq 2:(n+1):n$ in the presence of a catalyst at temperatures of from 130 to 160°C, is described by Schreurs et al.[118]

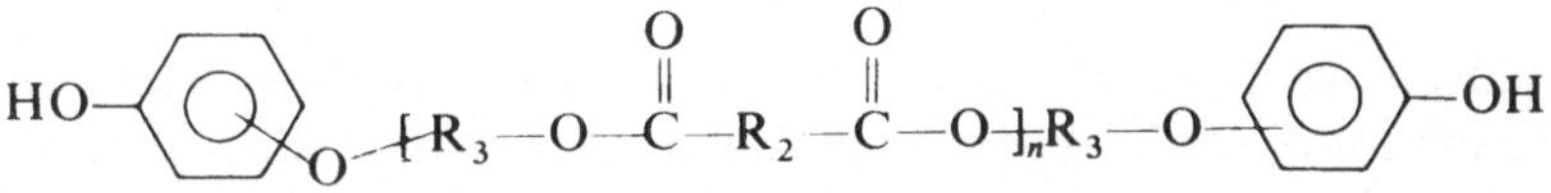

$$HO-\langle O \rangle - O + R_3 - O - \overset{\overset{O}{\|}}{C} - R_2 - \overset{\overset{O}{\|}}{C} - O -]_n R_3 - O - \langle O \rangle - OH$$

$$n = 0.3 \text{ to } 2,$$
$$R_2 = \text{hydrocarbon radical of dicarboxylic acid;}$$
$$R_3 = \text{divalent radical derived from epoxy resin}$$
$$\text{such as DGEBA}$$

110

Phenolic curing agent (ref. 118)

Shay and Griffith[119] describe a phenol-modified aminotriazine curing agent **113**, obtained by reacting phenol, **111**, or a substituted phenol (1.8 to 2.2 moles) with an alkoxymethyl aminotriazine, **112** (1.0 mole), for example trimethoxymethylaminotriazine, to yield a solid having a softening point

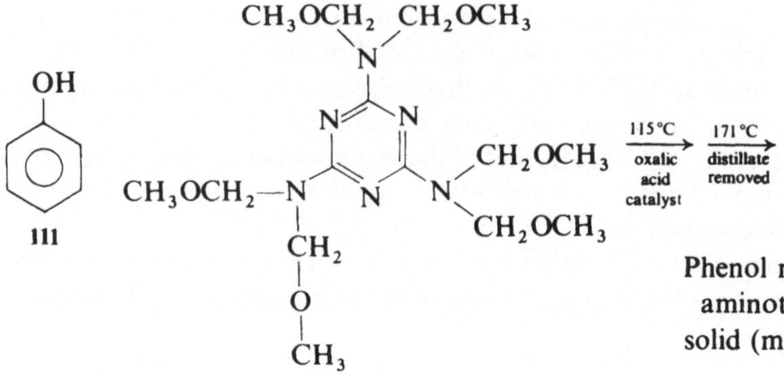

Phenol modified aminotriazine
solid (mp 116 °C)

112   **113**

Phenol modified amonitriazine (ref. 119)

of about (38 °C) 100 °F. The modified triazine is used at about 5 to 20 weight percent with an epoxy resin and cured at about 204 °C to form a film having good adhesion, and the proper combination of flexibility, hardness and gloss.

Alvino, Hammill and Seidel[120] describe the preparation of N-cyano-N'-hydroxymethylguanidine (**114**) from dicyandiamide and formaldehyde, and utilisation of this material as a curing agent with DGEBA to produce a high solids 'C' stage resin material useful as an impregnating composition.

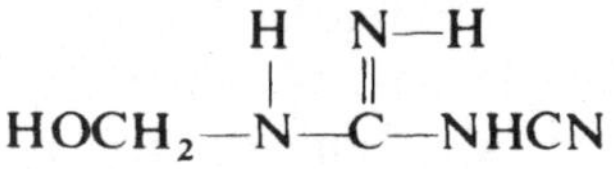

114

Hydroxymethyl dicyandiamide (ref. 120)

The advantage of this system over dicyandiamide is that three to four times more solvent would be needed to obtain dissolution if dicyandiamide were used, and the solids contents would only range from 64 to 67 %, compared with about 86 % for the formulation containing hydroxymethyl derivative **114**. Schulze, Zimmerman and Waddill[121] describe N,N'-polyoxy-alkylenebis(pyrolidinone-3-carboxylic acid) **115** compounds as epoxy

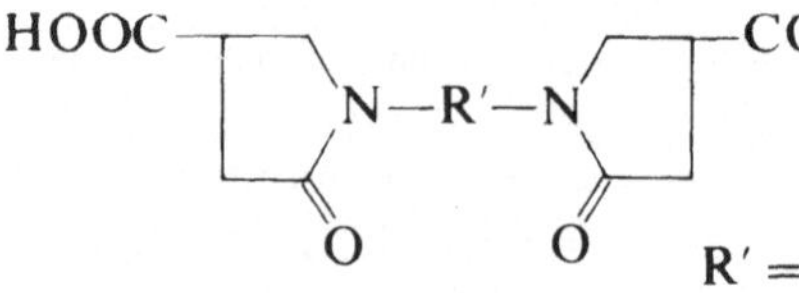

R' = polyoxyalkylene

115

N,N'-polyoxyalkylenebis(pyrolidinone-3-carboxylic acid) (ref. 121)

TABLE 5.70

DGEBA (EEW ~ 90) EPOXY RESIN CURED BY A POLYAROMATIC AMINE CONTAINING $2 \cdot 9$ wt % N,N'-POLYOXYPROPYLENE (PYROLLIDINONE-3-CARBOXYLIC ACID) ACCELERATOR[121]

Izod impact strength, ft-lb/in (Nm/m)	0·61 (32·6)
Tensile strength, MPa	64·1
Tensile modulus, GPa	2·78
Elongation at break, %	5·1
Flexural strength, MPa	10·3
Flexural modulus, GPa	2·90
Shore D hardness, 0–10 sec	86–84
HDT, °C, 1·82 MPa/0·45 MPa	44/48

curing agent accelerators. R' is a polyoxypropylene residue, of molecular weight varying from 200 to 2030, having terminal carbons to which the nitrogen atoms are bonded. Properties of a DGEBA (EEW ~ 190) resin (100 g), cured with 30 g of an aromatic polyamine (condensation product of formaldehyde and aniline) and 20 g of a 20 wt % solution of **115** (R' ≃ 400), in benzyl alcohol at 25 °C for seven days are listed in Table 5.70. When the curing of DGEBA was carried out in the absence of the accelerator, the cured resin was so brittle that it was not possible to cut samples for tests from the castings.

The diacetyl ester of catechol (**116**) and its ring substituted derivatives have been shown[122] to exhibit enhanced characteristics as accelerators in

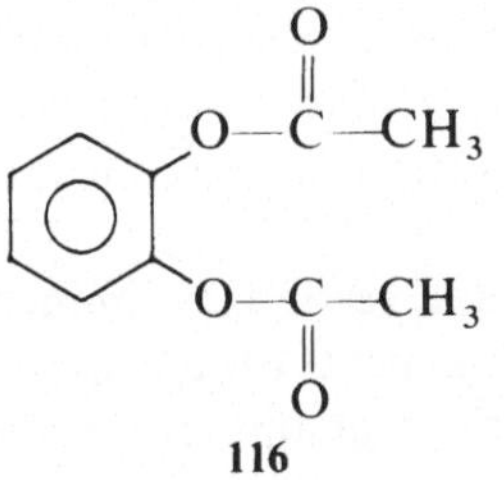

116

Diacetyl catechol (ref. 122)

the cure of epoxy resins, where utilised in combination with a cure initiator such as oxygen titanate or zirconate. The shelf-life stability at 50 °C was increased eight-fold compared with 4-chlorophenol or 4-chlorophenyl-acetate accelerated systems.

Gutekunst, Lohse, and Schmidt[123] describe a curing agent mixture consisting of polyglycidylether, a curing agent 2,6-dihydroxytoluene, containing an amine accelerator, such as 1-methylimidazole, tetramethyl-ammonium chloride or sodium ethylate. The mixture is used as a casting, laminating, or impregnating resin, or as a resin adhesive. The mixture is cured from 120 to 220 °C, and yields a product with high impact strength, good heat stability compared to amine cured epoxies, and a high stability to hydrolysis compared to anhydride cured epoxies.

Foscante *et al.*[124] describe an aminosilane (**117**) curing agent composition for epoxy resins, in which the aminosilane has at least 0·1

$$Y{-}Si{-}(OX)_3$$

117

Trisubstituted aminosilanes (ref. 124)

equivalent of amine per oxirane equivalent, where Y is a $H(NH-R)_a$ group, a is an integer from 2 to 6, and R is a difunctional radical, independently selected from the group consisting of: alkylene, arylene, dialkylarylene, alkoxyalkylene, cycloalkylene radicals, and $X = $ alkyl, hydroxyalkyl, alkoxyalkyl and hydroxyalkoxyalkyl radicals containing less than 16 carbon atoms. The aminosilane is mixed with the epoxy resin (DGEBA, EEW ~ 185) and amine curing agents in a ketone solvent. The ketone reacts with the amine to form a ketimine, which generates water. This water and atmospheric water cause hydrolytic condensation of the aminosilane, while simultaneously atmospheric moisture hydrolyses the ketimine, thereby causing the major epoxy–amine reaction. These simultaneous reactions, polymerisation of the silane and epoxy resin, as well as the amine–silane–epoxy reaction, bring about an interpenetrating polymer network of polysiloxane and epoxy resin. The interpenetrating network can also be formed in the absence of ketone. In this case moisture from the atmosphere or addition of water generates the interpenetrating network of polymers.

Cyanamides of primary and secondary amines, compounds **118** to **125** have been evaluated as curing agents for epoxy resins.[125,126] No properties of epoxy resins cured with cyanamides of the primary amines were reported, but some properties of N,N'-dicyanopiperazine-cured DGEBA (EEW ~ 185) epoxy were given.[126] These are as follows: HTD, 108 °C; flexural strength, 89·7 MPa; flexural modulus, 2·9 GPa; tensile strength, 77·9 MPa; tensile modulus, 2·9 GPa.

Cure of an epoxy novalac resin (EEW ~ 180) with N,N'-dicyano-piperazine at 200 °C/14 h yielded a tough, hard material having a glass transition temperature, T_g, of 208 °C.[126]

The chlorinated diamine **126**, when used as a curing agent for DGEBA

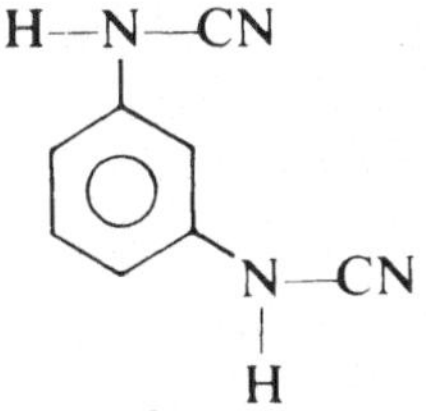

118

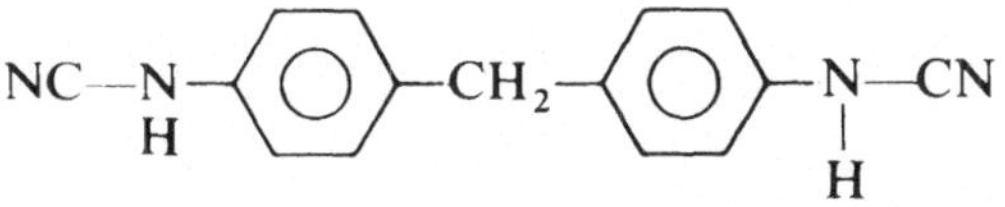

119

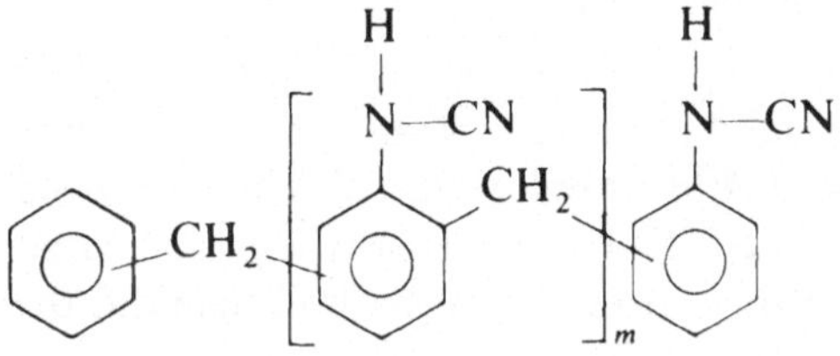

120

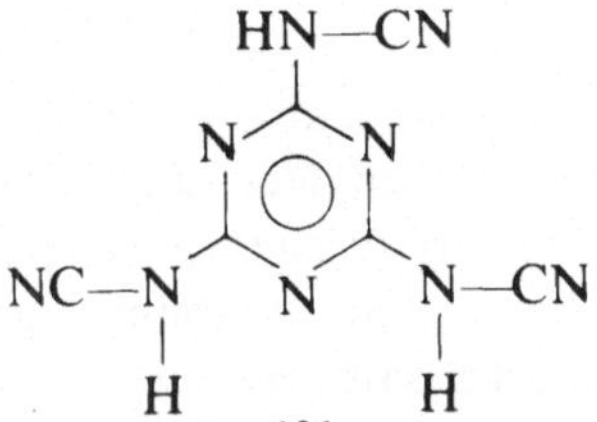

121

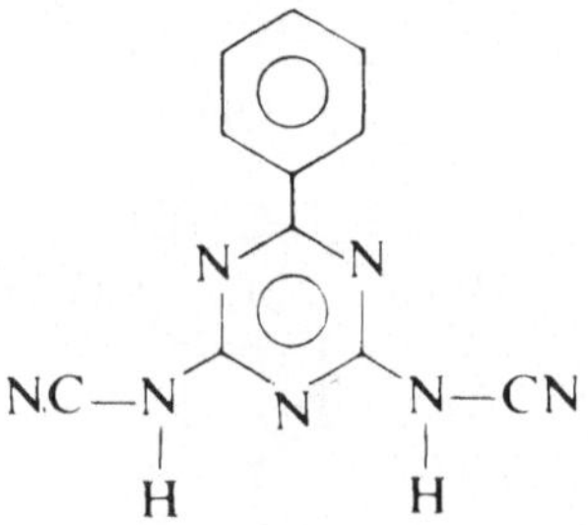

122

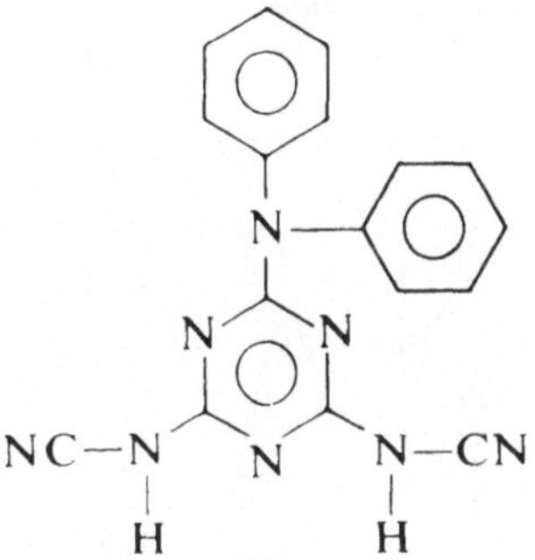

123

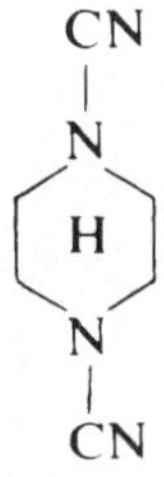

124

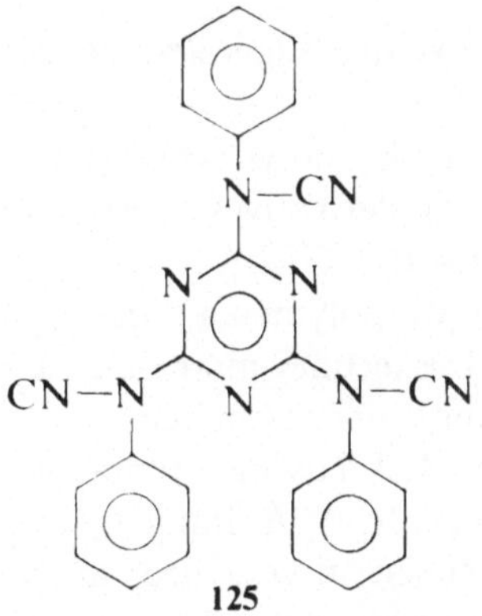

125

Cyanamides of primary and secondary amines (refs 125, 126)

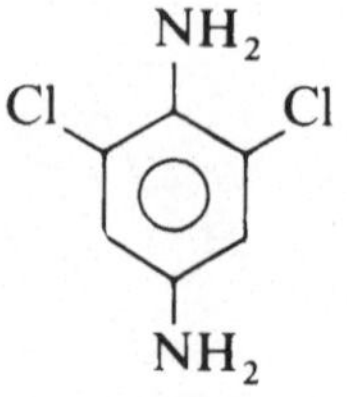

126

2,6-Dichloro-1,4-diaminobenzene (ref. 127)

(EEW $\sim$ 185), and other epoxy resins, produced resins with higher glass transition temperature (T_g values), and improved moisture resistance over the p-phenylene diamine-(Emporal 40-) cured resin.[127] The improved moisture resistance was based on the fact that chlorinated diamine cured epoxy resins exhibited higher T_g values after 24 h water boil than did the control samples.

Fluorinated aromatic anhydrides **127** and **128** have been suggested[128] as curing agents for epoxy resins. No properties of the epoxy resins cured with these materials were described.

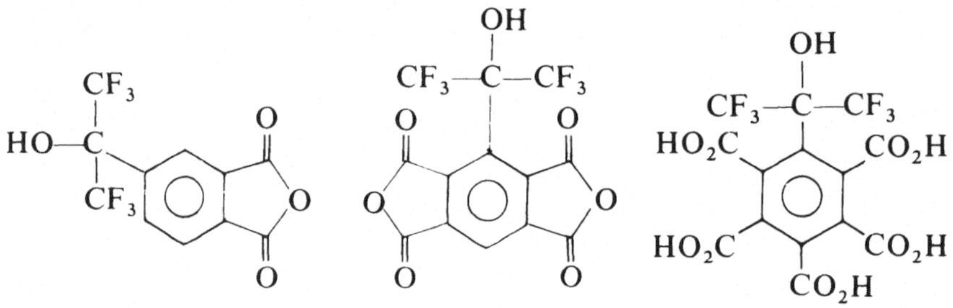

127   **128**   **129**

Fluorinated anhydrides and carboxylic acid (ref. 128)

The polymercaptans **130** are suggested as curing agents for epoxy resins.[129]

The polymercaptans of the polyoxypropylated pentaerythritol and polyoxypropylated sorbitol derivatives, designated **130a** and **130f** respectively, were evaluated with DGEBA (EEW $\sim$ 190). The results, shown in Table 5.71, indicate that the polymercaptan-cured epoxies exhibit longer working mix times than the tertiary-amine-cured epoxy, with equivalent hardness and improved moisture resistance.

Polyethers **131**, **132** and **133**, having aminoether side chains have been claimed as epoxy curing agents.[130] A film, cast from DGEBA and one of the polyether (aminoether) derivatives, cured at room temperature within 24 h to a tough film, free of blushing. The condensation product **134** of

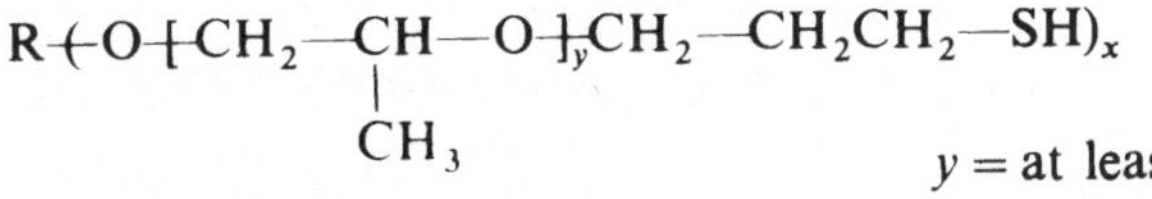

$$R\left(O\left[CH_2-\underset{\underset{CH_3}{|}}{CH}-O\right]_y CH_2-CH_2CH_2-SH\right)_x$$

$y =$ at least 1
$x =$ at least 2

a, R = pentaerythritol
b, R = dipentaerythritol
c, R = glycerol
d, R = propylene glycol
e, R = trimethylolpropane
f, |R = sorbitol

130

Polymercaptans (ref. 129)

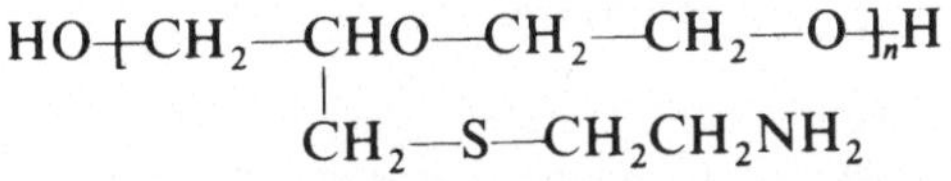

$$HO\left[CH_2-\underset{\underset{CH_2-S-CH_2CH_2NH_2}{|}}{CHO}-CH_2-CH_2-O\right]_n H$$

$n > 6000$
$n > 2500$

131

$$HO\left[CH_2-\underset{\underset{\underset{OH}{|}}{CH_2-SCH_2CHCH_2N^+(CH_3)Cl^-}}{CHO}-CH_2-CH_2-O\right]_n H$$

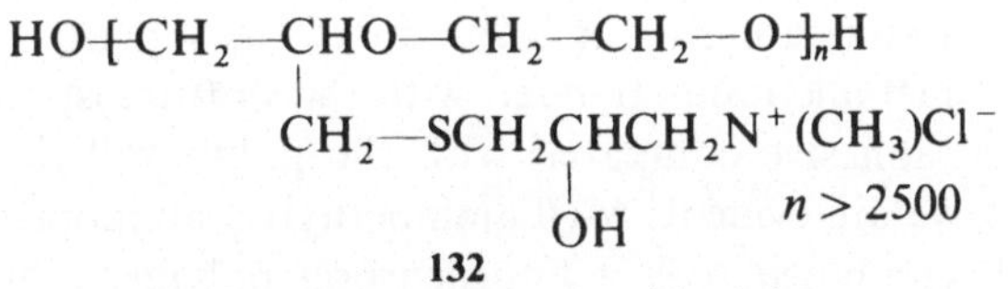

$n > 2500$

132

Polyether with aminothioether side chains

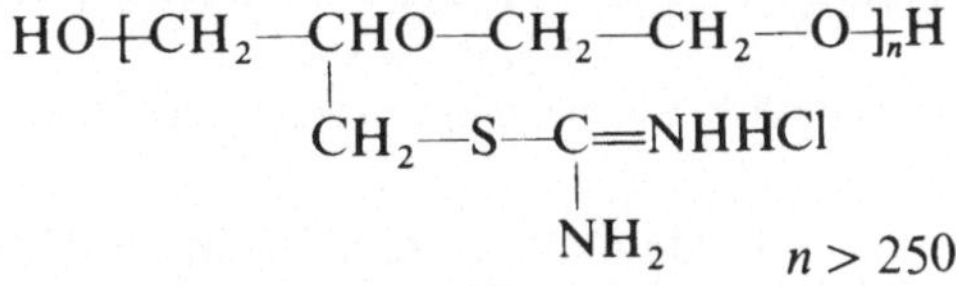

$$HO\left[CH_2-\underset{\underset{\underset{NH_2}{|}}{CH_2-S-C=NHHCl}}{CHO}-CH_2-CH_2-O\right]_n H$$

$n > 2500$

133

Substituted polyaminothioethers (ref. 130)

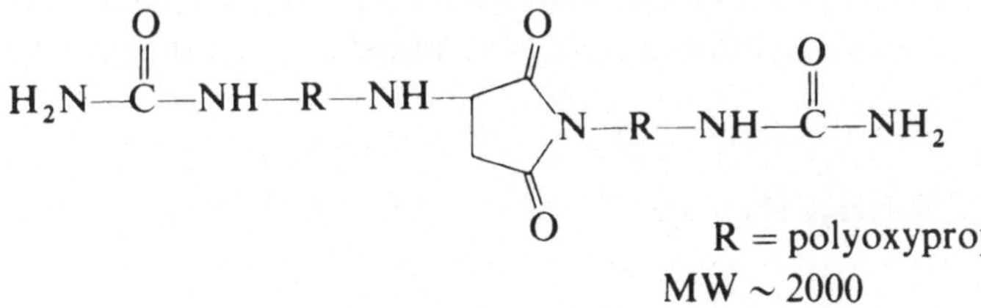

$$H_2N-\underset{\underset{O}{\|}}{C}-NH-R-NH-\left[\text{succinimide ring}\right]N-R-NH-\underset{\underset{O}{\|}}{C}-NH_2$$

R = polyoxypropylenediradial
MW ~ 2000

134

Maleic anhydride polyoxyalkylene diamine adduct (ref. 131)

TABLE 5.71
PROPERTIES OF POLYMERCAPTAN CURED[a] DGEBA (EEW ~ 190)[129]

Curing agent	Room temperature gel time (min)	Shore 'D' hardness		Water absorption[b] (g/cm^2)
		Initial	After 8 weeks	
EH 30 Tertiary amine	5·5	81	76	0·011
Mercaptan 129a	37	81	79	0·006
Mercaptan 129f	40	80	79	0·006

[a] Cured at 24 °C for 7 days.
[b] After immersion in distilled water at room temperature for 8 weeks, expressed in terms of weight gain per cm^2 of surface.

maleic anhydride polyoxypropylene diamine, and urea, was found to be an effective curing agent for epoxy resins such as DGEBA (EEW ~ 185).[131] Adhesive properties of metallic joints bonded with epoxy resin DGEBA (EEW ~ 185) cured with blends of the bis(urea)succinimide and poly-oxyalkylene amines are given in Table 5.72. The superior tensile lap shear strengths of metallic joints bonded with the bis(urea)succinimide/amine-cured epoxy adhesive compared with the polyoxyalkylene-amine-cured epoxy adhesive are evident. N-(β-aminoethyl)-γ-aminopropyltrimethoxy-silane (135) combined with a fluorocarbon polymer (Viton A), carbon

$$(CH_3O)_3Si-CH_2CH_2CH_2NHCH_2CH_2NH_2$$
135
N-(β-aminoethyl)-γ-aminopropyltrimethoxysilane (ref. 132)

black, light magnesium oxide and an epoxy resin DGEBA (EEW ~ 190) is reported[132] to function as a room-temperature curable elastomeric adhesive, coating and sealant.

Hirshmann and Russo[133] demonstrated that an organothiostannoic acid anhydride 136, such as butylthiostannoic acid anhydride, can function as

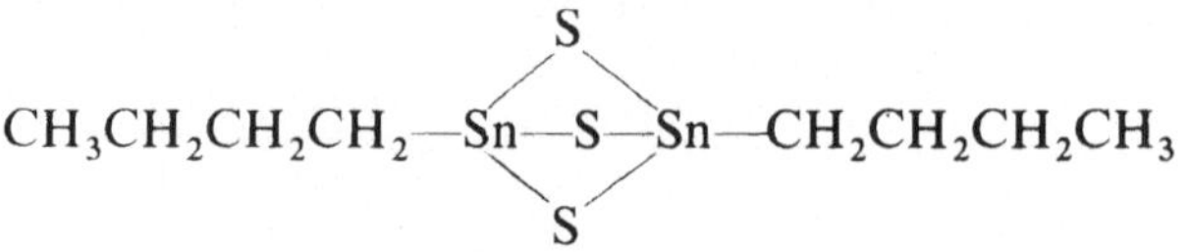

$$CH_3CH_2CH_2CH_2-Sn-S-Sn-CH_2CH_2CH_2CH_3$$

136
Butylthiostannoic acid anhydride (ref. 133)

TABLE 5.72

ADHESIVE PROPERTIES OF EPOXY RESINS[a] CURED[b] WITH BLENDS OF BIS(THIO) SUCCINIMIDE (133) AND POLYOXYALKYLENEAMINES[131]

	A	B	C	D	E	F	G	H
Composition								
DGEBA (EEW)	100	100	100	100	100	100	100	100
Succinimide derivative where R $\simeq$ 2000, phr	0	2	5	20	0	10	0	10
Polyoxypropylene diamine MW ~ 230, phr	30	30	30	30	—	—	—	—
Polyoxypropylene diamine MW ~ 400 phr	—	—	—	—	50	50	—	—
Polyoxypropylene triamine MW ~ 403 phr	—	—	—	—	—	—	45	45
Mechanical properties								
Tensile lap shear strength of bonded metallic joints, MPa	6·76	24·6	27·9	24·4	17·2	23·1	9·6	27·5
Peel strength, N/cm	13·3	17·5	26·0	47·4	—	—	—	—

[a] DGEBA (EEW ~ 185).
[b] Cured 7 days at room temperature.

an accelerator with nadic® methyl anhydride (NA) in curing epoxy resins. For example, 4 parts of NA with 1 part of butylthiostannoic acid anhydride, heated between 145 and 171 °C for 1·5 h, yields a homogeneous milk-white suspension which, when mixed (20 g) with DGEBA (EEW ~ 178) (30 g) and heated to 130 °C, caused gelation of the resin within 20 min. The material was a clear, hard, non-deformable solid at room temperature. By comparison, DGEBA cured with methyl nadic anhydride containing about 2·4 wt % benzyldimethylamine accelerator, heated at 130 °C for 40 min, yielded a material which was still soft at room temperature.

Bernhagen and Springer[134] report that a novel polycycloaliphatic aminoalcohol 137, can be considered as a curing agent for epoxy resins, but no properties of epoxy resins cured with this amine were given.

$$H_2NCH_2 \diagdown \diagup CH_2OH$$

137

Polycyclic aliphatic aminoalcohol (ref. 134)

Waddill[135] showed that a mixture of N-aminoethylpiperazine and triethanolamine is capable of accelerating the cure of an epoxy resin and an amine at ambient or elevated temperatures. Epoxy resin coatings and films can be prepared in less time than required without the accelerator. Waddill and Moss[136] also showed that an oligomeric poly(ethylenepiperazine) is an effective accelerator to cure a DGEBA/dicyandiamide resin mixture, decreasing the gel time at 120 °C from 180 min without accelerator to 20 min with accelerator. The DGEBA/dicyandiamide/accelerator mixture is stable at room temperature, but DGEBA/dicyandiamide/dimethyl-benzylamine mixtures are not.

Simon[137] reported that a corrosion resistant epoxy resin adhesive or coating can be prepared by a room temperature reaction of a mixture of chromium trioxide, a polyamide (with an amine value of 230–400) and water with a standard DEGBA (EEW ~ 185).

The product of a commercial esterified anhydride curing agent (namely the diol ester of trimellitic anhydride) and 2-ethoxyethanol was prepared by heating the reactants at 150 °C for 1 h in a 2-methoxyethylacetate solvent. The clear solution of acid ester alcohol was found to be stable in DGEBA (EEW ~ 190) at 40 °C for six weeks, with only a small increase in viscosity.[138] The esterified mixture is claimed to be useful as a curing agent. For example, a solution of DGEBA/acid ester alcohol, spread on to a tinplate to form a coating, and then cured at 200 °C for 12 min, was found to have good adhesion and methylethylketone solvent resistance. The butylglycolate ester of 3,3′,4,4′-benzophenone-tetracarboxylic acid dianhydride was found to produce films of similar quality.

REFERENCES

1. LEE, H. and NEVILLE, K., *Handbook of Epoxy Resins*, McGraw-Hill, Inc., New York, 1967.
2. SUZUKI, H. and INOUE, T., *Purasuchikkusu*, **23**(11), 87–92 (1972). *Chem. Abs.*, **78**, 168 25μ (1973).
3. KAMAN, T., *Shikizai Kyokaishi*, **47**(1), 2–11 (1974).
4. DiSTASIO, J. I., *Epoxy Resin Technology Developments since 1979*, Noyes Data Corporation (1982).
5. Uniroyal Chemical Data Sheet, Tonox, Tonox LC, and Tonox 60/40, ASP-4418 Uniroyal Chemical, PWA Uniroyal, Inc., Naugatuck, CT 06770.
6. Ciba-Geigy Corporation Data Sheet, Hardener HY 932, Resins Department, Ardsley, NY 10502.
7. Jefferson Chemical Co., Technical Data Sheet, Jeffamine® Poly(oxyethylene) diamines, Houston, Texas 77052.
8. Jefferson Chemical Co., Technical Data Sheet, Jeffamine®, Poly(oxypropylene) amines, Houston, Texas 77052.
9. Ciba-Geigy Corporation Data Sheet, Hardener HY 940, Resins Department, Ardsley, NY 10502.
10. Celanese Data Sheet, Solid phenolic curing agent for epoxy resins—Epicure 8451, Celanese Specialty Resins, Louisville, Kentucky.
11. FIC Corporation, Resins Division Data Sheet, 333 Market Street, San Francisco, California 94105.
12. Sherwin Williams Chemicals, MXDA and 1,3-BAC, Technical Bulletin 169, Sherwin Williams Company, 11541 Champlain Avenue, Chicago, Illinois 60628.
13. Milliken Chemicals, Technical Bulletins. Inman, South Carolina 29349.
14. McCRACKEN, J. H. and SCHULZ, J. C., US Patent 3,078,279, Feb. 19, 1963.
15. DuPont Co., Idea intermediates brochure, E-18333 5/29. DuPont Co., Wilmington, Delaware 19898.
16. Tokyo Chemical Industry Co., Ltd, 3-9-4 Nibonbashi-Honcho, Chuo-ku, Tokyo, Japan.
17. Schering AG Industrie Chemikalien, Technical Data Sheet on Euredur® 370, Postfach 15, Waldstrasse 14, D-4619 Bergkamen, Westf. Berlin, W. Germany (1976).
18. Humko Chemical Technical Data Sheet AMN-553 on Kemamine® DD-3680 dimer amine, Humko Chem. Div., Witco Chem. Corp., PO Box 125, Memphis, TN 38101.
19. General Mills Chemicals, Inc., Technical Data Sheet on versamine dimeryl diamine, Minneapolis, Minn. 55435.

20. Pacific Anchor Chemical Corp., Technical Data Sheets, 1145 South Tenth Street, Richmond, CA 94804.
21. CHIAO, T. J. and MOORE, R. L., 29th Annual Tech. Conf., 1974. Reinforced Plastics/Composites Institute, SPI, Inc., Section 16-B, 1–7.
22. CHIAO, T. J., JESOP, E. S. and NEWEY, H. A., *SAMPE Quart.*, April 1975, 38–43.
23. RINDE, J. A., CHIU, I., MONES, E. T. and NEWEY, H. A., *SAMPE Quart.*, January 1980, 22–30.
24. Veba-Chemie Aktiengellschaft, UK Patent Specification 1,568,183, May 29 (1980).
25. DAUM, G., German Patent 1,911,683 (1970).
26. Archer Daniels Co., Japanese Patent 41-15436 (1966).
27. Teijin Co., Japanese Patent 46-4659 (1971).
28. Ajinomoto Co., Japanese Patent 45-32155 (1970).
29. Chiba Co., Japanese Patent 46-24543 (1971).
30. SMITH, S. and HUBIN, A. J., US Patent 3,644,567, February 22 (1972).
31. LI, T. T., MINO, G. and KAIZERMAN, S., US Patent 3,531,527, September 29 (1970).
32. Union Carbide Co., Japanese Patents 41-12516 (1966) and 45-12864 (1970).
33. JONES, G. D. and TERFERTILLER, N. R., US Patent 3,703,553, November 21 (1972).
34. Ciba-Geigy Co., German Patents 2,113,884 and 2,113,885 (1971).
35. MARKOWITZ, M., *155th ACS Meeting, Org. Coating & Plastic Chemistry*, **28**(1), 392 (1968).
36. MEYER, R. V., KREUDER, H. J. and DeCLEUR, E., European Patent Application, EP 0051787, May 19 (1982).
37. KLUGER, E. W. and SU, TIEN-KUEI, US Patent 4,330,660, May 18 (1982).
38. RIEW, C. K., *Rubber Chemistry and Technology*, **54**(2), 374–402 (1981).
39. RIEW, C. K., ROWE, E. H. and SIEBERT, A. R., *Toughness and Brittleness of Plastics*, eds D. R. Deanin and A. M. Cragnola, Advances in Chemistry Series, **154** ACS, 1976, p. 326.
40. MINATO, I., SHIBATA, K. and FURUOYA, I., European Patent Application, EP 00053366, May 9 (1982).
41. FLOYD, D. E., US Patent 4,126,640, November 21 (1978).
42. DiBENEDETTO, M. and GANNON, J. A., US Patent 4,348,505, September 7 (1982).
43. DiBENEDETTO, M. and GANNON, J. A., European Patent 0044816, January 27 (1982).
44. KING, J. J., SELLERS, R. F. and CASTONGUAY, R. N., US Patent 4,330,659, May 18 (1982).
45. DANTE, M. F., US Patent 4,310,695, January 12 (1982).
46. DANTE, M. F. and ALLEN, R. A., US Patent 4,316,003, February 16 (1982).
47. DANTE, M. F., US Patent 4,322,321, March 30 (1982).
48. FOSCANTE, R. E., GYSEGEN, A. P. and MARTINICH, P. J., US Patent 4,229,563, October 21 (1980).
49. GRAVER, R. B., *J. Paint Technology*, **42**(540), 37–41 (1970).
50. GREENLEE, S. O., CROCKER, G. J. and WEIDNER, C. L., *J. Paint Technology*, **42**(540), 31–6 (1970).

51. GREENLEE, S. O., WEIDNER, C. L. and CROCKER, G. J., US Patent 3,288,766, November 29 (1966).
52. BILOW, N., *J. Appl. Poly. Chem.*, **12**, 175–90 (1968).
53. BLAHAK, J. and PREIS, L., UK Patent Specification 1,475,688, June 1 (1977).
54. SERAFINI, T. T., DELVIGS, P. and VANNUCCI, R. D., US Patent 4,244,857, January 13 (1981).
55. SERAFINI, T. T., DELVIGS, P. and VANNUCCI, R. D., *National SAMPE Tech. Conf., Boston, MA*, SAMPE, **11**, 564 (1979).
56. SCOLA, D. A. and PATER, R. H., NASA Final Report NASA-CR-165229, February 1, 1981.
57. SCOLA, D. A. and PATER, R. H., *National SAMPE Tech. Conf., Mt. Pocono, PA*, SAMPE, **13**, 487 (1981).
58. SCOLA, D. A., *Polymer Composites*, **4**(3) 154–61 (1983).
59. GIBBS, H. H., 28th Ann. Tech. Conf. Reinforced Plastics/Composites, SPI, Inc., Section 2-D (1973).
60. GIBBS, H. H., 29th Ann. Tech. Conf. Reinforced Plastics/Composites SPI, Inc., Section 11-D (1974).
61. CHAMIS, C. C. and SINCLAIR, J. H., *Exper. Mech.*, September 1977, 339.
62. TARNORUTKSII, M. M. and TIKHONENKO, M. I., *Polymer Mechanics*, No. 3, 564–6 (1967). *Chem. Abs.*, **68**, 105728a (1968).
63. MARQUIS, E. T. and WADDILL, H. G., US Patent 4,162,358, July 24 (1979).
64. Ciba Co., Japanese Patent 40-15970 (1965).
65. Furukawa Electric Co., Japanese Patents 45-28892 and 45-28879 (1970).
66. Toyo Rayon, Japanese Patent 46-24009 (1971).
67. BRADSHAW, J. S. and STEVENS, M. P., *J. Appl. Poly. Sci.*, **10**, 1809–12 (1966).
68. SCHULZ, J. G., European Patent Application, EP 0013785, August 6 (1980).
69. SCHULZ, J. G., US Patent 4,150,040, April 17 (1979).
70. PENCZEK, P. and STANIAK, H., *Plast. Kaut.*, **17**(4), 259–60 (1970). *Chem. Abs.*, **72**, 122275a (1970).
71. BELETSKAYA, T. V., MOSHINSKII, L. YA. and STETSYUK, M. F., *Plast. Massy*, **6**, 23–5 (1973). *Chem. Abs.*, **79**, 92883x (1973).
72. KVITA, V., DARMS, R., GREBER, G., MEINDL, H. and ROTH, M. (Ciba-Geigy AG), Swiss Patent 598,255, April 28 (1978). *Chem. Abs.*, **89**, 111365e (1978).
73. KVITA, V., DARMS, R., GREBER, G., MEINDL, H. and ROTH, H. (Ciba-Geigy AG), Swiss Patent 598,253, 28 April (1978). *Chem. Abs.*, **89**, 111366f (1978).
74. NIINO, H., NOGUCHI, S. and TAZUKI, S., *J. Appl. Poly. Sci.*, **27**, 2361–8 (1982).
75. CULBERTSON, B. M., SEDOR, E. A. and SLAGEL, R. C., *Macromolecules*, **1**, 254 (1968).
76. MCKILLIP, W. J. and SLAGEL, R. C., *Can. J. Chem.*, **45**, 2619 (1967).
77. HOLM, R. T., *J. Paint Technology*, **39**, 385–8 (1967).
78. SMITH, J. D. B., *J. Appl. Poly. Sci.*, **26**, 979–86 (1981).
79. STARK, C. J., US Patent 4,307,212, December 22 (1981).
80. VINCENT, H. L., FRYE, C. L. and OPPLIGER, P. E., *Advances in Chemistry Series*, **92**, 164–72 (1970).
81. CHIAO, W. B., UK Patent Application, GB 2,075,512A, November 18 (1981).
82. ELDIN, S. H., STOCKINGER, F., LOHSE, F. and RIHS, G., *J. Appl. Poly. Sci.*, **26**, 3609–22 (1981).

83. Ube Industries Ltd, Japan Kokai Tokkyo Koho JP 82-61,020, 13 April (1982).
84. Thom, K. F., US Patent 4,101,514, July 18 (1978).
85. Sawa, N., Nomato, T. and Suzuki, T., Japan Kokai, 77,111,570, September 19 (1977). *Chem. Abs.*, **88**, 89670 (1978).
86. Crivello, J. V., US Patent 4,230,814, October 28 (1980).
87. Crivello, J. V., Canadian Patent 1,118,550, February 16 (1982).
88. Crivello, J. V., Canadian Patent 1,118,936, February 23 (1982).
89. Crivello, J. V., US Patent 4,216,288, August 5 (1980).
90. Crivello, J. V., US Patent 4,225,691, September 30 (1980).
91. Crivello, J. V., US Patent 4,058,401, November 15 (1977).
92. Payne, W. L., US Patent 3,412,046, November 19 (1968).
93. Lutz, M. A., European Patent Application, EP 0051966, May 19 (1982).
94. Waddill, H. G., US Patent 4,264,758, April 28 (1981).
95. Smith, J. D. B., US Patent 4,273,914, June 16 (1981).
96. Laudise, M. A., Farone, E. R. and Higgins, W. A., *ACS Div. Org. Coatings & Plastics Chem.*, **28**(2), 203–9 (1968).
97. Tulimowsku, Z., *Polimery*, **18**(7), 350–2 (1973). *Chem. Abs.*, **80**, 27823i (1974).
98. Lyapichev, V. E. and Tuzhikov, O. I., *Funkts. Org. Seodin-Polim*, 57–61 (1972). *Chem. Abs.*, **81**, 78643d (1974).
99. Gulieva, K. A., Mustafaev, A. M. and Gadzhiev, T., *Plast. Massy*, **11**, 70 (1975). *Chem. Abs.*, **84**, 74985f (1976).
100. Anon., *Res. Disclosure*, **179**, 130 (1979). *Chem. Abs.*, **90**, 169510 (1979).
101. Gardiner, R. A., *Proc. Water-Borne Higher Solids Coating Symp.*, **5**(1), 105–33 (1978).
102. Matynia, J., *Polimery*, **25**(6–7), 227–30 (1980). *Chem. Abs.*, **94**, 84918g (1981).
103. Sakakibara, Y. and Okabe, E., Japan Kokai Application 76-29539, March 18 (1979). *Chem. Abs.*, **88**, 106246s (1978).
104. Popov, L. K., Tishchenko, M. A., Denisova, T. V. and Lapitskio, V. A., USSR 635,093, November 30 (1978). *Chem. Abs.*, **90**, 122529h (1979).
105. Popov, L. K., Ushakova, M. B., Lapitskii, V. A., Kozova, N. N., Volnykh, J. V. and Artemov, V. N., USSR 696,008, December 5 (1979). *Chem. Abs.*, **92**, 59769k (1980).
106. Tanaka, G. and Suzuki, H., Urano, T., Japan Kokai Tokkyo Koho 79-110,298, August 29 (1979). *Chem. Abs.*, **92**, 42871q (1980).
107. Takai, Y., Yamazaki, K., Yamamoto, J. and Shibahara, Y., Japan Kokai Tokkyo Koho 80-27365, February 27 (1980). *Chem. Abs.*, **93**, 9644r (1980).
108. Kluger, E. W. and Su, Tien-Kuei, US Patent 4,226,737, October 7 (1980).
109. Weisner, I., Czeck 184,973. August 15 (1980). *Chem. Abs.*, **95**, 8318s (1981).
110. Matsushita Electric Works Ltd, Japan Kokai Tokkyo Koho JP 81-118,415, September 17 (1981).
111. Matsushita Electric Works Ltd, Japan Kokai Tokkyo Koho JP 81-129214, October 9 (1981).
112. Gude, F., Neubold, K. Reimer, H. and Dormann, G., European Patent Application, EP 0044030, January 20 (1982).

113. FAKUDA, B., NAMBA, S. and NABESHIMA, H., *Nippon Secchaku Kyokai Shi*, 1(1), 24–8 (1965). *Chem. Abs.*, 67, 74090j (1967).
114. RABKINA, A. E. and POTEKHINA, E. S., *Novoe Obl. Sin. Otverzhdeniya Issled. Epoksidnykh Smol. Leningrad. Dom. Nauch. Tekh. Propag. SB Statei 1967*, 20–3. *Chem. Abs.*, 68, 115359v (1968).
115. BILOW, N., MURPHY, R. F. and PATTERSON, W. J., *J. App. Poly. Sci.*, 11(11), 2109–20 (1967).
116. TAKAHASHI, A., SHIMIZU, R., SHIMAZAKI, T. and WAJIMA, M., Japan Kokai 78-34897, March 31 (1978). *Chem. Abs.*, 89, 111373f (1978).
117. BEITCHMAN, B. D., JEFFERSON, D. E. and SEYMOUR, J. P., US Patent 4,250,293, February 10 (1981).
118. SCHREURS, G. C. M., RAUDENBUSCH, W. TH. and VISSER, T. N., US Patent 4,214,068, July 22 (1980).
119. SHAY, G. D. and GRIFFITH, J. H., US Patent 4,189,421, February 19 (1981).
120. ALVINO, W. M., HAMMILL, J. L. and SEIDEL, M. P., US Patent 4,327,143, April 27 (1982).
121. SCHULZE, H., ZIMMERMAN, R. L. and WADDILL, H. G., US Patent 4,332,720, June 1 (1982).
122. STARK, C. J., US Patent 4,307,213, December 22 (1981).
123. GUTEKUNST, F., LOHSE, R. and SCHMIDT, R., US Patent 4,216,304, August 5 (1980).
124. FOSCANTE, R. E., GYSEGEM, A. P., MARTIVICH, P. J. and LAW, G. H., US Patent 4,250,074, February 10 (1981).
125. SELTZER, R. and DiPRIMA, J., US Patent 4,168,364, September 18 (1979).
126. SELTZER, R., US Patent 4,140,658, February 20 (1979).
127. SUSMAN, S. E., US Patent 4,069,204, January 17 (1978).
128. GRIFFITH, J. R. and O'REAR, J. G., US Patent 4,045,408, August 30 (1977).
129. HARRIS, R. L. and GOBLE, P. H., US Patent 4,092,293, May 30 (1978).
130. HICKNER, R. A. and FARBER, H. A., US Patent 4,188,340, February 12 (1980).
131. WADDILL, H. G. and SCHULZE, H., US Patent 4,147,857, April 3 (1979).
132. ALLEN, C. M. and HINKLIEFF, I. R., European Patent Application, EP 00053002, June 2 (1982).
133. HIRSHMANN, J. L. and RUSSO, R. V., US Patent 4,120,875, October 17 (1978).
134. BERNHAGEN, W. and SPRINGER, H., European Patent Application, EP 0049,889, April 21 (1982).
135. WADDILL, H. G., US Patent 4,195,153, March 25 (1980).
136. WADDILL, H. G. and MOSS, P. H., US Patent 4,188,474, February 12 (1980).
137. SIMON, E., US Patent 4,042,544, August 16 (1977).
138. FALKENBURG, H. R., KRAUSE, S. and McGUINESS, R. C., European Patent Application, EP 0050,940, May 5 (1982).

INDEX

Absorbance subtraction of IR spectra,
    56–7, 75
    examples, 69, 70–3, 78, 79, 80, 81
Accelerators, 33, 166
    aluminium complexes, 243, 244
    aminimides, 227, 229, 230
    diacetyl catechol, 253
    metal acetylacetonates, 231–3, 234,
        235
    piperazine mixtures, 260
    pyrrolidinones, 252–3
    silicones, 234–5
Acetylacetonates, 231–3, 234, 235
Acrylonitrile–butadiene copolymers,
    191–4, 195
Adduct curing agents, 168–9, 247–8
    amine–epoxy, 170, 184–5, 188,
        189–91
    thioalcohol–epoxy, 204–7
Adhesion in failure processes, 160–1
Adhesivity of epoxy compositions,
    229, 237, 239, 258–9
Aerospace industry, 2–3, 14, 22
Aircraft industry, 2, 22, 27–8, 127, 128
Aluminium
    complexes, 243, 244
    compression strength of filled resins,
        136

Aluminium—*contd.*
    compression testing of laminate,
        139–40
    heat dissipation, 111, 114–15, 117
Amidines, 190
Amine curing agents, 166–96, 247,
        248, 254–6
    aliphatic and alicyclic, 33, 169–70,
        171, 172–96, 250, 256–8, 260
    aromatic, 162–72, 207–22, 248,
        250
    bisimide, 209–19
    epoxy adducts, 196–204, 244
    heterocyclic, 182–5, 188
    polyether, 33, 172–5, 256–8
Aminimides, 225–9, 230
Aminoalcohols, 260
Aminoesters, 208–9, 210
N-(2-Aminoethyl)piperazine, 191–3
3-Aminophenyl-4-chloro-3-amino-
        benzene sulphonate, 248
Anelastic effects, 106, 109–110
Anhydride curing agents, 172, 222–5,
        246, 256, 258, 260–1
    cured epoxy resins, 229, 230, 233,
        234, 235
    dianhydrides, 170, 222–5
Anisotropy, 142

Arsenate curing agents, 241–2
Automotive industry, 14

Benzophenonetetracarboxylic acid
    dianhydride, 170
Benzyldimethylamine, 230
2,4-Bis(p-aminobenzyl)aniline, 170
Bis(4-aminocyclohexyl)methane, 169,
    197, 198
Bis(N,N'-4-aminopiperidine) curing
    agents, 182–5, 188
N,N'-Bis(3-aminopropyl) dimer
    diamine, 171
2,2-Bis(epoxypropoxyphenyl)propane,
    207–8
Bisimide amine curing agents, 209–19
Bis(2-methyl-3-aminocyclohexyl)-
    methane, 169
Bis(monomethyl)cyclohexane, 168–9,
    194
Bisphenol A amine salts, 251
Bisphenol A epoxy resins
  degradation studies, 76, 77
  Raman spectra, 82
  solid polyamide curing agent, 167
  strength of carbon fibre composites,
    35
  see also Bisphenol A diglycidyl
    ether resins
Bisphenol A diglycidyl ether resins
  DSC scan, 41–3
  functional group stability, 78, 79
  IR spectra, 62–8
  one-part, 228–9, 233–5
  toughened, 192–5
  treated with curing agents
    aliphatic and alicyclic amines, 169,
      182, 185, 187, 189–90, 191,
      194–6
    amine adducts, 197–204, 248
    aminosilane, 254
    anhydride, 223–4, 236, 258–60
    complex salts, 241–6, 248, 249
    diacetone acrylamide, 247
    imidazoles, 239–40
    ketimines, 230, 232
    piperazines, 237, 238, 254

Bisphenol A diglycidyl ether
    resins—contd.
    piperidines, 184–5, 188
    polyamines, 249, 250
    polyether diamines, 33, 173–5, 178,
      179, 256–9
    polymercaptan, 256, 258
    polysiloxane diamines, 189
    pyrrolidinone accelerated, 253
Bisphenol F diglycidyl ether (DGEBF)
    resins
  cured with polyether triamine, 175,
    180–2, 183, 184
  IR spectra, 62–8
Bisphenol S resins, 34
Boron fibre epoxy composites, 143,
    155
  compression strength, 141, 143
  damage and fatigue, 114–15, 117,
    126, 151
  failure, 155
Boron trifluoride, 247, 228–9
Buckling of fibres, 137, 138, 139, 155–9
Butadiene–acrylonitrile copolymers,
    191–4, 195
1,4-Butadienediol diglycidyl ether
    resin, 242
Butylstannoic acid anhydride, 258–60

Calorimetric analysis, 37–43
Carbon fibre
  continuous (long), 7–23
  epoxy composites
    cured with bisimide amines,
      217–19
    effect of filler, 41
    failure, 154–5
    strength, 35, 142, 154–5
    thermography, 111, 114, 121, 125
  hybrid composites, 147
    compression strength, 149–51
  laminates
    compression strength and testing,
      140, 141, 143–4, 146
    fatigue, 151, 152
  short, 2, 3, 4–7
  thermoplastic composites, 1–29

Celanese compression test, 140–1
Charge transfer complexes, 246
Chemical resistance of epoxy resins, 199, 200, 205–7, 232
    see also Solvent sensitivity
Coatings, 23–5, 91–3
Compression properties
    composites
        failure processes, 153–61
        fatigue, 151–3
        hybrid, 147–51
    epoxy resins, 133, 134, 135–6, 153
        amine–epoxy adduct cured, 202
        bisimide amine cured, 215, 216, 219
        polyether amine cured, 181
    factors affecting, 131–2, 137
    polyester resins, 133–6, 153
    tests, 133, 137–42
    thermoplastic resins, 27, 136
    thermoset resins, 27
Continuous fibre reinforcement, 7–23
Crosslinking in epoxy matrices, 59–62, 69–75
Curing agents, 165–261
    adducts
        amine–epoxy, 196–204
        thioalcohol–epoxy, 204–7
    amines
        aliphatic and alicyclic, 33, 169–70, 171, 172–96, 250, 256–8, 260
        aromatic, 166–72, 207–25
        heterocyclic, 182–5, 188
    compositions, 249–54, 260
    cyanamides, 254, 255
    latent, 225–41, 251
    phenolic, 167, 172
    salts and complexes, 241–6, 248, 249
Curing process
    characterisation and control, 31–2, 37–49, 68 9, 123 5
    degree of cure, 39–41, 84, 133–5
    dielectric analysis, 83
    epoxy resins
        degradation, effect on, 77–8
        FTIR studies, 68–9
        polyether amine cured, 175, 177, 178

Curing process—*contd.*
    epoxy resins—*contd.*
        strength and state of cure, 133–5
        temperature of cure, 33, 37
Cyanamides, 254, 255
N-Cyano-*N*′-hydroxymethylguanidine, 252

Damage development, 107, 117, 121, 125–8
Defects, *see* Flaws
Degradation of epoxy resins
    IR spectra, 62–8, 75–9, 84
Degree of cure, 39–41, 84, 133–5
Delamination, 119–21, 122
Diacetone acrylamide, 246–7
Diacetylcatechol, 253
1,2-Diaminocyclohexane, 169–70, 249
    adducts, 190–1, 197, 198
4,4′-Diaminodiphenylmethane, 207
Diaminodiphenylsulphone
    3,3′-, 168
    4,4′-, 166, 168, 197–8, 200, 201, 207
Dianhydride curing agents, 170, 222–5
2,6-Dichloro-1,4-diaminobenzene, 254–6
Dicyandiamide, 237, 238, 252
Dielectric analysis, 82–4
Diethylenetriamine, 197
Diethyl-4,4′-methylenedianiline
    2,2′-, 167–8
    3,3′-, 166, 225
Difference spectra, 56–7, 75
    examples, 69, 70–3, 78, 79, 80, 81
Differential scanning calorimetry (DSC), 37–8, 75
    baseline problems, 41–3
    degree of cure measurement, 39–41, 84
Differential thermal analysis (DTA), 37, 43
Diffuse reflectance spectroscopy, 58–9
Diglycidyl ether of bisphenol A (DGEBA), *see* Bisphenol A diglycidyl ether
Dimeryl diamine, 171
2,5-Dimethyl-2,5-hexanediamine, 34, 182, 183, 185

Dimethyl 4-hydroxy-3,5-
    dimethoxyphenylsulphonium
    hexafluoroarsenate, 241–2
Dodecamethylenediammonium
    succinate, 239
Dynamic mechanical tests, 43–8
    dynamic spring analysis, 45

Elastic after-effect, 109–10
Electrical properties
    carbon fibre thermoplastic
        composites, 5, 26
    epoxy resins
        aminimide cured, 228, 230
        dielectric analysis, 82–4
        one-part, 228
        metal acetylacetonate cured, 233,
            234
Emissivity, 95–7, 116–17
EPON 828, 59, 60, 62, 70–3
Epoxy resins, 32
    compression strength, 133, 135–6
    curing agents, 165–261
    dielectric analysis, 82–4
    failure, 153
    fibre reinforced
        compression strength of
            laminates, 143–4
        mechanical properties, 26–7, 35–7
        quality control, 69–75
        rôle of matrix, 31–7
    IR and Raman spectra, 59–68
    IR degradation studies, 75–9
    mechanical properties, 133–6
    see also individual components, e.g.
        Bisphenol A diglycidyl ether
3,4-Epoxycyclohexylmethyl-3,4-
    epoxycyclohexane carboxylate,
    239–40, 242–4, 246
2-Ethyl-4-imidazole, 228 9

Factor analysis, 54–5
    example, 72, 73
Failure
    envelopes, 146–7
        examples, 148, 152

Failure—contd.
    mechanisms in compression, 138,
        139, 153–61
    fibre, 153–5
    fibre–matrix interface, 159–61
    matrix, 155–9
Fatigue
    compression, in, 151–3
    heat patterns, 115, 117, 127
    testing, 128
Fibre
    failure, 153–5
        buckling, 137, 138, 139, 155–9
    loading in thermoplastics, 25–6
Fibreglass, see Glass fibre
Flaws
    internal free surfaces, with, 108, 114
    monitoring, 107, 117, 125–7
    thermographic detection, 89–90,
        99–105, 107–8, 117–22
Fluorinated aromatic anhydrides, 256
Fluorinated polymers, 19
Fourier transform IR spectroscopy,
    49–54, 56–9, 75–9, 84

Gelation of epoxy resins, 43–8
    gel time determination, 47–8
Glass fibre
    failure processes, 154, 157–9
    fatigue, 151–2
    finishes, 23
    epoxy composites
        compression strength, 142, 143
        cure kinetics, 41
        FTIR studies, 68, 71, 73, 74, 75
        thermography, 114, 121, 123–5,
            127–8
    hybrid composites, 147, 149–51
    polyester composites, 135, 136, 138,
        142, 143
    thermography, 111
    thermoplastic composites, 2, 3, 5
Glass transition temperature
    epoxy resins, 44, 48, 191, 207, 214,
        216
    thermoplastics, 19, 136
Graphite, see Carbon

Heat
  deflection temperature, 34–5, 202
  distortion temperature, 35–7, 169,
    184, 222
  patterns
    production, 101–8
    steady-state, 104
    transient, 104–5, 117–19
    used to monitor damage
      development, 125–7
Hybrid composites, 147–51
Hydantoin-based resins, 35
Hydrolysis of epoxy resins, 80, 81

Ideal material, 132
Imidazoles, 170–1, 228–9, 239–41,
    250–1
  2-phenyl-2-imidazoline salts, 249
Imperfections, 132–3
  see also Damage development, Flaws
Infrared emission, 93–101
Infrared spectroscopy, 48–9, 54–7, 84
  epoxy resins, 59–68
  Fourier transform, 49–54, 56–9,
    75–9, 84
Internal reflection spectroscopy, 57–8
  example, 76, 77
Iodosobenzene diacetate, 242
Isoimidylphthalic anhydrides, 224–5
Isophoronediamine, 169–70, 197
Isophthalic ester resin, 133–5
ITTRI compression test, 141

Ketimines, 229–31, 232
Kevlar
  epoxy laminate, 143
  failure processes, 154, 157–9
  fatigue of composites, 152
  hybrid composites, 147, 148–9, 150

Laminates
  compression testing, 139–41, 142,
    143–6
  fatigue and moisture, 151–3
  hybrid, 147

Laminates—contd.
  thermography, 117, 119, 122–4
Least-squares analysis, 55–6, 74–5
  examples, 69, 72–5
Liquid
  crystalline polymers, 19, 27
  crystals, 92–3
Long fibre reinforcement, 7–23

Matrix failure processes, 155–61
Mechanical properties
  carbon fibre thermoplastic
      composites, 14
    fibre loading, effect of, 25–6
    nylon, 15–17
    poly(amide-imide), 11–12
    polyether etherketone, 20–1
    polyether sulphone, 11
    polyphenyl sulphone laminate,
      18–19
    polysulphone laminate, 4, 9–11
  epoxy resins
    aminimide cured, 228, 230
    bisimide amine cured, 215–19
    carbon fibre composites, 35–7
    diacetone acrylamide cured, 247
    mercaptan epoxy adduct cured,
      204–5
    one-part, 228
    polyether amine cured, 174, 175,
      179, 183, 184, 185
    salicylate cured, 244–6
    siliconate accelerated, 236, 237
    toughened, 193–5
    xylylenediamine cured, 220–2
  thermoplastic and thermoset
      matrices compared, 26–7
  see also Compression properties,
      Toughness
Melt impregnation, 12–21
Menthanediamine, 169
Mercaptans, 204–7, 256, 258
4,4'-Methylenedianiline, 166, 168
4-Methylhexahydrophthalic anhydride,
    172
Methyliminobispropylamine, 244–6
Michelson interferometer, 49–51

Mixing techniques, 32
Modulus of elasticity, *see*
    Compression properties
Moisture
  absorption by epoxy resins
    amine–epoxy adduct cured, 198,
      203
    bisimide amine cured, 213–14, 215,
      217
    FTIR study, 79–81
    polyether triamine cured, 184
  effect
    fatigue, on, 152–3
    strength, on, 146, 217
  thermographic detection, 123
Monitoring of flaws and damage, 107,
    117, 121, 125–7

Nadic methyl anhydride accelerators,
    243, 260
  crosslinking spectra, 70–3
  cured epoxy resin, 223
  IR and Raman spectra, 59, 61, 62
Novolac epoxy resins
  curing agents
    amine–epoxy adducts, 197
    N,N'-dicyanopiperazine, 254
    hexafluoroarsenate, 242
    onion salts, 243
    titanium complex adduct, 249
  degradation, 76, 77–9
  DSC scan, 41–3
Nylon, 14
  carbon fibre composites, 2, 3, 5, 6,
    7, 15–17
  finishes, 24
  curing agents, 250

Onion salts, 242–3
4,4'-Oxydianiline, 207

Phenolic curing agents, 167, 172, 251
  adducts, 198, 203–4
  diacetylcatechol, 253
Phenylenediamine, 166, 207

Phosphines, 247
Phosphor coatings, 91–2
Photodegradation of epoxy resins,
    62–8, 76–9
Photon-effect IR detectors, 100–1
Phthalic anhydrides, 172, 222–5, 246,
    256
Piperazines, 236–7, 254, 260
Piperidines, 182–5, 188, 194
Plastic deformation, 113–14
Polyamide curing agent, 167
Poly(amide-imide) resin, 8, 11–12
Polyamine curing agents, 166–83,
    185–90, 194–6, 207–8, 222,
    248–50
  bisphenol A salts, 251
  diethylenetriamine complex, 246–7
  epoxy adducts, 170, 184, 189–94,
    197, 201–4
Polybutyl terephthalate, 24
Polycarbonates, 6, 8
Polyester etherketone (PEEK) carbon
    fibre composites, 19–21, 22, 23,
    26–7
Polyester resins, 133–6
  failure processes, 153
  fibre composites
    compression properties, 143,
      148–9, 150
    finishes, 24
    isophthalic, 133–5
Polyether amines, 33, 172–5, 256–8
Polyether sulphones, 8, 11, 26
Polyimides, 8, 35
Polymercaptans, 256, 258
Polymethylene dianilines, 166, 225
N,N'-Polyoxyalkylenebis-
    (pyrrolidinone-3-carboxylic
    acid), 252–3
Polyoxyethylene diamines, 172–3
Polyoxypropylene diamines, 173–83,
    238–9, 240
Polyphenylsulphone carbon fibre
    composites, 9–11, 16–19
  solvent impregnated prepregs, 8,
    9–11
  see also Polysulphones, Sulphone
    epoxy resins

Polysiloxane diamines, 187, 189
Polysulphide dianhydrides, 224
Polysulphones
  carbon fibre composites, 3–4
  solvent impregnated prepregs, 8–9, 11
  solvent sensitivity, 3, 8, 9–11, 13
  see also Polyphenylsulphone carbon fibre composites, Sulphone epoxy resins
Poly(thioxyalkanoic acids), 205–7
Polyxylylenepolyamine, 220–2
Pot-life of epoxy resins
  aminimide cured, 227–8
  ketimine cured, 229
  polyamine cured, 167, 169
Prepregs
  epoxy
    compression testing, 147
    dielectric analysis, 83
    quality control, 69–71
    storage, 33–4, 41
    TBA, 45
  impregnation of thermoplastic, 7–8, 13, 24–5
Pultrusions
  compression testing, 137–8, 143
  failure processes, 154, 157–61
  hybrid, 148–9
  thermographic process control, 125
Pyrrolidones, 238–9, 240

Quality control by FTIR, 69–75, 84

Raman spectroscopy, 59–62, 82
Reflectivity, 116–17
Repair of composites, 34
Resonance, 106–8, 113, 118, 119–20

Salicylates, 244–6
Sampling techniques for FTIR, 57–9, 69
Sandwich beam, 139–40, 141
Shelf-life of epoxy resin prepregs, 41

Shelf-life of epoxy resin prepregs
  —contd.
  curing agents, with
    diamine, 182
    latent, 233, 234, 236, 238–40
    phenolic, 167
  diacetylcatechol accelerator, with, 253
  fibre reinforced, 33–4
Short fibre reinforcement, 2, 3, 4–7
Silanes, 244, 250, 254, 258
Silica filled resins, 41, 136
Siliconates, 235–6, 237
Silicone accelerators, 234–5
Solvent
  impregnation, 7–12
  sensitivity
    polyether etherketone, 20–1
    polysulphones, 3, 8, 9–11, 13
Sporting goods, 2, 7
Steel
  fibre composites, 154
  heat dissipation, 111
Stress
  epoxy resins
    carbon fibre composites, 35–7, 144
    moisture stability, effect on, 80, 81
  fields, 134–5
  heat generation, effect on, 106, 108–9, 111–13, 114–15, 122–3
  SPATE system, 122
  yield stress, 134–5
Sulphone epoxy resins, 35
  see also Polysulphones, Polyphenylsulphone carbon fibre composites
Sulphonium salts, 241–3

Temperature measurement, 90–101
4,4'-(N,N'-Tetraglycidyl)methylene-dianiline epoxy resin, 197–8, 201, 202, 213–19
2,2,6,6-Tetramethyl-4-aminopiperidine adducts, 184–5, 188
Thermal conductivity, 110–11, 115–16
Thermal degradation of epoxy resins, 62–8, 78, 79

Thermal expansion
  bisimide amine-cured epoxy resins,
    213
  carbon fibre composites, 5, 25–6
  dependence of thermoelastic
    temperature change, 122
Thermography, 89–128
  active, 90, 105–8, 120, 121, 123–4
  heat pattern production, 101–8
  passive, 90, 102–5, 119
  thermal pulse video, 121–2
Thermoplastics, 136
  carbon fibre composites, 1–29
Thermoset matrix systems, 26–7
Thioalcohol–epoxy adducts, 204–7
Torsional braid analysis (TBA), 44–5,
  46, 84
Torsional impregnated cloth analysis
  (TICA), 45–6
Toughness
  epoxy
    ATBN-toughened, 192–4
    bisimide amine cured, 209
    improvement, 35–7
  hybrid composites, 147
  thermoplastics and thermosets
    compared, 3–4, 26
Toxicity
  aminimides, 226, 227
  polyamines, 166–7, 246–7
Triazines, 251–2
Tris(dimethylaminomethyl)phenyl
  2-ethylhexanoic salt, 228

Ultraviolet spectra of epoxy resins,
  76–9

Vibrothermography, 106, 107–8,
  108–9, 123–4
Video-thermographic systems, 100,
  101, 128
  thermal pulse video thermography,
    121–2
4-Vinylcyclohexene dioxide resins, 243
Viscoelastic dissipation, 105–8, 113,
  114–15
Viscoelastic methods
  dielectric analysis, 82–3
  gel time determination, 47–8
Viscosity of epoxy resins, 32, 43, 175,
  177, 180, 181
Vitrification of epoxy resins, 44–5

Water, see Moisture

Xylylenediamine, 168–9, 194, 197,
  220–2, 225
  ketimines, in, 230, 231

Zinc pyrrolidone carboxylic acid
  poly(oxypropylidenediamine) salt,
    238–9, 240